The Quantum Dissidents

Olival Freire Junior

The Quantum Dissidents

Rebuilding the Foundations of Quantum Mechanics (1950-1990)

With a Foreword by Silvan S. Schweber

Springer

Olival Freire Junior
Instituto de Física – UFBa Campus de Ondina
Salvador
Brazil

ISBN 978-3-662-51181-7 ISBN 978-3-662-44662-1 (eBook)
DOI 10.1007/978-3-662-44662-1
Springer Heidelberg New York Dordrecht London

Printed on acid-free paper

Springer is part of Springer Science+Business Media (www.springer.com)

To Olival and Antonieta, my parents,
in memory.
To Vitor, Fátima, Inês, and Silvana,
with love.

Preface

Olival Freire's *The quantum dissidents – Rebuilding the foundations of quantum mechanics 1950-1990* is a compelling, important book. It is also a remarkable book. At one level it is a richly documented history of how the foundations of quantum mechanics were formulated and variously interpreted from 1925 until the 1990s. Special emphasis is given to the developments from the 1950s on, and two threads are initially followed that eventually combine. The first has as its point of departure, the interpretation of the mathematical formalism of quantum mechanics that David Bohm and Hugh Everett formulated in the early 1950s. Bohm's was a deterministic interpretation in contrast to the conventional probabilistic one, and Everett's became known as a "many world" formulation of quantum mechanics. Their interpretations differed radically from those by the founding fathers, in particular the ones formulated by Werner Heisenberg, by Wolfgang Pauli, and by Niels Bohr, that became amalgamated and loosely referred to as the Copenhagen interpretation. Freire begins the second thread with Eugene Wigner's post-World War II critical analysis of John von Neumann's formulation of the measurement process as framed in his *Mathematische Grundlagen der Quantenmechanik* in 1932. The two threads became intertwined as foundational issues assumed greater legitimacy in the late 1950s. A new phase opened in the early 1960s when John Bell showed how to quantitatively address the quantum weirdness exhibited by entanglement and non-locality, and John Clauser and Abner Shimony indicated how to translate these insights into executable experiments. Alain Aspect's definitive experiments in the early 1980s confirmed the validity of quantum mechanics and corroborated what John Archibald Wheeler had said regarding delayed choice experiments, namely that "no phenomenon is a phenomenon until it is an observed phenomenon." Research on the foundations of quantum mechanics became highly regarded by the community after Aspect's experiments. The subsequent refinements of these experiments made them critically relevant to computer science and helped establish the field of quantum information, one of whose aims is to revolutionize computing, and another is to make the transmission of information absolutely secure and thereby revolutionizing cryptography. All these developments are beautifully expounded by Freire.

If *The quantum dissidents* contained only its detailed, internalist, presentation of the history of how the foundations of quantum mechanics became differently interpreted, this would already be a most impressive accomplishment by virtue of the command and synthesis of the huge amount of materials Freire had gathered and made use of: personal interviews, the American Physical Society's Center for the History of Physics' as well as other interviews, biographies, documents from numerous archives, correspondences, published articles and books, unpublished notes and papers, annotations to papers, ... And equally impressive is the fact that Freire explains all the physics he presents in a readily accessible, accurate, clear, succinct fashion. For example, we learn what the measurement problem is, how it became a foundational issue, and why by virtue of the extreme fineness of the level structure of a ***macroscopic*** body when described quantum mechanically, its interactions with its surrounding can never be neglected and that it can never be considered a closed system. This is the basis of the decoherence mechanism that Dieter Zeh, Wojciech Zurek, and others have introduced in order to explain how definitive pointer readings come about in the quantum mechanical description of the measurement process. Today, by virtue of these advances a complete quantum mechanical description of the measurement process is almost at hand.

But Freire wanted his presentation to be more than *a longue durée* internalist narration of the history of the changes in the conceptualization of the foundations of quantum mechanics brought about by the investigations of various theorists who dissented from the orthodox view. He wanted to understand why *investigating* foundational questions regarding quantum mechanics was actively discouraged until the 1960s. And in addition to answers to questions such as: "What were the factors that led these "dissenters" to choose issues from the foundations of quantum mechanics as research themes? What issues did each one of them come to grips with? What were the favorable factors, and what were the obstacles to their activities? And to what extent did they succeed in their endeavor?" Freire wanted to know in what ways the political and cultural contexts made the change possible, and in what ways these contexts—as well as ideology and metaphysics—were reflected in the interpretations given.

Considering the founding fathers of quantum mechanics—Heisenberg, Dirac, Pauli, Schrödinger, Bohr—as being "off-scale" was part of a creation myth and contributed to the belief that all foundational problems had been answered by the Copenhagen interpretation. Similarly, von Neumann, whose axiomatization of quantum mechanics made rigorous mathematical statements regarding the formalism possible, was deemed off-scale among the then off-scale mathematicians. His proof of the impossibility of introducing hidden variables was assumed flawless and went unchallenged until Bell—who was trying to understand the consistency of Bohm's deterministic interpretation of quantum mechanics with a particle's position and momentum considered hidden variables—discovered an invalid assumption in von Neumann's "proof." Interestingly, the mistake had been detected in the mid-1930s by Grete Hermann, but because she was primarily a mathematician interested in philosophical problems and perhaps because she was a woman, her finding went unnoticed by the physics community. In any case, physicists during the 1930s were fully occupied successfully extending the boundaries of the applicability of quantum mechanics to solid state and nuclear physics, and exploring its validity at ever smaller distances.

After World War II, the plethora of new precision instruments that became off-shelf equipment in the laboratory, the success of the renormalization program in quantum electrodynamics, masers, lasers, transistors, and PDP computers opened up new worlds in "table-top" physics. And ever more powerful accelerators did the same in high energy physics. In the United States, the one country whose home grounds had not been devastated by the war, worrying about the foundations of quantum mechanics—when the latter had been responsible for successfully designing an atomic bomb during the war—seemed misguided given all the concrete problems that were being successfully addressed using the conventional interpretation of quantum mechanics to get measurable numbers out. Furthermore, philosophizing had always been looked at askance in the United States and positivistic pragmatism flourished there after it was introduced by Charles Sanders Pierce and William James in the last third of the nineteenth century.

But two new factors altered the postwar political and social contexts of the physics community in the United States. One was the Cold War and the concomitant McCarthyism; the other was the large increase of its physics community—from some 3,000 before the war to over 8,000 after the war— the number of theorists among them and the new status accorded to them. Freire sensitively conveys the consequences of the Cold War and of McCarthyism in his narration of how and why David Bohm formulated his particular interpretation of quantum mechanics. Likewise, the paternalism that bound the physics community and the power it had vested in Bohr and his apostle, Leon Rosenfeld, are clearly described when Freire tells the story of Hugh Everett and of the reception of his "relative state" formulation of quantum mechanics. Similarly, the crucial importance of the political and cultural contexts is convincingly rendered when Freire analyzes the ways the civil rights movement, the Vietnam war, and the student upheavals transformed what had been deemed good physics and helped bring center stage foundational issues in quantum mechanics in the early 1970s.

One of the outstanding features of the book is its weaving together of professional, cultural, and political contexts with the personal and individual. We thus get short, incisive biographies of the principal actors, their family background, the institutions they were educated in, their mentors and thesis advisors, the universities they became associated with, the resources they could draw on, the encouragement and support they received from colleagues at their home institution, and from the wider physics community. And these presentations are supplemented by sociological insights gleaned from various sources: Pierre Bourdieu on habitus and various forms of capital, the strong program of the sociology of scientific knowledge, Timothy Lenoir, David Kaiser,... In the final chapter of the book, Freire makes use of prosopography to characterize the two dozen or so courageous physicists who were primarily responsible for effecting the dramatic changes in the conceptualization of quantum mechanics, the ones he calls the "quantum dissidents." They belonged to different generations, but they all had integrity, were self-confident, and they all shared the belief "that issues in foundations of quantum mechanics were worthy enough to be pursued as part of a professional career in physics, and that denying this was a dogmatic attitude. This was the main feature of their dissidence, as most physicists at the time disagreed with this." One other feature stands out as a result of Freire's analysis. The decisive changes came about by virtue of what a few in that group had done: Bell, Shimony, Clauser, Aspect. The changes were engendered by the actions of ***individuals*** making use of the resources of the collectivity they were part of. The seminal paper of John Bell, John Clauser, Abner Shimony, Michael Horne, and Richard Holt seems to be the exception. But it turns out to have resulted from pooling together into one paper the conclusions Bell, Clauser, and Shimony had reached independently. They did so in order to maximize its impact.

Commendations similar to the above can be made regarding Freire's discussion of philosophical issues. One of the central concerns of the book is explaining how come the same mathematical structure can support so many different physical interpretations. When explaining why this is so, Freire introduces the reader to the Quine-Duhem thesis regarding the under-determination of theories, to concerns with realism, to the equivalence of various mathematical formulations, to what constitutes deterministic or probabilistic explanations, to when are explanations causal, and much else. And Freire always does so simply, concisely and without ostentation.

I would characterize the book as exemplifying what the successful synthesis of the history, sociology, and philosophy of science can accomplish. I can give *The quantum dissidents – Rebuilding the foundations of quantum mechanics 1950-1990* no higher compliment than to say that anyone aspiring to become a physicist would become a better one by reading it.

Waltham, MA Silvan S. Schweber

Archives and Abbreviations

Aage Bohr Papers, Niels Bohr Archive, Copenhagen—ABP
Abner Shimony Papers, ASP.2009.02, Archives of Scientific Philosophy, Special Collections Department, University of Pittsburgh—ASP
Archives for the History of Quantum Physics, American Philosophical Society, Philadelphia, PA—AHQP
Archives of the Italian Physical Society, Bologna—ASIF
Archivio Occhialini, Università degli studi, Milan—AO
Arquivos do CNPq, Museu de Astronomia, Rio de Janeiro—AC
Bohr Scientific Correspondence (BSC–AHQP)
Costas Papaliolios Papers—[Accession 14811], Harvard University Archives—CPP
Guido Beck Papers, Centro Brasileiro de Pesquisas Físicas, Rio de Janeiro—GBP
David Bohm Papers, Birkbeck College, University of London—BP
Eugene Wigner Papers, Manuscripts Division, Department of Rare Books and Special Collections, Princeton University Library—WigP
Everett Papers—American Institute of Physics, College Park, MD—EP
Everett Papers in possession of Mark Everett, available at http://hdl.handle.net/10575/1060—ME
Henry Margenau Papers, Manuscripts and Archives, Yale University Library—MP
John Clauser Papers. Clauser's Personal Archive—JCP
Klaus S. Tausk Personal Archive, São Paulo—KST
Lancelot L. Whyte Papers, Department of Special Collections, Boston University, Boston—LWP
Léon Rosenfeld Papers, Niels Bohr Archive, Copenhagen—RP
John von Neumann Papers, Library of Congress, Washington, DC—JvNP
John Wheeler Papers, American Philosophical Society, Philadelphia, PA—WP
Niels Bohr Library & Archives, American Institute of Physics, College Park, MD, USA—AIP

Norbert Wiener Papers, MC022, Institute Archive, MIT, Cambridge, MA—NWP
Pipkin Papers [Accession 12802], Harvard University Archives—PP
Princeton University Library (Seeley G. Mudd Manuscript Library, Graduate Alumni Records)—GAR
Thomas Kuhn Papers, MC240, Institute Archive, MIT, Cambridge, MA—TKP

Contents

Chapter 1
Dissidents and the Second Quantum Revolution

Abstract The second quantum revolution, which may lead to a major technological breakthrough in science and technology with the creation of quantum computers, was the term coined by the French physicist Alain Aspect to describe changes in physics, the beginnings of which date back to the 1960s. To flesh out the new term he brought together two different threads. The first one embraced the emergence of the awareness of the importance of a new physical effect, entanglement. This refers to the quantum description of a composite system which is not reducible to the sum of its parts. It started a conceptual revolution, including the perspective of building quantum computers with calculation power exponentially greater than the best computers of today. The second thread derives from physicists' ability to isolate, control, and observe single quantum systems such as electrons, photons, neutrons and atoms. Finally these threads merged into the creation of a new field of research entitled quantum information. In Aspect's formulation, found in his introduction to John Bell's papers (Speakable and unspeakable in quantum mechanics: collected papers on quantum philosophy. Cambridge University Press, 2004), he posited two quantum revolutions taking place in the twentieth century. The first one, in the first half of the century, created the scientific theory that describes the behavior of atoms, radiation, and their interactions. The second one occurred in the second half and is still evolving, as the promise of quantum computers remains unaccomplished. This book deals with the origins of this alleged second revolution—from the early 1950s to the mid-1990s—and is a historical account of the context and intellectual aspects that arose from the renewal of research on the foundations of quantum physics. It roughly covers the period from the 1950s, when this research gained momentum with the appearance of new interpretations for the mathematical formalism of this physical theory, to the early 1990s, when research on these foundations was established as a promising topic on the agenda of research in physics. As "quantum information" became a new field of research in the middle of the 1990s, this narrative ends when quantum information as a blossoming field of research starts. This book can thus be regarded as a prehistory of quantum information.

O. Freire Junior, *The Quantum Dissidents*, DOI 10.1007/978-3-662-44662-1_1

Quantum theory is an exemplary case in the history of physics in that the success of its predictions and explanations coexisted with profound doubts about the soundness of its foundations. However, analogous doubts had appeared with major physical theories such as Newtonian mechanics and thermodynamics. Devices such as lasers and transistors, which dramatically changed science, technology, and society in the second half of the twentieth century, were based on quantum theory. Strange as it may seem, the number of scientists who called for the need to scrutinize its foundations grew over the same period. This recalls the Pascalian view that a broad scope of knowledge leads to restricted certainty about its foundations. Thus, it is legitimate to ask questions such as, "Does their instrumental effectiveness stand on the rock of secure concepts or the sand of unresolved fundamentals?" (Briggs et al. 2013). Physicists were troubled at the existence of different interpretations for the theory's mathematical formalism. Indeed, some of them thought that the theory's foundations were insufficiently established for the next stage in the development of physics. John Bell, one of the most distinguished physicists to work on these issues, used to state that quantum mechanics is "rotten," using Hamlet's famous line in an oblique reference to the father of the standard interpretation of this theory, the Danish physicist Niels Bohr (Gottfried 1991). The doubts about the foundations of quantum theory have become one of the most compelling controversies in the history of science, comparable either to that which pitted Newtonians against Cartesians at the dawn of modern physics or the supporters of energeticism against those of atomism in the late nineteenth century. Since most of the research on the foundations of quantum physics in the second half of the twentieth century was intertwined with controversy that roiled about those foundations, our account focuses on both of these aspects, which we refer to as "the quantum controversy."

The quantum controversy, therefore, drew a divide between those who thought that there was nothing further to be researched in the foundations of the theory after they were set by its founding fathers, such as Niels Bohr, Werner Heisenberg, Wolfgang Pauli, Max Born, Pascual Jordan, Paul Dirac, and John von Neumann, and those, mostly from a younger generation, who committed their professional careers to investigating such themes. Indeed, until the late 1970s, research on alternative interpretations of quantum mechanics was not considered "real physics" by many; even the existence of such a controversy was a controversial position. This is why Léon Rosenfeld, for example, objected to the use of the term "Copenhagen interpretation" because it could mean the validity of a diversity of interpretations (Ballentine 1987, p. 786; Freire Jr. 2005, p. 28). Let us now illustrate this view with two recent testimonies. The French physicist Franck Laloë published in 2001 a paper provocatively titled "Do we really understand quantum mechanics? Strange correlations, paradoxes, and theorems." The good reception of the paper led him to enlarge it into a full book (Laloë 2012). Laloë (2001, p. 656) gives us the following account:

> Until about 20 years ago, probably as a result of the famous discussions between Bohr, Einstein, Schrödinger, Heisenberg, Pauli, de Broglie, and others [...], most physicists seemed to consider that "Bohr was right and proved his opponents to be wrong," even if this was expressed with more nuance. In other words, the majority of physicists thought that the so-called "Copenhagen interpretation" had clearly emerged from the infancy of quantum mechanics as the only sensible attitude for good scientists.

"Most physicists" here requires clarification. The question was not the existence of a majority of physicists consciously adopting the complementarity view or von Neumann's presentation. Indeed, complementarity itself never was part of the training of physicists, being absent from most textbooks (Kragh 1999, p. 211). However, the received view among physicists was that foundational issues were already solved by the founding fathers of quantum physics and one did not need to spend time reading the papers where such problems were already solved. References to this tacit knowledge will appear many times through this book. The second testimony to illustrate this view is given by Christopher Gerry and Kimberley Bruno, who wrote "The Quantum Divide," intended for a wider audience than professional physicists. They told us the following anecdote (Gerry and Bruno 2013, p. 172):

> Some years ago, the senior author of this book (CCG) gave a talk about Bell's inequalities, and in the audience was a retired professor who had once been a post-doctoral research associate at Bohr's institute. After the talk he informed the audience that there was nothing of importance in the Bell inequalities, and that Bohr had already solved all the problems of quantum mechanics.

This "all foundational issues are solved" approach to the foundations of quantum physics was, however, challenged by other physicists who thought that these issues were worth pursuing as part of a professional career in physics. By doing so, the latter were questioning the very definition of what good physics was and challenging the established distribution of scientific capital, to use Bourdieu's notion of scientific fields (Bourdieu 1975). I have called them, in this sense, "quantum dissidents," a borrowing from the notion of dissidence in politics and religion. They include David Bohm, Jean-Pierre Vigier, Hugh Everett, John Bell, John Clauser, Abner Shimony, Heinz Dieter Zeh, Bernard d'Espagnat, Anthony Leggett, Franco Selleri, GianCarlos Ghirardi, Anton Zeilinger, and Alain Aspect, along with some physicists from the old guard of quantum mechanics, such as Louis de Broglie and Eugene Wigner.

In the early stages of this controversy the debate was restricted to theoretical arguments. Bohm, Everett, and de Broglie in the early 1950s, as well as Wigner and Shimony in the early 1960s, could not have imagined how to move the debate into the laboratory. The absence of experiments led a physicist, Albert Messiah, to say in his influential textbook that "the controversy has finally reached a point where it can no longer be decided by any further experimental observations; it henceforth belongs to the philosophy of science, rather than to the domain of physical science proper" (Messiah 1961, p. 48). Such a conclusion had clear-cut professional implications; it meant the controversy was not a professional matter for physicists, particularly for those new to the profession. The perception that the beginnings of the quantum controversy was a philosophical controversy survived in later accounts by the new protagonists of this research. In 1974, the historian Max Jammer wrote a comprehensive book on the history of interpretations of quantum mechanics entitled *The Philosophy of Quantum Mechanics* (Jammer 1974). As late as 1999, the physicist Anton Zeilinger recalled: "most work on the foundations of quantum

technology, it becomes now feasible to perform many experiments which could only be conceived in theoreticians' brain before" and "[this conference] was organized [...] for the purpose of reviewing fundamental concepts of quantum mechanics with the aid of experimental means made available by recent technological advancements." These were the opening words at the International Symposia on Foundations of Quantum Mechanics in the Light of New Technology held in Tokyo in 1983 and 1986 (Nakajima 1983, 1987). Similar ideas were expressed by American and European physicists leading research on these topics (Greenberger 1986; Haroche 2004). Historian Joan Bromberg has exploited this answer furthermore. After noticing that historiography so far had been focusing on themes such as Marxism and alternatives to the Copenhagen interpretation, David Bohm's causal interpretation, and the inception of Bell's theorem, she emphasized "one lead that historians have yet to pursue is constant reference that working physicists make to the role of new instrumentation" (Bromberg 2008, p. 327).[3] This kind of explanation tends to be the received view on the subject not only due to the bulk of materials concerning Bell's inequalities and experiments on them in the last two decades but also for the impressive technical improvements, particularly from the 1980s on, which enabled the manipulation of single quantum systems. Furthermore, this view is akin to the description of changes in physical sciences in which only theory and experiments could play a role. It explains the changes in the quantum controversy mainly as a consequence of the role played by experiments in physics. It may be a kind of experimental determinism, heir to technological determinism. However, is it the only or even the most interesting explanation?

This book explores an alternative perspective about the changes in the research on foundations of quantum mechanics from the 1950s on. There was a slowly developing change in the perceptions of the physicists concerning the foundations of physics, both as a controversial subject and a field of research. New institutional and professional opportunities related to the subject were created, even before the first experiments on Bell's theorem had taken place. This change happened during the 1950s and 1960s, and it can explain the elaboration and the positive reception the Bell's theorem experiments obtained. Experiments on Bell's theorem certainly increased the speed of that change and later other factors played their role. However, and this is crucial to our point, even after the first experiments on Bell's theorem began to be carried out, professional stigma against the physicists who were working on these experiments remained, as demonstrated by John Clauser's case and John Bell and Alain Aspect's concerns throughout the 1970s. Explaining changes in physics based only on theory and experiment as driving factors does not harmonize with the survival of professional stigma against a topic of research despite the performance of successful experiments. It begs for a wider kind of

[3] Bromberg's own contributions go in that direction. Bromberg (2008) deals with Wheeler's delayed-choice and Vigier's one-way experiments, while Bromberg (2006) exploits the relation between "device" and "fundamental" physics considering the case of quantum optics and Scully's works.

explanation. Indeed, it is not enough to have experiments for work to be considered good physics; it is necessary that many other physicists consider such experiments to be relevant. It is certainly true that the existence of technical improvements and real experiments were influential in the emergence and consolidation of the research on the foundations of quantum physics. It was an effective driver if you restrict your analysis to the 1980s, but it is a particularly limited explanation if one considers the whole transformation happening since the early 1950s. In addition, as we will see, even in the 1980s traces can be found of professional and cultural prejudice against research on these topics. Indeed, diverse factors may have played their roles in the evolving controversy over the foundations of this theory. Among these factors, it is worth considering philosophical and ideological issues, professional biases, generational and cultural changes, and the diversity of the social and professional environments in which physics was practiced throughout the century. In addition to this, there were conceptual and theoretical breakthroughs, technical innovations, *Gedankenexperiments* and factual experimental feats as well as technological expectations.

Let us illustrate the diversity of factors driving the change in the intellectual and professional landscape of the foundations of quantum theory after World War II. While the first round of experiments were concerned with Bell's theorem in the early 1970s, other theoretical issues were pressing physicists, both before and during the surge of interest on Bell's theorem. In the 1950s alternative interpretations of quantum physics were formulated by David Bohm and Hugh Everett. Bohm conjectured about different predictions at what he called the subquantum level but none of them at that time consistently considered the experimental implications of their proposals. While Bohm's interpretation was influential in motivating John Bell for his later work and its experimental implications, Everett's proposal never had and possibly never will have experimental predictions other than those of usual quantum physics. And yet it has been influential for its heuristic capabilities. Furthermore, after a decade without attracting the attention of experts, Everett's approach was revived by Bryce DeWitt. He was motivated by the problem of the marriage between quantum physics and general relativity, a domain which even today is far from experimental or observational concerns but has increasingly been attractive to physicists. Another pressing theoretical issue was the analysis of the quantum measurement processes, which led Eugene Wigner to diagnose the existence of a quantum measurement problem in the early 1960s. When discussions on the measurement problem became acute, pitting Wigner against Léon Rosenfeld, there was no perspective of experiments to enlighten the debate. Finally, when a problem related to the measurement problem—the transition from the quantum to the classical behavior—gained momentum among physicists in the early 1980s, it was not immediately driven by possible experiments, although a little later it did enter the lab. From this we can conclude that there was an agenda of theoretical problems in the foundations of quantum mechanics driving research after World War II. This agenda did not have any immediate bearing on experiments as these came later. In addition, there was the increasing philosophical discomfort with the instrumentalistic overtones related to the standard views of quantum mechanics.

The history of the quantum controversy may provide a window on the relationships between physics and its broader contexts. In the early 1950s, for example, Cold War tensions inevitably framed this debate both in the East and in the West. McCarthyism was a major factor shaping the career of Bohm and made him perhaps the most notable American scientist to choose exile in the last century. As philosophical themes such as determinism and realism were at stake, it comes as no surprise that ideological trends, such as the Soviet Zhdanovism, were influential in raising criticism against the standard interpretation of quantum theory. Thus as early as 1974 the historian of physics Max Jammer suggested "the extent to which this process [decline of influence of the complementarity interpretation] was fomented and supported by social-cultural movements and political factors such as the growing interest in Marxist ideology in the West deserves to be investigated" (Jammer 1974, p. 250). As we have argued elsewhere (Freire Jr. 2011b), Marxist criticisms contributed to the decline in the influence of the complementarity interpretation, even though there were Marxist physicists on both sides of the dispute, both pro and contra the complementarity view, as we will see throughout this book, particularly in Chaps. 2, 4, and 5. This tension was diluted in the late 1950s, but we can find traces of it later in the 1960s, as we will see in Chaps. 5 and 6. Resonance between physics practice and wider cultural trends were not limited to the ideological issues concerning Marxism. The surge of interest in foundations of quantum physics around 1970 was not out of tune with wider political and cultural changes that marked the times. Opposition to the Vietnam War and the cultural and political unrest of the late 1960s echoed in the decision of the Italian Physical Society to dedicate the 1970 issue of its traditional Varenna Summer School to the foundations of quantum mechanics and its 1972 issue to the history of physics in the twentieth century and its social implications. The former was the first major scientific gathering entirely dedicated to the foundations of quantum physics after World War II. Echoes from that unrest may also be found in John Clauser's shift from high precision measurements in astrophysics towards foundations as well as in the opening of the magazine *Physics Today* to the debates on the diverse interpretations of quantum physics. In the same vein, historian David Kaiser in his book *How the Hippies Saved Physics* (Kaiser 2012) has convincingly argued that cultural trends inspired by the counter culture and based on the West Coast of the U.S. were influential in supporting some research on the foundational issues and provoking the physics establishment to produce one of the key results related to quantum information, the no-cloning theorem.

Last but not least, the threads of this story are also intertwined with technical developments such as lasers, photo-detectors, optical fibers, and computers; scientific breakthroughs such as the manipulation of quantum single systems, particularly photons; the flourishing of new disciplines, such as quantum optics; theoretical breakthroughs, such as the concepts of entanglement and decoherence; and the trade of skills between applied and foundational research. Thus the challenge for historians dealing with research on the quantum foundations is integrating such a diversity of factors into a single narrative. Indeed, bringing together the diversity of factors shaping science is the ultimate goal of historians of science. However, not all factors prevail at the same time; in fact, in each diachronic slice of this history the workings of only a few can be found. The job of the historian is therefore to disentangle the roles played by each factor in each local and temporal context. That is what we have tried to do throughout this book. In the final chapter I present a synopsis of the diverse factors which have played a role in each context.

1.2 Strategy and Historiographical Issues

My strategy to build a narrative on the research on foundations of quantum mechanics after 1950 was to follow people, issues, and their relevant contexts. It was a choice inspired in the dictum of the historian Marc Bloch: the historian is like the ogre of fairy tales, "he knows that wherever he catches the scent of human flesh, there his quarry lies" (Bloch 1953, p. 28). In doing this I deal with figures who attracted public attention well beyond physics, such as David Bohm, Hugh Everett and John Bell. However, this is not a story of *great men*. Bohm and Everett were not considered such by their fellow physicists at the time; their reputations developed later. Alongside great physicists, many of our characters are ordinary physicists who collaborated to develop research on foundations and in some cases also suffered professional prejudice. Some of these rank-and-file physicists can also be classified as anti-heroes, bearing the burden of the prejudices of the times, as was the case of Klaus Tausk, mentioned in Chap. 5. In addition to physicists, some characters in the quantum controversy were well-known philosophers, such as Karl Popper, Thomas Kuhn and Paul Feyerabend. Since I only became fascinated by this topic in the late 1980s when it had already become a regular field for research in physics, and the field of quantum information was beginning to blossom, it was important to avoid the sins of anachronism or the Whig interpretation of history (Kragh 1987, pp. 89–107). This choice of strategy was an antidote to these temptations. Another strategy was to ask the same questions to different people in order to allow me to build a collective biography of the scholars who worked on a theme they thought worthy of research. In doing this I was inspired in the historiographical method of prosopography (Stone 1971; Kragh 1987, pp. 174–181), although I did not follow this method strictly as I only used the biographical data in a qualitative manner. Chapter 9 represents my attempt to synthesize the collective biography. Thus, archival sources, oral histories, published papers, dynamics of science citations, and dialogue with the secondary literature relevant to the subject were the tools used throughout the research.

The subject of the book also required a dialogue with some theoretical issues, in addition to those presented in the previous paragraph. Our narrative is a story of disciplinary change and power distribution in an established scientific field, physics in this case. Issues at stake included the value of the research on foundational issues and to what extent quantum mechanics could be applied to other areas in physics. If these issues were addressed by the physics community there would be a rearrangement in terms of professional recognition. Thus it almost naturally invites contributions from Pierre Bourdieu, as already mentioned. Let us use two quotations from Bourdieu's seminal paper on scientific capital as a form of symbolic capital. For the French sociologist, "the 'pure' universe of even the 'purest' science is a social field like any other, with its distribution of power and its monopolies, its struggles and strategies, interests and profits, but it is a field in which all these *invariants* take on specific forms" (Bourdieu 1975, p. 19). Most of the quantum controversy may be read as a story of struggles for power and monopolies, as will be evident throughout the book. Bourdieu also noted that "in the struggle in which every agent must engage in order to force recognition of the value of his products and his own authority [...], what is at stake is in fact the power to impose the definition of science (i.e. the delimitation of the field of the problems, methods and theories that may be regarded as scientific)" (Bourdieu 1975, p. 23). Bohm and Everett fought to maintain that they were doing good physics instead of metaphysics, philosophy, or, as some critics saw it, just pointless reasoning. We will see that some American physicists doubted if what Clauser was doing was "real physics." Bourdieusian lenses are fruitful not only in these cases but also in a number of other episodes in our narrative.

Bourdieu's distinction between two kinds of professional strategies, either succession or subversion, a choice young scientists in particular need to make when they enter into the profession, may be helpful for our analysis. According to his words (Bourdieu 1975, pp. 30–31),

> Depending on the position they occupy in the structure of the field (and also, no doubt, on secondary variables such as their social trajectory, which governs their assessment of their chances), the 'new entrants' may find themselves orientated either towards the risk-free investments of succession strategies, which are guaranteed to bring them, at the end of a predictable career the profits awaiting those who realise the official ideal of scientific excellence through limited innovations within authorised limits; or towards subversion strategies, infinitely more costly and more hazardous investments which will not bring them the profits accruing to the holders of the monopoly of scientific legitimacy unless they can achieve a complete redefinition of the principles legitimating domination.

Many of the physicists who appear in our story chose the subversion strategy. The sociologist Trevor Pinch was the first to look for Bourdieu's contributions while analyzing the dispute between Bohm and von Neumann around the validity of von Neumann's proof against the possibility of existence of hidden variables compatible with quantum mechanics (Pinch 1977). I exploited Pinch's suggestion further. Everett's case, with his attempt to provide a new interpretation of quantum theory that should be the natural presentation of its mathematical formalism, thus displacing both Bohr's and von Neumann's views, fits in the subversion strategy. While he had meaningful capital to bid this game—a doctoral thesis at Princeton under John Wheeler—he did not succeed, at least in the short term, as he did not achieve a "complete redefinition of the principles legitimating" which was considered the right interpretation of quantum physics. Short term may be too much time for a singular career. Everett chose to leave physics and academia for a profession using mathematics in the U.S. defense system.

My narrative also made use of a number of other contributions and readings from sociology, history, and philosophy. Timothy Lenoir's book on the cultural production of scientific disciplines was influential due to the diversity of factors he mobilized to discuss how disciplines are created and how they evolve (Lenoir 1997). He also used Bourdieu's conceptual framework to make sense of the dynamics of the birth and change of disciplines. Lenoir (1997, p. 12) argues that "one of the objectives of disciplinary struggles is to rechart the boundaries of the field, to legitimate and consecrate new combinations of assets with cultural prestige and authority, to revalue a form of capital previously considered 'impure,' and to secure that valuation through an institutionalized structure." Among the examples Lenoir used to illustrate his point are current "efforts to legitimate computational mathematics as a field of mathematics on a par with traditional mathematical disciplines," and "the consecration of science fiction as a literary genre admissible within academic departments of literature." The move of the foundations of quantum physics from a fringe position to the mainstream of physics seems to me another illustration of the disciplinary shifts Lenoir has studied. Furthermore, for Lenoir (1997, p. 19), "ideology has a crucial role to play in this process." It is not "negatively valued in [his] account." Quantum controversy is a case where the scientific disputes are loaded with philosophical and ideological commitments, and this has not been an obstacle to its cognitive development.

The controversy over the interpretation and foundations of quantum physics is thus an exemplary case of science as a cultural production, which demands that "understanding science as a cultural activity…means learning to identify and to interpret the complicated and particular collection of shared actions, values, signs, beliefs and practices by which groups of scientists make sense of their daily lives and work" (Galison and Warwick 1998). While I have studied each case or episode as rooted in its local contexts, thus attentive to study the history of science as the study of science at work, as a practice, the whole story presented in this book had to deal with a diversity of local settings in order to make this narrative intelligible. Some of the places featured are Princeton, São Paulo, Copenhagen, London, Paris, Boston, Berkeley, Heidelberg, Moscow, Geneva, Varenna, Vienna, College Park, and Bari, making the story truly international. I also had to appeal to history, tout court, and not only the history of science in order to make sense of backgrounds such as the Cold War, McCarthyism, Zhdanovism, Marxism, and the cultural and political unrest of the late 1960s.[4] As philosophical themes popped up from time to time I could not, nor did I want to, be insensitive to the literature of the philosophy of science. Readings from Michel Paty (1989, 1999, 2000), Abner Shimony (1993), and Ian Hacking (1983) have been most influential in my own work as I coped with the philosophical dimension of the quantum controversy and in particular with the

[4] On this subject it is impossible to acknowledge all the readings which were influential for my work, but at least I should cite the following ones: Schrecker (1986), Wang (1999), Graham (1972), Graham (1987), Gaddis (2005), and Hobsbawm (1994, 1982).

constraints that the very practice of physics in the twentieth century forced upon realism in science.

Finally, insofar as this history is also a history of a scientific controversy, I benefited from the attention science studies scholars have dedicated to controversies (McMullin 1987; Collins and Pinch 1993, 1998a, b).[5] Bruno Latour emphasized this interest to the point of basing his first rule of method on them: "We study science in action and not ready made science or technology; to do so, we either arrive before the facts and machines are blackboxed or we follow the controversies that reopen them" (Latour 1987, p. 258). However, I must admit that I was attracted to the controversy over the quantum at a time when I was not familiar either with the literature on scientific controversies, in particular, or with science studies. Later, when I read Paul Forman's paper on "Weimar Culture, Causality, and Quantum Theory, 1918–1927" (Forman 1971), I was impressed both by its historiographical power as well as by the fact that such a controversy had been revived in the 1950s and was still alive. One of the pleasures I had while working on this subject was to establish a connection between Bohm's proposal of the causal interpretation, in the early 1950s, and Forman's motivation for writing his Weimar Culture 1971 paper (Freire Jr. 2011a). Still on controversies and the history of science, the idea that science develops most of the time through consensus among its practitioners, as once suggested by T. S. Kuhn (1970) with the idea of shared paradigms in normal science, is challenged by the story of the controversy over the foundations of quantum theory (Freire Jr. 2014). Indeed, the history of the foundations of quantum theory has not been a history of shared paradigm; instead it has been a matter of permanent dispute among physics practitioners.

This book is not a comprehensive history of the research on the foundations of quantum theory in the second half of the twentieth century.[6] It is an attempt to make sense of how a topic once on the fringes of physics moved to its mainstream. Thus I chose contents, people, cases, and disputes which were in my view the most influential in this move. Thus many interpretations from the gamut of quantum interpretations do not appear, or appear only incidentally, in this story. The same holds for subjects such as quantum logics or axiomatics. Other subjects such as quantum gravitation and quantum optics were touched upon insofar as they directly contributed to the recognition of foundations of quantum physics as a worthy theme of research. Those two subjects reveal fascinating stories in themselves and their historiographical treatment is only beginning to be done, as one can see in the following works: Bromberg (2006), Silva and Freire (2013), Silva (2013), and Hartz (2013).

I present now a brief outline of what appears in each chapter where our history unfolds. Chapters 2 and 3 are dedicated to the most notorious cases of quantum dissidence in the 1950s, David Bohm and Hugh Everett with their alternative

---

[5] See also "Controversies", the special issue of *Science in Context* 11(2) (1998), 147–325.

[6] For an updated and comprehensive review of the conceptual issues in foundations of quantum mechanics, see the book *Do we really understand quantum mechanics*?, by Franck Laloë (2012).

interpretations of quantum physics. In Chap. 4 we move to the 1960s and the dispute among Eugene Wigner, Léon Rosenfeld, and others concerning the existence of a problem in quantum measurement. The dispute led to splitting the dominant orthodoxy in quantum mechanics. Chapter 5 is dedicated to the case of the lesser-known physicist, Klaus Tausk, who moved to foundations in the mid-1960s and whose career was subsequently mutilated. In Chap. 6 we explore how the political and cultural unrest of the late 1960s helped reconfigure the agenda of research in physics. Chapter 7 is dedicated to John Bell, his seminal theorem about the conflict between quantum theory and any local realist theories, and the early experiments on this theorem. Prominent figures include John Clauser, Abner Shimony, Edward Fry, and Alain Aspect, in addition to Bell himself. The chapter covers the period from the mid-1960s, when this theorem appeared, to the early 1980s, when Aspect's experimental results were favorably received among physicists around the world. Chapter 8 summarizes the acceleration of foundational research in the 1980s leading ultimately to the emergence of a new research field, quantum information. Finally, while the previous chapters are case studies, locally grounded, in Chap. 9, I build a collective biography of physicists who worked on foundations of quantum physics from the early 1950s to the early 1990s. I conclude by arguing that most of them can be rightly referred to as what I have called quantum dissidents.

Acknowledgements I am indebted to the following colleagues who read and commented on parts of this book: David Kaiser, Joan Bromberg, Jeffrey Bub, Osvaldo Pessoa Jr., Thiago Hartz, Indianara Silva, George Musser, Saulo Carneiro, Jose Perillan, and Ileana Greca. I am also indebted to Angela Lahee, for editorial assistance, and to Stefano Osnaghi, Osvaldo Pessoa Jr., Fabio Freitas, and Alexis De Greiff, who were co-authors of papers I have used in this book. I acknowledge Denise Key and Shaun Akhtar for their help revising the English. For sure, deficiencies in the text are my entire responsibility. The research leading to this book was made possible due to the encouragement of a number of colleagues and students. When I began to consider writing this book I was encouraged by Sam Schweber, Michel Paty, Olivier Darrigol, Dieter Hoffmann, and Cathryn Carson. I am thankful for their support. In addition to these colleagues, I am indebted to other colleagues and students who read some of the previous papers, or commented upon them after my talks; thus I am indebted to Paul Forman, Abner Shimony, Alain Aspect, Anja Jacobsen, Alexei Kojevnikov, Cristoph Lehner, Jan Lacki, Finn Aaserud, Aurino Ribeiro, José Fernando Rocha, Nelson Studart, Jean-Jacques Szczeciniarz, Jean Eisenstaedt, Martha-Cecilia Bustamante, Michael Kiessling, Harvey Brown, Albin Volte, Michael Stöltzner, Peter Byrne, Cássio Vieira, Ana Maria Ribeiro de Andrade, Ademir Santana, Frederik Santos, Virgile Besson, Climerio Silva, Robert Robinson, Mayane Nóbrega, and Wilson Bispo.

This research would not have been possible without the aid from the following institutions and agencies: CNPq, the Brazilian federal agency for research, for the continuous funding and support and the distinction as its fellow; the Brazilian agencies CAPES and FAPESB; the American Institute of Physics and American Philosophical Society for funding and aid; the General Physics department at Universidade Federal da Bahia, for my leaves of absence; Dibner Institute for the History of Science and Technology, the institution which supported me with a Senior Fellowship in 2004–2005; Université Paris 7, where I stayed as a visiting researcher for some months in 2004 and 2012; MIT, Harvard University, and University of Maryland for the stays at those institutions as a guest researcher in 2005, 2009, and 2014. I am also indebted to the following archives for allowing me to consult their collections and authorizing citations from those sources: Niels Bohr Archive, Copenhagen (Aage Bohr Papers and Léon Rosenfeld Papers); Archivio Occhialini,

Università degli studi, Milan; Arquivos do CNPq, Museu de Astronomia, Rio de Janeiro; Guido Beck Papers, Centro Brasileiro de Pesquisas Físicas, Rio de Janeiro; David Bohm Papers, Birkbeck College, University of London; Lancelot L. Whyte Papers, Department of Special Collections, Boston University; John von Neumann Papers, Library of Congress, Washington; John Wheeler Papers, American Philosophical Society, Philadelphia; Center for History of Physics, American Institute of Physics, College Park, Maryland (Archives for the History of Quantum Physics, Bohr Scientific Correspondence, Physics Today Collection, and all the oral histories used in this book); Thomas Kuhn Papers, Institute Archive, MIT; and Archives of the Italian Physical Society, Bologna.

Finally I am thankful to the following journals for allowing me to republish papers or fragments of papers originally published in their journals: *Studies in History and Philosophy of Modern Physics*, *Historical Studies in the Physical and Biological Sciences*, *Foundations of Physics*, and *Physics in Perspective*.

References

Ballentine, L.E.: Resource letter IQM2: foundations of quantum mechanics since the Bell inequalities. Am. J. Phys. **55**, 785–792 (1987)

Bell, J.S.: Speakable and unspeakable in quantum mechanics : collected papers on quantum philosophy. Cambridge University Press, Cambridge (2004). With an introduction by Alain Aspect

Bloch, M.: The Historian's Craft. Vintage Books, New York (1953)

Bourdieu, P.: Specificity of scientific field and social conditions of progress of reason. Soc. Sci. Inf. **14**, 19–47 (1975)

Briggs, G.A.D., Butterfield, J.N., Zeilinger, A.: The Oxford questions on the foundations of quantum physics. Proc. R. Soc. A Math. Phys. Eng. Sci. **469**, 20130299 (2013)

Bromberg, J.L.: Device physics vis-à-vis fundamental physics in Cold War America: the case of quantum optics. ISIS **97**, 237–259 (2006)

Bromberg, J.L.: New instruments and the meaning of quantum mechanics. Hist. Stud. Nat. Sci. **38**, 325–352 (2008)

Collins, H.M., Pinch, T.J.: The golem: what everyone should know about science. Cambridge University Press, Cambridge (1993)

Collins, H.M., Pinch, T.J.: The golem at large: what you should know about technology. Cambridge University Press, Cambridge (1998a)

Collins, H.M., Pinch, T.J.: The golem: what you should know about science. Cambridge University Press, Cambridge (1998b)

Forman, P.: Weimar culture, causality, and quantum theory, 1918-1927: adaptation by german physicists and mathematicians to a hostile intellectual environment. Hist. Stud. Phys. Sci. **3**, 1–115 (1971)

Freire Jr., O.: Science and exile: David Bohm, the cold war, and a new interpretation of quantum mechanics. Hist. Stud. Phys. Biol. Sci. **36**, 1–34 (2005)

Freire Jr., O.: Causality in physics and in the history of physics: a comparison of Bohm's and Forman's papers. In: Carson, C., Kojevnikov, A., Trischler, H. (eds.) Weimar culture and quantum mechanics: selected papers by Paul Forman and contemporary perspectives on the forman thesis. Imperial College & World Scientific, London (2011a)

Freire Jr., O.: On the connections between the dialectical materialism and the controversy on the quanta. Jahrbuch Für Europäihce Wissenschaftskultur **6**, 195–210 (2011b)

Freire Jr., O.: On the influence of science milestones on the history and philosophy of science. In: Blum, A., Gavroglu, K., Renn, J. (eds.) Towards a history of the history of science: 50 years since "Structure". Edition Open Access, Berlin (2014)

Gaddis, J.L.: The Cold War: A New History. Penguin Press, New York (2005)
Galison, P., Warwick, A.: Introduction: cultures of theory. Stud. Hist. Philos. Mod. Phys. **29B**, 287–294 (1998)
Gerry, C.C., Bruno, K.M.: The quantum divide: why Schrödinger's cat is either dead or alive. Oxford University Press, Oxford (2013)
Gottfried, K.: Does quantum mechanics carry the seeds of its own destruction? Phys. World, 34–40 (1991)
Graham, L.R.: Science and Philosophy in the Soviet Union. Knopf, New York (1972)
Graham, L.R.: Science, Philosophy, and Human Behavior in the Soviet Union. Columbia University Press, New York (1987)
Greenberger, D. (ed.): New Techniques and Ideas in Quantum Measurement Theory, vol. 480. Annals of the New York Academy of Sciences, New York (1986)
Hacking, I.: Representing and Intervening: Introductory Topics in the Philosophy of Natural Science. Cambridge University Press, Cambridge (1983)
Haroche, S.: Physique Quantique. Collège de France - Fayard, Paris (2004)
Hartz, T.: As heterodoxias quânticas e o olhar do historiador: uma história dos usos dos argumentos de Niels Bohr acerca da medição de campos quânticos (1930-1970). PhD dissertation, Universidade Federal da Bahia and Universidade Estadual de Feira de Santana (2013)
Hobsbawm, E.J.: The History of Marxism. Indiana University Press, Bloomington (1982)
Hobsbawm, E.J.: The Age of Extremes. The Short Twentieth Century 1914-1991. Penguin, London (1994)
Jammer, M.: The Philosophy of Quantum Mechanics—The Interpretations of Quantum Mechanics in Historical Perspective. Wiley, New York (1974)
Kaiser, D.: How the Hippies Saved Physics: Science, Counterculture, and the Quantum Revival. W. W. Norton, New York (2012)
Kragh, H.: An Introduction to the Historiography of Science. Cambridge University Press, Cambridge (1987)
Kragh, H.: Quantum Generations : A History of Physics in the Twentieth Century. Princeton University Press, Princeton, NJ (1999)
Kuhn, T.S.: The Structure of Scientific Revolutions. The University of Chicago Press, Chicago (1970)
Laloë, F.: Do we really understand quantum mechanics? Strange correlations, paradoxes, and theorems. Am. J. Phys. **69**(6), 655–701 (2001)
Laloë, F.: Do We Really Understand Quantum Mechanics? Cambridge University Press, New York (2012)
Latour, B.: Science in Action: How to Follow Scientists and Engineers Through Society. Harvard University Press, Cambridge, MA (1987)
Lenoir, T.: Instituting Science: The Cultural Production of Scientific Disciplines. Stanford University Press, Stanford, CA (1997)
Ma, X.S., Herbst, T., Scheidl, T., Wang, D.Q., Kropatschek, S., Naylor, W., Wittmann, B., Mech, A., Kofler, J., Anisimova, E., Makarov, V., Jennewein, T., Ursin, R., Zeilinger, A.: Quantum teleportation over 143 kilometres using active feed-forward. Nature **489**, 269–273 (2012)
McMullin, E.: Scientific controversy and its termination. In: Engelhardt, H.T., Caplan, A.L. (eds.) Scientific Controversies: Case Studies in the Resolution and Closure of Disputes in Science and Technology. Cambridge University Press, Cambridge (1987)
Messiah, A.: Quantum Mechanics. North Holland, Amsterdam (1961)
Nakajima, S.: Opening address. In: Nakajima, S. (ed.) International Symposium on Foundations of Quantum Mechanics in the Light of New Technology. Physical Society of Japan, Tokyo (1983)
Nakajima, S.: Preface. In: Nakajima, S. (ed.) The 2nd International Symposium on Foundations of Quantum Mechanics in the Light of New Technology. Physical Society of Japan, Tokyo (1987)
Paty, M.: La Matière dérobée l'appropriation critique de l'objet de la physique contemporaine. Éd. des Archives Contemporaines, Paris (1989)

Paty, M.: Are quantum systems physical objects with physical properties? Eur. J. Phys. **20**, 373–388 (1999)
Paty, M.: Interpretations and significations in quantum physics. Rev. Int. Philos. **54**, 199–242 (2000)
Pinch, T.: What does a proof do if it does not prove? A study of the social conditions and metaphysical divisions leading to David Bohm and John von Neumann failing to communicate in quantum physics. In: Mendelsohn, E., Weingart, P., Whitley, R. (eds.) The Social Production of Scientific Knowledge. Reidel, Dordrecht (1977)
Schlosshauer, M., Kofler, J., Zeilinger, A.: A snapshot of foundational attitudes toward quantum mechanics. Stud. Hist. Philos. Mod. Phys. **44**, 222–230 (2013)
Schrecker, E.: No Ivory Tower: McCarthyism and the Universities. Oxford University Press, New York (1986)
Shimony, A.: Search for a Naturalistic World View. Cambridge University Press, Cambridge (1993). 2 vols
Silva, I., Freire, O.: The Concept of the Photon in Question: The Controversy Surrounding the HBT Effect circa 1956—1958. Hist. Stud. Nat. Sci. **43**, 453–491 (2013)
Silva, I.: Uma história do conceito de fóton na segunda metade do século XX: para além de histórias do modelo bola de bilhar. PhD dissertation, Universidade Federal da Bahia and Universidade Estadual de Feira de Santana (2013)
Stone, L.: Prosopography. Daedalus, **100**(1), 46–79 (1971)
Wang, J.: American Science in an Age of Anxiety : Scientists, Anticommunism, and the Cold War. University of North Carolina Press, Chapel Hill, NC (1999)
Zeilinger, A.: Experiment and the foundations of quantum physics. Rev. Mod. Phys. **71**, S288–S297 (1999)

Chapter 2
Challenging the Monocracy of the Copenhagen School

Abstract Quantum mechanics—the physical theory for atoms, radiation and their interaction—was developed in the first quarter of the twentieth century. This was accompanied by a quarrel, with philosophical overtones, on its interpretation. Bohr called it complementarity and later it was labeled the Copenhagen interpretation. Complementarity spurred a debate among giants such as Bohr and Einstein. In 1952 David Bohm made the boldest challenge to this interpretation suggesting instead a causal interpretation. The proposal was harshly criticized by most commentators and supported by just a few. Bohm had joined the Communist Party in 1943 while at Berkeley and was caught in the witch-hunt of the McCarthyism era. He opted for a life of exile in Brazil, then Israel, and eventually England. His passport was confiscated by U.S. officials and his citizenship was revoked. While some Soviet scholars criticized the complementarity interpretation as idealistic, and thus bourgeois, they did not endorse Bohm's endeavor to recover determinism, which frustrated Bohm. At the forefront of this battle it was two giants who quarreled: Bohm and Rosenfeld. Both gifted physicists and dedicated Marxists, they nonetheless disagreed about how to interpret physics and its philosophical lessons. Bohm promised to generalize his approach for the relativistic domain, but this was not fulfilled. In the late 1950s, Bohm experienced a major intellectual change. He broke with Marxism, abandoned the causal interpretation, moved towards Eastern thinkers and began a long-standing project of reforming physics along the themes of order and wholeness.

2.1 Interpretation of Quantum Theory Before David Bohm[1]

The inception of quantum physics, between 1925 and 1927, and its early origins dating back to 1900, along with the debates about its interpretation, are some of the topics better exploited in the literature concerning the history of physics of the

[1] In this chapter I draw from some of my previous works, in particular (Freire Jr. 1999, 2005, 2011a, b).

O. Freire Junior, *The Quantum Dissidents*, DOI 10.1007/978-3-662-44662-1_2

twentieth century. As it is impossible to give a fair account of that history in a few introductory lines, I will only refer to some milestones of the debates before Bohm entered the scene in the early 1950s. Quantum physics began with old phenomena concerning electromagnetic radiation and its interaction with matter, and a few new ideas, such as the quantum of action and the granularity of light. Its development was marked by a close relationship between novel ideas and precision measurements of new and old phenomena, which led to a completely new theoretical landscape between 1925 and 1927. This landscape needed to be interpreted in terms of physics in order to make sense of its abstract mathematical formalism. This was divisive for physicists, creating what the philosopher Karl Popper (Popper and Bartley 1982) would later call "the schism in Physics," a controversy now more than 8 decades old and only comparable to the one that pitted Newtonians against Cartesians at the dawn of modern physics.

One of the poles of the controversy was the line of interpretation developed by the Danish physicist Niels Bohr, which he christened "the complementarity view." In general, he considered complementarity's main features to be the following: Max Born's quantum probabilistic descriptions are non-reducible to deterministic descriptions; quantum jumps are intrinsic to quantum descriptions; the means of observation play a prominent role, i.e. quantum phenomena should consider both the system and the observation devices; classic concepts, such as wave and particles, are used in a complementary manner, i.e. not jointly in the same experiments but in mutually exclusive ways; discreteness of the physical magnitude action is a fundamental feature of nature; and quantum theory is considered a complete theory, i.e. not superseded by other theories dealing with phenomena concerning radiation and its interaction with matter.

Bohr's thoughts were presented embedded in philosophical considerations concerning the role of ordinary language in guaranteeing objectivity to research in physics. First considered "obscure" by many, Bohr's philosophy has been scrutinized by philosophers in recent decades. Shoulder-to-shoulder with Bohr were some of the creators of quantum theory, such as Werner Heisenberg, Max Born, Pascual Jordan, and Wolfgang Pauli, each with subtle differences in their interpretations. However, not all the founding fathers of quantum physics aligned themselves with Bohr's complementarity. Albert Einstein was initially skeptical about the consistency of the theory, and later about its completeness, producing several *Gedankenexperiments* to reveal shortcomings in the theory; Louis de Broglie tried to maintain determinism and the images of wave and particles, suggesting a model of particles being piloted by waves; and Erwin Schrödinger did not accept quantum jumps and pleaded for a wave representation of quantum phenomena.

In parallel to more philosophical and conceptual debates, John von Neumann looked for a rigorous presentation of the quantum mathematical formalism in an attempt to replace the coexisting, equivalent presentations of the quantum theory in terms of matrixes, wave functions, and algebras. The Hungarian-born mathematician described all these previous presentations as special cases of a more general mathematical framework, namely that of Hilbert's space vectors. Von Neumann's presentation would leave a lasting imprint on the debates on the foundations of quantum physics as many of these debates took von Neumann's as the orthodox presentation of quantum theory. In particular, he formalized what quantum physicists call "the reduction of the wave packet" to describe measurement processes as an independent axiom in quantum theory (technically he described it as the workings of a projection operator). It meant that measurements were not ruled by the Schrödinger equation which now describes only the evolution of quantum states before measurement processes. Later, this formulation would be the standard introduction to the intractable measurement problem, the problem of the process through which superposition of quantum states, the most fundamental quantum feature, disappears during measurements. In addition, he provided a theorem prohibiting the enlargement of quantum theory with additional or "hidden" variables. As we will see, one of David Bohm's first achievements with his causal interpretation was to create a practical counterexample to this rule.

The controversy on the interpretation and foundations of quantum physics, "quantum controversy" as shorthand, was a heated topic among physicists in 1927, particularly at the Fifth Solvay International Conference on Electrons and Photons held in Brussels in October of that year (Institut International de Physique Solvay 1928; Bacciagaluppi and Valentini 2009). In the following years the debates between Einstein and Bohr on this subject attracted the attention of physicists and images and text featuring the two giants quarreling are now iconic in the culture of physics. However, as time went by, physicists tended to attribute a less important role to the controversy as more mundane subjects occupied their agenda. Applications and extensions of quantum physics and mainly nuclear physics dominated the scene in the 1930s.

Picture 2.1 Bohr and Einstein's debate portrayed in the artistic imagination—Sculpture in Park Muzeon, Moscow. Photo by Climério P. da Silva Neto

During the war more practical efforts, mainly related to the building of the atomic bomb and radar, absorbed physicists' energies, at least in the US. After the war, old themes such as the discovery of new particles, fixing the machinery of quantum electrodynamics, and a renewed approach to solid state caught the attention of the physicists. Indeed, since the middle of the 1930s, the typical physicist either considered foundational issues to be off the main agenda of physics or thought that they had already been solved by Niels Bohr and his close companions. As reminded by the Danish physicist Christian Møller, assistant of Niels Bohr in Copenhagen, "although we listened to hundreds and hundreds of talks about these things, and we were interested in it, I don't think, except Rosenfeld perhaps, that any of us were spending so much time with this thing . . . When you are young it is more interesting to attack definite problems. I mean this was so general, nearly philosophical."[2] Those were the times the historian of physics Max Jammer

[2] C. Møller, interviewed by T.S Kuhn, 29 July 1963, Archives for the History of Quantum Physics (hereafter AHQP), American Philosophical Society, Philadelphia, PA, cited in Jacobsen (2012, p. 55).

referred to as the "almost unchallenged monocracy of the Copenhagen school in the philosophy of quantum mechanics." Although research on the foundations of quantum theory was not a priority in the early 1950s, interest in the subject was far from dead. Indeed the first salvos in the new battles were fired by Niels Bohr reviewing their disagreements in the paper ("Discussion with Einstein on Epistemological Problems in Atomic Physics") for the volume of "The Library of Living Philosophers" edited by P. A. Schilpp to honor Einstein (Bohr 1949; Einstein 1949; Schilpp and Einstein 1949). Bohr's paper and Einstein's reply played a role in arousing the dormant debate. Einstein himself continued to voice his discomfort with quantum physics and new critics of the complementarity view were appearing among Soviet physicists and philosophers.[3] All these criticisms influenced David Bohm, as we will see.

2.2 Bohm's Causal Interpretation of Quantum Mechanics

In early July 1951, the American physicist David Bohm, from Princeton University, submitted a lengthy paper entitled "A suggested interpretation of the quantum theory in terms of 'hidden' variables" to the prestigious journal *Physical Review*. The paper was organized in two parts, both published in early 1952, and the technical title hid its far-reaching philosophical implications (Bohm 1952b). Soon both David Bohm and his critics were using "causal interpretation" to label his approach to quantum theory, clarifying Bohm's ambition to restore a kind of determinism analogous to that of classical mechanics (Bohm 1952a, 1953a; Bohm and Vigier 1954). Unlike the early critics of quantum mechanics, Bohm did not just express hopes of going back to a causal description for atomic phenomena. In fact, he built a model for his approach which assumed that an object like an electron is a particle with a well defined path, which means it is simultaneously well defined in both position and momentum. It is noteworthy that in quantum theory it is precisely the impossibility of such a simultaneous determination which breaks with the classical determinism, while in classical mechanics the possibility of that simultaneous definition assures the classical deterministic description. Bohm's work had philosophical implications as a consequence of its physical assumptions. According to him, this interpretation "provides a broader conceptual framework than the usual interpretation, because it makes possible a precise and continuous description of all processes, even at the atomic level." More explicitly, he stated that,

> This alternative interpretation permits us to conceive of each individual system as being in a precisely definable state, whose changes with time are determined by definite laws, analogous to (but not identical with) the classical equations of motion. Quantum-

[3] For the debates before 1950 see Jammer (1974), for Jammer's quotation, see his p. 250. For the Soviet critics see Graham (1987).

mechanical probabilities are regarded (like their counterparts in classical statistical mechanics) as only a practical necessity and not as a manifestation of an inherent lack of complete determination in the properties of matter at the quantum level. (Bohm 1952b, p. 166)

Fully aware of the philosophical implications of his proposal, Bohm concluded the paper by criticizing the usual interpretation of quantum mechanics on philosophical grounds. He accused "the development of the usual interpretation" of quantum theory of being "guided to a considerable extent by the principle of not postulating the possible existence of entities which cannot now be observed," and remarked that "the history of scientific research is full of examples in which it was very fruitful indeed to assume that certain objects or elements might be real, long before any procedures were known which would permit them to be observed directly," the case of the atomistic hypothesis being the best historical example. Bohm also noted that this principle derived from the "general philosophical point of view known during the nineteenth century as 'positivism' or 'empiricism.'" Then he explained to his readers that "a leading nineteenth-century exponent of the positivist view was Mach." While conceding that "modern positivists appear to have retreated from this extreme position," he stated that this position was still reflected "in the philosophical point of view adopted by a large number of modern theoretical physicists." Apart from this philosophical digression, the philosophical implications of Bohm's proposal concerned not only the recovery of determinism as a mode of description of physical phenomena, but also the adoption of a realist point of view toward physical theories, both discarded by the complementarity view.[4]

Later in his career, Bohm (1987, p. 33) emphasized that recovering determinism was not his main motivation and that his major dissatisfaction was that "the theory could not go beyond the phenomena or appearances." Building an ontology to explain phenomena would become a permanent goal in Bohm's research with determinism pushed down on his agenda. However, in the 1950s Bohm and the debate triggered by his proposal did indeed promote the recovery of determinism.

To illustrate the strength of the attachment of Bohm and his collaborators to the philosophical priority of causality, we can make reference to the work he and Jean-Pierre Vigier did in 1954, changing Bohm's original model slightly. In this work, they embedded the electron in a fluid undergoing "very irregular and effectively random fluctuation" in its motion (Bohm and Vigier 1954). While these fluctuations could be explained by either a deterministic or a stochastic description, Bohm and Vigier framed them into the causal interpretation approach, giving their paper the title "Model of the causal interpretation of quantum theory in terms of a fluid with irregular fluctuations."

[4] Bohm (Bohm 1952b, pp. 188–189). Bohm's reference to Ernst Mach, criticizing the positivist view, is a shibboleth of his Marxist background, a feature we will return to later, as this reference gained currency among Marxists in the first half of the twentieth century following the diffusion of *Materialism and Empirio-criticism* (Lenin 1947).

Bohm not only suggested a new conceptual and philosophical framework. He also raised the stakes by suggesting his approach could be fruitful in new domains of physics, promising that "modifications can quite easily be formulated in such a way that their effects are insignificant in the atomic domain [. . .] but of crucial importance in the domain of dimensions of the order of 10^{-13} cm." Bohm was indeed referring to intra-nuclear distances, an area in which there was a proliferation of discoveries of new particles requiring the development of new methods in quantum field theories. Bohm's promises, however, were as appealing as vague, saying that "it is thus entirely possible that some of the modifications describable in terms of our suggested alternative interpretation, but not in terms of the usual interpretation, may be needed for a more thorough understanding of phenomena associated with very small distances." The promise of fulfillment of such an expectation was then postponed: "we shall not, however, actually develop such modifications in any detail in these papers" (Bohm 1952b, p. 166).

In Bohm's original model, electrons suffer physical influences both from potentials, such as electromagnetic potentials, and from a new potential resulting from mathematical manipulations of the Schrödinger equation, which Bohm labeled the "quantum potential." Technically this new potential arises when one exploits analogies between the Schrödinger equation of quantum theory and the Hamilton-Jacobi equation of classical mechanics. To make a clear comparison, let us take an electron with well defined positions described by a function of the form $\psi = R \exp(iS/\hbar)$, which must satisfy the Schrödinger equation, and let us call $R(\mathbf{x})^2 = P(\mathbf{x})$. After some mathematical manipulations we get Eqs. (2.1) and (2.2) resulting from Bohm's approach.

$$\partial P/\partial t + \nabla(P\nabla S/m) = 0 \tag{2.1}$$

$$\partial S/\partial t + (\nabla S)^2/2m + V + U = 0 \tag{2.2}$$

where

$$U = -(\hbar^2/2m)(\nabla^2 R/R)$$

Bohm then further exploited these analogies by suggesting that electrons have a well defined momentum $\mathbf{p} = \nabla S(\mathbf{x})$. The same analogies suggest that the "extra" term U in Eq. (2.2) may be interpreted as the action of a "quantum potential" on electrons, in addition to the potentials known from classical physics, such as electromagnetic potentials. In addition, according to this model, Eq. (2.1) is a continuity equation, and Bohm suggests that we take $P = |\psi(x)|^2$, where ψ is the solution of the Schrödinger equation, to assure the conservation of the probability density of an ensemble of particle positions. As remarked by Max Jammer (1988, p. 693), "Bohm interprets [P] as the probability of the particle's *being* at the position defined by the argument x of $\psi(x)$ and not, as Born conceived it, as the probability of *finding* the particle at that position if performing a suitable measurement." Bohm's model of electrons has well defined positions as well as momenta;

thus, they have continuous and well defined trajectories. These *p*'s and *x*'s are the hidden variables in Bohm's models. They are "hidden" when compared to standard quantum mechanics as Heisenberg's uncertainty relations forbid the simultaneous precise definitions of positions and momenta. Later on, however, the physicist John Bell, a supporter of Bohm's proposal, would consider Bohm had been unhappy choosing the term "hidden variables." Bell would remark that complementarity is the interpretation which hides either of the complementary variables as they could not be considered images of the phenomena (Bell 2004, p. 201).[5] In order to get models which were able to produce the same results as quantum mechanics, Bohm needed to ascribe well defined positions and momenta to the measurement devices too. Thus, from the Hamiltonian (kinetic plus potential energies) of the coupling between such devices and the micro systems, observable results could be predicted. Bohm used these models to carry out detailed calculations of a number of different problems, for instance, stationary states, transitions between stationary states (including scattering problems), the Einstein-Podolsky-Rosen *Gedankenexperiment*, and photoelectric and Compton effects. To achieve results compatible with those from quantum mechanics, Bohm modeled light as electromagnetic waves. In all these problems he found the results predicted by the usual mathematical formalism of quantum theory (Jammer 1988; Bohm 1952b, p. 183).

Bohm's achievement was not a minor one. He was able to build an approach to quantum theory leading to the same predictions as usual quantum mechanics and develop the first alternative interpretation to the dominance of the complementarity view among physicists. This empirical equivalence was dependent on adopting hidden variables in the system and in the measurement device, which was an improvement on his initial approach. This was done in reacting to criticisms made by Wolfgang Pauli. In fact, with this improvement Bohm's approach became superior to an earlier and analogous approach that had been suggested by Louis de Broglie in 1927, then entitled the "pilot-wave approach" (Bohm 1952b, pp. 191–193). This earlier approach was unknown to Bohm until he received Pauli's criticisms, as we will see later. To be more precise, Bohm's approach was equivalent to non-relativistic quantum mechanics as his electron model, for instance, did not have "spin."

That Bohm's approach was unable to deal with relativistic systems is clear from the equation of the quantum potential. Indeed, it is enough to take a system with two electrons to see that the quantum potential tells us that an interaction could propagate from one electron to the other instantaneously without any time dependency. This would not have been considered a major flaw when Bohm published his papers if one recalls that in the historical process of the creation of quantum physics, non-relativistic equations came first and relativistic generalizations a little

[5] According Bell's words, "absurdly, such theories are known as 'hidden variable' theories. Absurdly, for there is not in the wavefunction that one finds an image of the visible world, and the results of experiments, but in the complementary 'hidden'(!) variables." I am thankful to Michael Kiessling for calling my attention to Bell's remarks.

later. At any rate, critics would ask Bohm for these generalizations and Bohm would promise that they were under way.

Bohm's papers also raised other philosophical and technical issues. Empirically equivalent to the standard quantum mechanics, Bohm's would be a nice example of what philosophers call the underdetermination of theories by empirical data. According to the philosopher Paul Feyerabend, after Bohm's work, "it follows that neither experience nor mathematics can help if a decision is to be made between wave mechanics and an alternative theory which agrees with it in all those points where the latter has been found to be empirically successful" (Feyerabend 1960, p. 325). This philosophical thesis, also called the Duhem-Quine thesis, a reference to the scientist and philosopher Pierre Duhem and the philosopher Willard Van Orman Quine, was well set in logical terms but it was, and it is, at least unpleasant for physicists to realize that some of their best theories are not the only possible description of phenomena.[6] Finally, Bohm's approach was a practical example showing that something was wrong with von Neumann's mathematical proof against the possibility of introducing hidden variables in quantum mechanics. Bohm was fully aware of this in his approach, making it explicit in his papers, and he would attentively follow von Neumann's reactions to his proposals (Bohm 1952b, pp. 187–188). Finally, as we have already noted, Bohm did not refuse the philosophical debate implied by his proposals as he not only defended his approach with both technical and conceptual arguments, but also accused supporters of the standard interpretation of being the twentieth-century equivalents of the anti-atomists in the nineteenth century.

2.3 Backgrounds of Bohm's Causal Interpretation

Before analyzing the reception of Bohm's proposal by his fellow physicists, let us go back to see how Bohm evolved towards this interpretation of quantum mechanics. In addition, let us flesh out our history by considering the life and environment of the person involved. Bohm's proposal of a causal interpretation for quantum physics was a surprising move both on the physics scene at the time and in his own professional career. He had been awarded his PhD during World War II at Berkeley under the supervision of Robert Oppenheimer, who was then already involved in the Manhattan Project. His dissertation was dedicated to the subject of scattering in proton deuteron collisions, which was a sensitive subject for the Manhattan Project; thus it was immediately classified. As Bohm had no clearance to present his dissertation and did not work on the atomic project due to his union activities and links with the Communist Party he could not defend his PhD dissertation. Bohm was then involved with the American Communist Party and the union activities of

[6] On the Duhem-Quine thesis, see (Harding 1976). On quantum mechanics as an illustration of this thesis, see Cushing (1994).

technical and scientific workers. It was the eve of the war and Bohm was attracted to the Communist ranks as it seemed to him that the USSR could be an essential force against the Nazis. In this move he was joined by a few of Oppenheimer's students, all of whom paid a high price for it after the war. Subsequently Bohm broke his organizational ties with the Communist Party while keeping the same ideological inclinations until the late 1950s. According to historian Alexei Kojevnikov (2002, p. 166), "Bohm severed his ties with the organized communist movement while remaining a convinced Marxist with a special interest in the philosophy of dialectical materialism."

The problem was solved with Oppenheimer testifying to the quality of Bohm's dissertation, which led Berkeley to grant him his doctoral degree. Still during the war, as a research fellow at Berkeley, he began to work with the Australian Harrie S. W. Massey on the problem of electrical currents passing through a gas in magnetic fields, an issue considered relevant for the enrichment of uranium and thus part of the Manhattan Project. The problem led them to study plasmas, but the process of enrichment did not prove useful for the war effort. After the war, hired by Princeton University, Bohm and his graduate student Eugene Gross resumed work on plasmas, developing the approach called "collective variables in classical plasmas." Then, with the graduate student David Pines, Bohm moved to study current in metals, elaborating a quantum approach to the phenomenon using the same collective variable resource he had successfully applied to the classical treatment of plasmas. His jointly-authored papers with Pines and Gross would become landmarks in this field.[7]

Bohm was then considered by elder fellow physicists to be one of the most promising American theoretical physicists of his generation—"probably Oppenheimer's best student at Berkeley" according to historian Sam Schweber—and it was in this capacity that he was one of the few to be invited to the 1947 Shelter Island conference, the first of a series of conferences held in the US dealing with topics such as high energy nuclear physics, new nuclear particles, and anomalies and procedures for fixing quantum electrodynamics.[8] The list of topics on which Bohm had worked until the late 1940s did not presage his move towards working on the foundations of quantum physics.

Clues that might shed light on the inception of his alternative interpretation of quantum theory come from his teaching duties at Princeton. Having been educated at Berkeley, where "Bohr was God and Oppie [Oppenheimer] was his prophet," according to Weinberg, one of Oppenheimer's students at Berkeley, Bohm's classes on quantum mechanics naturally reflected Bohr's views on this theory. From these classes his textbook *Quantum Theory* (Bohm 1951) emerged. A close inspection of

---

[7] On Bohm's biography, see Peat (1997) and Mullet (2008b). For an analysis of Bohm's works, see Kojevnikov (2002). Bohm and Gross (1949a, b), Bohm and Pines (1951, 1953), Pines and Bohm (1952). The fourth paper in the series was authored only by Pines (1953).

[8] Sam Schweber, "Bohm Memorial," Folder A.M., David Bohm Papers, Birkbeck College, University of London (hereafter BP), cited in Mullet (2008a, p. 40) and Mehra (1994, pp. 217–218).

this book, however, reveals how far Bohm was from being truly Bohrian. Indeed, Bohm's *Quantum Theory* is remarkable for its attempt to combine Niels Bohr's complementarity with Bohm's own kind of realism. While the former denied quantum theory the ambition of describing a world independent of measurements, the latter included an ontological description of the quantum world, referred to by Bohm as "an attempt to build a physical picture of the quantum nature of matter." The book is also noteworthy for his conceptual clarity and a few innovations, such as the reformulation of the EPR thought experiment using spin instead of position and momentum, which later became the standard formulation for EPR theory and experiments due to its mathematical simplicity. Bohm also included a treatment of the measurement process using random phases, which he would use later in his work on the causal interpretation.[9]

When the book came out, Bohm was already moving towards the elaboration of his causal interpretation. Later he would acknowledge at least two influences on his move: a discussion with Albert Einstein at the Institute of Advanced Studies in Princeton after the book was published and the reading of a paper by a Soviet physicist criticizing the complementarity view for its idealistic and subjectivist inclinations. As told by historian Max Jammer,

> Stimulated by his discussion with Einstein and influenced by an essay which, as he told the present author, was "written in English" and "probably by Blokhintsev or some other Russian theorist like Terletzkii," and which criticized Bohr's approach, Bohm began to study the possibility of introducing hidden variables. (Jammer 1974, p. 279)

Later Jammer (1988, p. 692) reiterated this story in a kind of Festschrift for Bohm's 70th birthday. Bohm never contested it. This information, however, raised a doubt, as Jammer himself noted. "Bohm [had] forgotten the exact title and author of this paper" and there was no paper either by Blokhintsev or by Terletzkii published in English before Bohm's shift to the causal interpretation (Jammer 1974, p. 279 footnote 63). Indeed, the papers by the Soviets criticizing complementarity published in Western languages appeared in French in 1952, while Bohm's shift to the causal interpretation occurred in 1951.[10] The riddle may be explained by free translations from the Soviet papers which may have circulated among Marxist intellectual circles in the West before their publication. Plausible as this explanation is, unfortunately, we do not have documentary evidence to support it. Furthermore, the statement on the influence of the Soviet views in the inception of the causal interpretation is crucial information given Bohm's Marxist beliefs and the role played by the criticisms from Soviet philosophers and physicists against the complementarity interpretation in the quantum controversy (Graham 1972, 1987). However, and again unfortunately, archival documents unearthed since then have not been able to reinforce Jammer's interesting clue. Indeed, most of Bohm's personal papers did not survive and he did not keep copies of his correspondence. Later on, a few letters from him to some friends and fellow physicists surfaced but

[9] For Weinberg's statement, see Mullet (2008a, p. 39).

[10] Blokhintsev (1952) and Terletsky (1952).

they are not enough to document the personal and intellectual environment at the time of his move.[11] Einstein's conversation with Bohm after the publication of *Quantum Theory* was the start of a relationship which would last until Einstein's death in 1955. Ironically, as it may seem, Einstein would support Bohm on several grounds except in defense of Bohm's approach to quantum mechanics. We will see some of their exchanges throughout this chapter.

2.3.1 Trapped in the Cold War Storm

During the 1950s David Bohm would fight his most important intellectual battle while pushing for the causal interpretation of quantum theory. That battle happened in extreme personal circumstances as he was trapped in a Cold War storm that made his story almost Kafkaesque. Back in 1949 he had been subpoenaed to appear before the HUAC (House Committee on Un-American Activities) where he was asked about his connections with the Communist Party. Bohm took the Fifth Amendment of the US Constitution (the right to refuse to answer a question because the response could be self-incriminating). In the anti-communist hysteria typical of Cold War times in the US, a period later called McCarthyism, he was indicted for contempt of Congress, arrested and then released on bail. In the following months the court would find him not guilty. Like Kafka's character in *The Trial*, Bohm never knew exactly what he was accused of. As the historian David Kaiser remarked, being a theoretical physicist with leftist inclinations in Cold War America was enough to mark anybody as a highly probable target of anticommunist hysteria. For the American laymen, the atomic bomb could be reduced to a single equation which could be passed on to the USSR. This would mean that the enemy would immediately possess the same weapons America had developed during World War II.[12] Meanwhile, Princeton University suspended his contract,

[11] The David Bohm Papers, deposited at Birkbeck College, University of London, reveal few documents from the period prior to his departure to Brazil at the end of 1951, when the papers on the causal interpretation had already been submitted for publication. After leaving the U.S., there is a meaningful correspondence with Einstein; Melba Philips, an American physicist and friend of Bohm; Hanna Loewy and Miriam Yevick, his friends. Most of the correspondence with Wolfgang Pauli, relevant for the period prior to his departure from the U.S. and after the completion of his paper in the causal interpretation, was recovered and published by Karl von Meyenn in the collection dedicated to Pauli's correspondence (Pauli and Meyenn 1996, 1999). More recently, a batch of letters between Bohm and the French astrophysicist Evry Schatzman was unearthed by Virgile Besson at Schatzman's papers, Observatoire de Paris. These letters corroborate the main points of our work. Furthermore, they weaken the possibility of Bohm's reading of Soviet papers while moving to build the causal interpretation. Indeed, he did not mention this in his letters to Schatzman while describing his work to obtain this interpretation.

[12] "The early years of the Cold War were not a pleasant time to be an intellectual in the United States, especially if he or she happened to have a past or present interest in the political left. [. . .]

prevented him from attending classes and using the university libraries, and in June 1951 did not renew his contract.[13]

Picture 2.2 David Bohm reading a newspaper; after refusing to testify whether or not he was a member of the Communist Party before the House Un-American Activities Committee. Library of Congress, New York World—Telegram and Sun Collection, courtesy AIP Emilio Segre Visual Archives

theoretical physicists emerged as the most consistently named whipping-boys of McCarthyism" (Kaiser 2005, p. 28).

[13] Historians have already set the record of most of this history. The cases of persecution towards Bohm and his colleagues at Berkeley, Bernard Peters, Joseph Weinberg, and Giovanni Rossi Lomanitz have been well charted by Shawn Mullet (2008a); Princeton's attitudes towards him were analyzed by Russell Olwell (1999); the anti-communist hysteria in American academia was studied by Ellen Schrecker (1986), Jessica Wang (1999), and David Kaiser (2005). Bohm's imprisonment and bail is also in Kojevnikov (2002, p. 181).

The intersection of Bohm's political persecution and his move towards a new interpretation of quantum theory has attracted the attention of historians. Christian Forstner has suggested that isolation from Princeton and the American community of physicists was influential in Bohm's abandoning the standard interpretation of quantum physics and adopting a heterodox interpretation. Unfortunately, as we have previously discussed, historians have to deal with the scant documentary evidence around the circumstances of these events in the crucial months in Bohm's life between his appearance at the HUAC in 1949, the publication of his book *Quantum Theory*, and the completion of his causal interpretation in the middle of 1951. As such, we may only deal with plausible conjectures in mapping the influences and motivations of his move toward the causal interpretation.[14]

2.3.2 Bohm, de Broglie, and Pauli: Conceptual Issues and Disputes About Priorities

Bohm was unaware of previous work by Louis de Broglie along analogous lines. What we may reconstruct about how Bohm reacted when informed of de Broglie's works sheds light on the kind of technical problems he needed to solve in order to make his proposal consistent. It is also illuminating regarding the disputes and alliances in the controversy over the foundations of quantum physics. Last but not least, as Wolfgang Pauli was one of the people to warn Bohm about de Broglie's works and as their exchange is one of the most relevant for the early debate on the causal interpretation, it is interesting to see their discussion in some detail. Before Bohm's papers appeared in print, Einstein and Pauli informed him that de Broglie had suggested a similar approach at the 1927 Solvay conference, which Bohm had not known about. Pauli had criticized de Broglie's approach when first proposed and de Broglie had reacted by giving up his idea. Now Bohm had to face the same objections. Pauli had argued that de Broglie's proposal fitted Max Born's probabilistic interpretation of the ψ function only for elastic collisions. In the case of inelastic scattering of particles by a rotator, a problem Enrico Fermi had solved in 1926, de Broglie's idea was incompatible with assigning stationary states to a rotator, before and after the scattering. Pauli had considered this failure intrinsic to de Broglie's picture of particles with definite trajectories in space-time, an approach de Broglie had called the "pilot wave", which means particles with well defined paths ruled by waves coming from the Schrödinger equation.[15]

[14] On the influences on Bohm's shift towards the causal interpretation, see Jammer (1974, 1988) and Forstner (2008).

[15] Einstein's remark is in Paty (1993). Bohm to Pauli, [Jul 1951], in Pauli and Meyenn (1996, pp. 343–345). Most of Pauli's letters to Bohm did not survive; we infer their contents from Bohm's replies. Bohm to Karl von Meyenn, 2 Dec 1983, ibid, on 345. Broglie's pilot wave and Pauli's criticisms are in (Institut International de Physique Solvay 1928, pp. 105–141 and 280–282). See also Bacciagaluppi and Valentini (2009).

Pauli addressed his criticisms toward a draft version, which Bohm corrected in consequence. This draft has not survived, but an indication of the corrections has. In response to Pauli's criticisms Bohm wrote: "I hope that this new copy will answer some of the objections to my previous manuscript . . . to sum up my answer to your criticisms . . . I believe that they were based on the excessively abstract assumptions of a plane wave of infinite extent for the electrons' Ψ function. As I point out in section 7 of paper I, if you had chosen an incident wave packet instead, then after the collision is over, the electron ends up in one of the outgoing wave packets, so that a stationary state is once more obtained." Initially Pauli did not read the second manuscript as he considered it too long, which angered Bohm. He rebuked Pauli: "If I write a paper so 'short' that you will read it, then I cannot answer all of your objections. If I answer all of your objections, then the paper will be too 'long' for you to read. I really think that it is your duty to read these papers carefully." As a precaution, he summarized his views and the improvements in letters[16]:

> In the second version of the paper, these objections are all answered in detail. The second version differs considerably from the first version. In particular, in the second version, I do not need to use "molecular chaos." You refer to this interpretation as de Broglie's. It is true that he suggested it first, but he gave it up because he came to the erroneous conclusion that it does not work. The essential new point that I have added is to show in detail (especially by working out the theory of measurement in paper II) that his interpretation leads to all of the results of the usual interpretation. Section 7 of paper I is also new [transitions between stationary states—the Franck-Hertz experiment], and gives a similar treatment to the more restricted problem of the interaction of two particles, showing that after the interaction is over, the hydrogen atom is left in a definite "quantum state" while the outgoing scattered particle has a corresponding definite value for its energy.

Eventually, Pauli analyzed Bohm's papers as well as the letters. Pauli conceded that Bohm's model was logically consistent, which was recognition of Bohm's work: "I do not see any longer the possibility of any logical contradiction as long as your results agree completely with those of the usual wave mechanics and as long as no means is given to measure the values of your hidden parameters both in the measuring apparatus and in the observed system." Pauli ended with a challenge, related to Bohm's promise of applying his approach to new domains such as high energy physics: "as far as the whole matter stands now, your 'extra wave-mechanical predictions' are still a check, which cannot be cashed." Pauli never ceased to oppose the hidden variable interpretation and would formulate new objections, as we will see later. For Bohm, however, Pauli's challenge now was less pressing than de Broglie's.[17]

Before 1927, Louis de Broglie had had the idea of a "double solution," in which the waves of Schrödinger's equation pilot the particles, which are singularities of the waves. Just before the meeting of the Solvay council on October 24–29, 1927 he gave up this idea because of its mathematical difficulties and presented his report to

[16] Bohm to Pauli, July 1951, Summer 1951, Oct 1951, 20 Nov 1951 (Pauli and Meyenn 1996, pp. 343–346, 389–394, and 429–462).

[17] Pauli to Bohm, 3 Dec 1951, plus an appendix, (Pauli and Meyenn 1996, pp. 436–441).

the meeting with just the "pilot wave" proposal. The particles were reduced to objects external to the theory. After the 1927 meeting he adhered to the complementarity interpretation. Bohm was right in remarking that de Broglie had not carried his ideas to their logical conclusion, but de Broglie surely had a share in the idea of hidden variables in quantum mechanics. Bohm resisted accepting this. He suggested the following interesting analogy which expresses the silent dispute about priorities between the two physicists, the young American and the elder Frenchman: "If one man finds a diamond and then throws it away because he falsely concludes that it is a valueless stone, and if this stone is later found by another man who recognize its true value, would you not say that the stone belongs to the second man? I think the same applies to this interpretation of the quantum theory."[18]

In the end Bohm adopted a diplomatic way, suggested by Pauli, to recognize de Broglie's contribution while maintaining the superiority of his own work: "I have changed the introduction of my paper so as to give due credit to de Broglie, and have stated that he gave up the theory too soon (as suggested in your letter)." In addition to changing the introduction, he added "a discussion of interpretations of the quantum theory proposed by de Broglie and Rosen" and rebutted Pauli's criticisms. By the time Bohm's papers appeared in print, de Broglie had returned to his old causal approach reviving the idea of "double solution" with his assistant Jean-Pierre Vigier. They would become the most important of Bohm's allies in the hidden-variable campaign.[19]

2.3.3 Exile in Brazil

Let us now return to the outcomes of the troubles Bohm was facing in the era of McCarthyism. After Princeton refused to renew his contract, Bohm realized that it was highly unlikely that he would get another job in American academia as the witch-hunt was growing in America. He therefore looked for opportunities abroad and left the US exiling himself for the rest of his life. He left the US for Brazil, Brazil for Israel, and Israel for the UK, where he finally settled. It was after unsuccessfully attempting to work in the UK, in particular Manchester, that Bohm considered the possibility of exile in Brazil. A small group of Brazilian physicists had graduated from Princeton, among them Jayme Tiomno, who had graduated under John Wheeler and Eugene Wigner in 1950; José Leite Lopes, who had studied under Wolfgang Pauli and Josef Jauch in 1946 and was named a Guggenheim Fellow in 1949; and Walter Schutzer who had completed a Master's degree in 1949. Bohm was one of the readers of Tiomno's doctoral dissertation and

[18] For the evolution of de Broglie's ideas, see Broglie (1956, pp. 115–143). Bohm to Pauli, Oct 1951, *op. cit.*

[19] Bohm to Pauli, 20 Nov 1951, *op. cit.* (Bohm 1952b, pp. 191–193).

served as the chairman of his dissertation committee when Wigner was away. Tiomno invited Bohm to the University of São Paulo. The appointment had the recommendation of Einstein and Oppenheimer and the support at the University of São Paulo of Abrahão de Moraes, then the head of the Physics Department, and Aroldo de Azevedo, an influential geographer. Later, to keep Bohm in his Brazilian position, de Moraes asked Einstein to send letters for an eventual promotion addressed to the highest administrative levels, including President Getúlio Vargas. Bohm arrived in Brazil on October 10, 1951 and would leave for Israel in early 1955.[20]

Bohm went to Brazil an innocent and, as soon as he arrived, he wrote optimistically to Einstein, "The university is rather disorganized, but this will cause no trouble in the study of theoretical physics. There are several good students here, with whom it will be good to work." Later, however, he expressed considerable dissatisfaction: "The country here is very poor and not as advanced technically as the U.S., nor is it as clean." "I am afraid that Brazil and I can never agree." "Brazil is an extremely backward and primitive country." One month after his arrival American officials confiscated his passport and told him that he could only retrieve it to return to his native country. This profoundly changed Bohm's fate and morale. He wrote to Einstein, "Now what alarms me about this is that I do not know what it means. The best possible interpretation is that they simply do not want me to leave Brazil, and the worst is that they are planning to carry me back because perhaps they are reopening this whole dirty business again. The uncertainty is certainly very disturbing, as it makes planning for the next few years very difficult." Bohm's stay in Brazil, without a passport, changed his mood; he wrote to Melba Phillips: "Ever since I lost the passport, I have been depressed and uneasy, particularly since I was counting very much on [a] trip to Europe as an antidote to all the problems that I have mentioned." Bohm's mood oscillated also depending on the reception of his ideas and the work he had done on them. In addition, his hopes were not modest. "If I can succeed in my general plan, physics can be put back on a basis much nearer to common sense than it has been for a long time." Once he wrote, "I gave two talks on the subject here, and aroused considerable enthusiasm among people like Tiomno, Schutzer, and Leal-Ferreira, who are assistants ... Tiomno has been trying to extend the results to the Dirac equation, and has shown some analogy with Einstein's field equations." And then, "I am becoming discouraged also because I lack contact with other people, and feel that there is a general lack of interest in new ideas among physicists throughout the world."[21]

[20] Albert Einstein to Patrick Blackett, 17 Apr 1951, Albert Einstein Archives. Jayme Tiomno, interviewed by the author, 4 Aug 2003. Record number 816/51 [microfilm], Archives of the Faculdade de Filosofia, Ciências e Letras, USP. Abrahão de Moraes did not need to use the letter to President Vargas, it is published in *Estudos avançados*, [São Paulo] 21 (1994).

[21] David Bohm to Albert Einstein, Nov 1951, BP (C.10–11). David Bohm to Hanna Loewy, 6 Oct 1953, BP (C.39). David Bohm to Albert Einstein, Dec 1951, BP (C.10–11). David Bohm to Albert Einstein, 3 Feb 1954. Albert Einstein Archives. David Bohm to Melba Phillips, n.d., BP (C.46–C.48). David Bohm to Melba Phillips, 28 June 1952; ibid., [w.d.], BP (C.46–C.48). David Bohm to Hanna Loewy, 6 Oct 1953, BP (C.39).

As we will see, in Brazil Bohm continued to work consistently on the causal interpretation, kept in contact with colleagues abroad, discussed his proposal with visitors from Europe and the United States, profited from collaboration with Brazilian physicists, and published results on the causal interpretation. Thus Bohm's activities in Brazil did not reflect the pessimistic views he expressed in some of the letters he wrote at the time. They tell us more about his personality and the context. That context was conditioned by political insecurity and by the adverse reception of his proposal among his fellow physicists, a subject we discuss in the following section. Bohm would have faced many of the obstacles that he faced in Brazil elsewhere in working on a causal interpretation. Furthermore, Bohm's double identity as a Marxist and a Jew was not a liability in Brazil; on the contrary, it probably garnered him support. Brazil had been a *terre d'accueil* for Jews since the beginning of the twentieth century and following the participation of the country in World War II with the Allies, the dictatorship called *Estado Novo* (1937–1939) ceded room to a democratic regime. While political liberties were limited from 1945 to 1964, they were enough for Communists to continue to play a role in Brazilian life. Examples are the writers Jorge Amado and Graciliano Ramos, the painter Cândido Portinari, the historian Caio Prado Jr., the physicist Mário Schönberg, and the architect Oscar Niemeyer.[22]

Moreover, Bohm arrived in Brazil at a propitious time for Brazilian physics. Cesare Lattes had participated in the discovery of cosmic-ray pions in 1947 in the UK, and in 1948 in the detection of artificially produced pions at Berkeley. These achievements resonated in Brazil, especially after the role of science in the war and the production of the first atomic bomb. An alliance among scientists, the military, businessmen, and politicians was developed so as to strengthen physics in Brazil. This led to the creation of the Centro Brasileiro de Pesquisas Físicas [CBPF] in Rio de Janeiro and, in the same year that Bohm arrived in Brazil, to the creation of the first federal agency exclusively dedicated to funding scientific research, CNPq. Bohm received several grants from CNPq to develop the causal interpretation. Visits to Brazil by Ralph Schiller and Mario Bunge, both his invitees, and visits by Jean-Pierre Vigier and Léon Rosenfeld were afforded by this agency. Most of the money Bohm received went to research on cosmic rays, a field under Bohm's responsibility at USP. Nevertheless, the board of CNPq explicitly supported the development of the causal interpretation. An indication of the interest of CNPq in the research appears in the report of Joquim Costa Ribeiro, physicist and the Scientific Director of the agency, on Bohm's application for funds for Vigier[23]:

[22] For more details on Bohm's stay in Brazil, see (Freire Jr. 2005, pp. 4–7 and 10–19). On Jews in Brazil, see Rattner (1977); on Brazilian communist intellectuals, see Rodrigues (1996, p. 412). During the 1930s, however there were some obstacles to Jews in Brazil, see Saidel and Plonski (1994).

[23] On Lattes's cosmic ray work, see Vieira and Videira (2014). On Brazilian physics in the early 1950s, see Andrade (1999), Brownell (1952). Costa Ribeiro's report is in Arquivos do CNPq (Records of the Conselho Diretor, 139th meeting, 25 Feb 1953), Museu de Astronomia, Rio de Janeiro.

> I call the attention of the Board to the interest of this subject. Prof. Bohm is today on the agenda of theoretical physics at an international level owing to his theory, which is a little revolutionary because it intends to restore to quantum mechanics the principle of determinism, which seems, in a certain way, to have been shaken by Heisenberg's principle. Prof. Bohm seems to have found one solution to this difficulty of modern physics, trying to reconcile quantum mechanics with the rigid determinism of classical physics. I am not speaking in detailed technical terms, but summarizing the issue. Bohm's theory has given rise to a great debate in Europe and United States, and Prof. Vigier has expressed his willingness to come to Brazil, mainly to meet the team of theoretical physicists and discuss the problem here. This seems to me to be a very prestigious thing for Brazil and our scientific community.

2.4 Critics and Supporters of the Causal Interpretation

Bohm's approach to quantum mechanics did not pass unnoticed, as revealed by research into archives containing correspondence and papers from the early 1950s. As a matter of fact, most of the physicists who reacted to the causal interpretation were downright hostile to it, while a few of them became strong supporters, and a number of others had mixed reactions. Let us try to chart the initial reception of the causal interpretation, as it is illuminating of the dominant climate at that time towards research on the foundations of quantum physics. Wolfgang Pauli and Léon Rosenfeld were the first to react, Pauli even while the papers were in draft, as we have seen. Pauli concentrated on the physical and epistemological aspects, Rosenfeld on the philosophical and ideological ones. As Rosenfeld explained his strategy to Pauli, "My own contribution to the anniversary volume [for de Broglie] has a different character. I deliberately put the discussion on the philosophical ground, because it seems to me that the root of evil is there rather than in physics." Let us first examine Pauli's reaction.[24]

After Bohm's papers appeared in print, Pauli advanced new criticisms, which Bohm knew of before their publication. Bohm was astonished: "I am surprised that Pauli has had the nerve to publicly come out in favor of such nonsense ... I certainly hope that he publishes his stuff, as it is so full of inconsistencies and errors that I can attack him from several different directions at once." Pauli had criticized the causal interpretation for not preserving the symmetry between position and momentum representations, expressed in the standard formalism by the possibility of changing basis in the vector space through unitary transformations. He had also complained that Bohm's approach had borrowed the meaning of Ψ from quantum theory. In a letter to Markus Fierz, Pauli raised the stakes on the philosophical grounds. He observed that Catholics and Communists depended on determinism to buttress their eschatological faiths, the former in the heaven to come, the latter in paradise on earth. These references were implicitly directed to Louis de Broglie on the one hand, and to Bohm and Vigier on the other. Pauli also

[24] Léon Rosenfeld to Wolfgang Pauli, 20 Mar 1952, in Pauli and Meyenn (1996).

warned his old friend Giuseppe Occhialini about "Bohm in São Paulo and his 'causal' quantum theory." Occhialini had worked in Brazil at USP during the 1930s and continued scientific collaboration there after the war. Pauli's substantive and persistent attack on Bohm's approach was based on two issues: Since it does not have "any effects on observable phenomena, neither directly nor indirectly ... the artificial asymmetry introduced in the treatment of the two variables of a canonically conjugated pair characterizes this form of theory as artificial metaphysics." And yet, "[if the] new parameters could give rise to empirically visible effects ... they will be in disagreement with the general character of our experiences, [and] in this case this type of theory loses its physical sense." Apparently, this criticism of Pauli echoed well among physicists. "Incidentally, Pauli has come up with an idea (in the presentation volume for de Broglie's 50th birthday) which slays Bohm not only philosophically but physically as well," wrote Max Born to Einstein.[25]

Among the physicists who supported the complementarity view Rosenfeld played a singular role as a vocal and harsh critic of the causal interpretation. This role should be framed, however, by considering the following issues. While he had been Niels Bohr's closest assistant for epistemological matters, as an adept of Marxism he saw the battle against the causal interpretation as part of the defense of what he considered to be the right relationship between Marxism and science. Indeed Rosenfeld was sensitive to criticisms against complementarity coming from the Marxist camp even before the appearance of the causal interpretation. It was the beginning of what the historian of science Loren Graham called "the age of the banishment of complementarity" in the USSR, as part of the *Zhdanovshchina*, "the most intense ideological campaign in the history of Soviet scholarship." As early as 1949, following criticisms appearing in the USSR, he wrote to Bohr, "[I am] just writing an article on 'Komplementaritet og modern Rationalisme' in order to clear up the various misunderstandings which arise when one tries to mix complementarity in all possible sorts of mysticism, whatever it is a question of idealism as with Eddington and others, or about the Russian Pseudo-Marxism. These many 'ismes' are surely too tedious to [you] but [I] feel that one cannot any longer content oneself by ignoring that nonsense."[26]

Rosenfeld went so far as to deny the very existence of a controversy on the interpretation of quantum physics, writing to Bohm, "I certainly shall not enter into any controversy with you or anybody else on the subject of complementarity, for the simple reason that there is not the slightest controversial point about it."

[25] Bohm to Beck [w/d], Guido Beck Papers, Centro Brasileiro de Pesquisas Físicas, Rio de Janeiro. Beck had reported to Bohm the content of Pauli's seminar in Paris, in 1952. The criticisms were published in Pauli's contribution to the Louis de Broglie Festschrift, see Pauli (1953). Pauli to Markus Fierz, 6 Jan 1952, in Pauli and Meyenn (1996, pp. 499–502); Pauli to Giuseppe Occhialini, [1951–1952]. Archivio Occhialini 5.1.14, Università degli studi, Milan. Max Born to Einstein, 26 Nov 1953, in (Einstein et al. 1971).

[26] For Rosenfeld's biography, see Jacobsen (2012). On the debates on the quantum theory in the former USSR, see Graham (1987, p. 325 and 328). Rosenfeld to Bohr, 31 May 1949, Bohr Scient. Corr, AHQP.

For Rosenfeld, complementarity was both a direct result of experience and an essential part of quantum theory. Since complementarity implied the abandonment of determinism, as it precludes the simultaneous definition of position and momentum, which is the basis of mechanical determinism, Rosenfeld saw the causal interpretation as a metaphysical regression, writing, "determinism has not escaped this fate [becoming an obstacle to progress]; the physicist who still clings to it, who shuts his eyes to the evidence of complementarity, exchanges (whether he likes it or not) the rational attitude of the scientist for that of the metaphysician." Every good Marxist should understand that. "The latter [metaphysician], as Engels aptly describes him, considers things 'in isolation, the one after the other and the one without the other,' as if they were 'fixed, rigid, given once for all.'" Rosenfeld believed that complementarity was a dialectical achievement that had to be defended not only against Bohm's criticisms but also against Soviet critics who blamed it for introducing idealism into physics. Rosenfeld's brand of Marxism was Western Marxism rather than the Soviet variety, to use the terms used by Perry Anderson. Thus Rosenfeld was orthodox in quantum mechanics and heterodox in Marxism.[27]

Rosenfeld mobilized colleagues wherever he could to take up the fight against the causal interpretation. He appealed to his professional connections as well as companions sharing ideological ties with Marxism. He pushed Frédéric Joliot-Curie—a Nobel prize winner and member of the French Communist Party—to oppose French Marxist critics of complementarity[28]; advised Pauline Yates—Secretary of the "Society for cultural relations between the peoples of the British Commonwealth and the USSR"—to withdraw her translation of a paper by Yakov Ilich Frenkel critical of complementarity from *Nature*; asked *Nature* not to publish a paper by Bohm entitled "A causal and continuous interpretation of the quantum

[27] Léon Rosenfeld to David Bohm, 30 May 1952, Léon Rosenfeld Papers, Niels Bohr Archive, Copenhagen (hereafter RP). In the French version of the paper, Rosenfeld (1953) emphasized the idea of complementarity resulting from experience, but in the English version, reacting to criticisms from Max Born, he attenuated his stand, changing "La relation de complémentarité comme donné de l'expérience" to "Complementarity and experience." On Born's criticism, see Freire Jr. and Lehner (2010). "But in any case the relation of complementarity is the first example of a precise dialectical scheme, whose formal structure has been accurately analysed by the logicians" (Rosenfeld 1953). For Western Marxism, see Anderson (1976).

[28] "Je crois mon devoir de vous signaler une situation que je considère comme très sérieuse et qui vous touche de près. Il s'agit de vos 'poulains' Vigier, Schatzman, Vassails e tutti quanti, tous jeunes gens intelligents et pleins du désir de bien faire. Malheureusement, pour le moment, ils sont bien malades. Ils se sont mis en tête qu'il fallait mordicus abattre la complémentarité et sauver le déterminisme." He did not succeed; Joliot diplomatically kept his distance from the battle. "Autant je suis d'accord avec leurs préoccupations concernant les grands principes de la physique moderne, autant je suis d'accord avec vous sur la nécessité d'en comprendre le sens exact et profond avant de se lancer dans des discussions avec des citations qui ne sont que des planages trahissant parfois leurs auteurs." Léon Rosenfeld to Frédéric Joliot-Curie, 6 Apr 1952; Joliot to Rosenfeld, 21 Apr 1952. RP. See also Pinault (2000, p. 508).

theory;" and advised publishers not to translate one of de Broglie's books dedicated to the causal interpretation into English.[29]

Rosenfeld's correspondence shows that his campaign had wide support, as testified by Denis Gabor, "I was much amused by the onslaught on David Bohm, with whom I had a long discussion on this subject in New York, in Sept. 51. Half a dozen of the most eminent scientists have got their knife into him. Great honour for somebody so young." Positive letters came from Abraham Pais, Robert Cohen, Vladimir Fock, Jean-Louis Destouches, Robert Havemann, and Adolf Grünbaum. Pais, who had been a student of Rosenfeld in Utrecht, wrote, "I find your piece about complementarity interesting and good ... I could not get very excited about Bohm. Of course it doesn't do any good, but (with the exception of Parisian reactions) it also doesn't do any harm. I find that Bohm wastes his energy and that it will harm him personally a lot because he is moving into the wrong direction—but he needs to realize this himself, he is a difficult person." Cohen, a young Marxist physicist, wrote, "I turn to you because my own reaction to the Bohm thing and to the pilot wave revival has been quite negative, while yet I share Professor Einstein and others' uneasiness at the orthodox situation." Fock, who was the most influential and vocal supporter of complementarity in the USSR, wrote complaining that "Bohm-Vigier illness" was so widespread. Havemann, a German Communist physical chemist, sent him a paper on quantum complementarity, and Rosenfeld replied, "I read with great interest your paper and I am glad to see that our ideas are, in their essential aspects, in agreement."[30]

Guido Beck and Eric Burhop took issue with Rosenfeld's rhetoric, however, and Lancelot L. Whyte challenged him publicly over his review of Bohm's later book *Causality and chance in modern physics*. Guido Beck, one-time assistant to Heisenberg who had fled to Brazil from the Nazis, did not share a belief in the causal interpretation, but defended Bohm against Léon Rosenfeld's criticisms and insisted Bohm should be encouraged to show what his approach could achieve. Rosenfeld was sensitive to Beck's remarks. In the English translation of the original French

[29] Pauline Yates to Léon Rosenfeld, 7 Feb 1952, 19 Feb 1952, RP. Rosenfeld succeeded, "the editors stopped work on this article." The paper had been submitted to *Nature* by Harrie S.W. Massey [with whom Bohm had worked in the Manhattan Project at Berkeley]. *Nature*'s editors to Léon Rosenfeld, 11 Mar 1952, RP. "Also I sent a brief article to Massey with the suggestion that he publish it in *Nature*." David Bohm to Miriam Yevick, n.d., BP. Bohm did not keep a copy of the unpublished paper, but there is a copy of it in Louis de Broglie Papers, Archives de l'Académie des sciences, Paris. Léon Rosenfeld, "Report on L. de Broglie, La théorie de la mesure en mécanique ondulatoire." n.d. RP. The book Rosenfeld advised against translating was Broglie (1957).

[30] Denis Gabor to Léon Rosenfeld, 7 Jan 1953; Abraham Pais to Léon Rosenfeld, 15 May [1952]; Robert Cohen to Léon Rosenfeld, 31 Jul 1953; Vladmir Fock to Léon Rosenfeld, 7 Apr 1956; all papers at *RP*. For Fock's criticism of Bohm's views, see Fock (1957). Jean-Louis Destouches to Léon Rosenfeld, 19 Dec 1951; Léon Rosenfeld to Robert Haveman, 7 Oct 1957; Haveman to Rosenfeld, 13 Sep 1957; Adolf Grünbaum to Léon Rosenfeld, 1 Feb 1956; 20 Apr 1957, 3 Oct 1957; Rosenfeld to Grünbaum, 14 Feb 1956; 21 May 1957; 11 Dec 1957. All letters are at RP. On Havemann, see Hoffmann (1999).

paper Rosenfeld deleted the comparison which had been criticized by Beck. The original expression is: "on comprend que le pionnier s'avançant dans un territoire inconnu ne trouve pas d'emblée la bonne route; on comprend moins qu'un touriste s'égare encore après que ce territoire a été levé et cartographié au vingt-millième." Burhop, who was at that time organizing a meeting among Rosenfeld and Marxist or left-wing physicists, such as John Bernal, Maurice Levy, Maurice Cornforth, and Cecil Powell to discuss Rosenfeld's article, also wrote: "Incidentally the only other comment I would offer on your article was I thought perhaps you were a little cruel to Bohm. Do you think you could spare the time to write to him? He is a young Marxist. . .being victimized for his political views in the U.S."[31]

Rosenfeld went to Brazil to discuss the epistemological problems of quantum mechanics. He offered a course on classical statistical mechanics in Rio de Janeiro, published papers in Portuguese on the epistemological lessons of quantum mechanics, and gave a talk in São Paulo on complementarity. Bohm met him and reported on their exchange to Aage Bohr: "Prof. Rosenfeld visited Brazil recently, and we had a rather hot and extended discussion in São Paulo following a seminar that he gave on the foundations of the quantum theory. However, I think that we both learned something from the seminar. Rosenfeld admitted to me afterwards that he could at least see that my point of view was a possible one, although he personally did not like it." Bohm and Rosenfeld would meet each other again at the conference held in Bristol in 1957 and dedicated to foundational issues in quantum mechanics.[32]

Werner Heisenberg criticized Bohm's approach as "ideological" while Max Born initially was not impressed.[33] It was Rosenfeld who brought to his attention this interpretation, which led Born to criticize it. "I have already written my Guthrie Lecture in rough draft and have done there just what you suggest, namely, I have included the other party who prefer particles, like Bohm and the Russians which you quote (I cannot read Russian and I take it from your article.)" The common front against the causal interpretation hid disagreements, usually in private, over tactics. Rosenfeld publicly criticized Heisenberg of leaning towards idealism. Pauli and Born privately criticized Rosenfeld's mixture of Marxism with complementarity. As part of their debate, Max Born sent Rosenfeld a ten-page typed text arguing

[31] Guido Beck to Léon Rosenfeld, 1 May 1952, RP. Rosenfeld to Beck, 9 Feb 1953; Bohm to Beck, 16 Sep 1952; 31 Dec 1952; 13 Apr 1953; 5 May 1953; 26 May 1953; Guido Beck Papers, Centro Brasileiro de Pesquisas Físicas, Rio de Janeiro. Rosenfeld (1953). Eric Burhop to Léon Rosenfeld, 5 May 1952, RP. Lancelot Whyte to Léon Rosenfeld, 8 Apr 1958; 14 Mar 1958; 22 Mar 1958; 27 June 1958; Rosenfeld to Whyte, 17 Mar 1958, RP. Rosenfeld to Whyte, 28 May 1958, is in Lancelot L. Whyte Papers, Department of Special Collections, Boston University. The disputed papers were Rosenfeld (1958) and Whyte (1958).

[32] Bohm to Aage Bohr, 13 Oct 1953, ABP; Rosenfeld (1954) and Rosenfeld (2005). For the Bristol conference's proceedings, see Körner (1957).

[33] Heisenberg's criticism was published in the widely read and translated *Physics and Philosophy* (Heisenberg 1958). However, Heisenberg did not pursue the combat. In the late 1950s, "[he] had written more than enough on the subject and had, he said, 'nothing new to say'" (Carson 2010, p. 92).

approach is to assume "pure chance" which may propagate from level to level, but which is never explained in any way, except in terms of itself.

2.4.1 Supporters

If the critics set the tone in the reception of Bohm's proposal, supporters were no less active, and included attempts to further develop the initial papers. The most important adherents came from France with Louis de Broglie, who reconverted to his early ideas of a deterministic description of quantum systems, and Jean-Pierre Vigier, his young assistant. The importance of de Broglie's support may be inferred from the fact that Rosenfeld and Pauli chose to criticize Bohm's approach in their contributions to the de Broglie *Festschrift*, while the French Nobel Prize laureate was cogitating about the implications of Bohm's papers. Eventually de Broglie abandoned the complementarity view in the quest for a causal interpretation of quantum physics.

The influence of de Broglie's reconversion to his earlier ideas can be seen in terms of the weight Rosenfeld attributed to it in a later letter to Niels Bohr: "This comedy of errors [the attempt to develop a 'theory of measurement' based on the 'causal interpretation' of quantum mechanics] would have passed unnoticed, as the minor incident in the course of scientific progress which it actually is, if it had not found powerful support in the person of L. de Broglie, who is now backing it with all his authority." In fact, de Broglie did not directly support Bohm's proposal, instead he pleaded for what he called the "double solution," which would remain as a mathematical suggestion and not a physical model for a causal interpretation. From 1953, through Vigier's visit to Bohm in Brazil, when their collaboration was already underway, de Broglie reminded Bohm of their differences: "You know our viewpoints are not entirely the same because I do not believe in the physical existence of the Ψ wave, which seems only to be the representation—rather subjective—of probabilities. By the way, when we have more than just one particle the Ψ wave must be represented in the configuration space with more than three dimensions and its non physical character appears then absolutely evident."[41]

Vigier brought momentum to the causal interpretation. He was influential among the French communists and in the Cold War times of the early 1950s he mobilized young Marxist physicists to work on the causal interpretation. With de Broglie and Vigier, the Institut Henri Poincaré became the world headquarters of the causal interpretation. A testimony from Jean-Louis Destouches reveals the isolation of complementarity in the French milieu: "The young people received with enthusiasm Bohm's work, which corresponds to the philosophical trends supporting their positions: Thomistic realism, Marxist determinism, Cartesian rationalism. I am

[41] Rosenfeld to Bohr, 21 Oct 1957, BSC, reel 31, AHQP, reel 31, cited in (Osnaghi et al. 2009, p. 101). Louis de Broglie to Bohm, 29 March 1953, Louis de Broglie Papers, Box 7, Archives de l'Académie des sciences, Paris.

almost the only one here to support Bohr's quantum interpretation."[42] Bohm also gathered support from the US, Argentina, and Brazil, through Hans Freistadt, Ralph Schiller, Mario Bunge, and Jayme Tiomno.[43]

Bohm considered the papers he wrote with Tiomno and Schiller and with Vigier to be the main achievements of the causal program in the early 1950s. With Vigier, Bohm answered Pauli's objection that he had included an arbitrary element in the causal interpretation, by using a ψ function that satisfied Schrödinger's equation. Bohm had tried to solve the issue by himself without success, while De Broglie and Vigier were cognizant of the problem in 1952. In 1954, Bohm and Vigier were able to prove that under certain general conditions any function could become a solution of the Schrödinger equation. To achieve this, they used an analogy between Bohm's approach and the hydrodynamic model suggested by Erwin Madelung in 1926, which embedded microscopic quantum particles in a subquantum medium with random fluctuations. Thus, the "molecular chaos", an idea Bohm had abandoned after his discussions with Pauli, came back into his work with Vigier.[44]

Jayme Tiomno had met Bohm at Princeton while he was doing his PhD under John Wheeler on weak interactions. Ralph Schiller had worked on gravitation in his PhD under the supervision of Peter Bergmann at Syracuse University and had gone to Brazil to be Bohm's research assistant. With Tiomno and Schiller, Bohm enlarged the scope of his model to include spin, although via analogy with Pauli's equation and not through a relativistic treatment of electrons. Tiomno, however, was not an adherent of the causal interpretation. He worked with Bohm looking for the consequences of extending Bohm's model to other fields of physics, but did not share its philosophical assumptions concerning causality. The Argentinian Mario Bunge, who had been supervised by Guido Beck at La Plata University, spent a year working with Bohm in Brazil, but nothing came of it. Bunge attacked the difficult problem of the "Bohmization" of relativistic quantum mechanics and the elimination of infinities in quantum electrodynamics. Bunge had studied physics in order to develop a better philosophy of the subject, later developing a successful career in

[42] "Les jeunes gens ont accueilli avec enthousiasme le travail de Bohm qui correspond à toutes les tendances philosophiques qui les animent: réalisme thomiste, déterminisme marxiste, rationalisme cartésien. Je suis donc maintenant à peu près le seul ici à soutenir encore l'interprétation quantique de Bohr." Jean-Louis Destouches to Léon Rosenfeld, 19 Dec 1951, RP.

[43] Freistadt worked both on the philosophical and technical aspects of the causal interpretation; on his activities on this subject in the context of American physics, see (Kaiser 2012, pp. 20–22). For Freistadt's works, see Freistadt (1953, 1955, 1957). Schiller, Bunge, and Tiomno worked with Bohm in Brazil and their cases are discussed in this chapter.

[44] For the role Bohm attributed to those papers, see Bohm (1981, pp. 114 and 118, notes 11 and 12), Bohm and Hiley (1993, p. 205), Pauli (1953) and Bohm (1953a); a simplified and shortened version of this paper was presented in New research techniques in physics (1954, pp. 187–198). "C'était aussi un des problèmes décisifs que Bohm n'avait pas traité dans ses papiers de 1952." Jean-Pierre Vigier, interviewed by the author, 27 Jan 1992 (Bohm and Vigier 1954, 1958; Broglie et al. 1963). A lacuna in the history of physics in the twentieth century—an analysis of the activities of the de Broglie-Vigier group—is now being filled by the works of Vals (2012) and Besson (2011).

the philosophy of science in Canada. In the mid-1960s, disenchanted with the hidden variable interpretation, he gave up on it, accepted indeterminism as part of physics theories, and focused his criticisms of quantum mechanics on the role played by observation in the complementarity view.[45]

The collaboration between Bohm and Vigier was aided by an irony typical of the Cold War. Had Bohm remained in the U.S., Vigier might not have been able to visit and work with him. Vigier had made a name for himself in the Communist Party in France and, as Jessica Wang has pointed out in writing about the "age of anxiety" in American history, "in addition to refusing passports to American scientists, the State Department also restricted the entry of foreign scientists with left-wing political ties into the United States . . . Scientists from France, where the Left was particularly strong, had an especially hard time. As much as 70–80 % of visa requests from French scientists were unduly delayed or refused." However, supporters who just applauded the causal interpretation on ideological grounds without trying to develop it did not help Bohm much; apparently, this was the case of French astrophysicist, and Marxist, Évry Schatzman.[46] After all, the causal interpretation needed to win the technical challenges promised by Bohm himself.

2.4.2 *Mixed Reactions*

Not all reactions were clear-cut criticisms or support. The contributions of two people—Einstein and Feynman—were especially meaningful for Bohm. Einstein, the iconic critic of complementarity, had influenced Bohm while at Princeton to see quantum theory as an incomplete theory. On political grounds, Einstein was an enduring supporter of Bohm against McCarthyism. When the causal interpretation came out, however, he did not support it. "Have you noticed that Bohm believes (as de Broglie did, by the way, 25 years ago) that he is able to interpret the quantum theory in deterministic terms? That way seems too cheap to me," was his comment in a letter to Max Born. Moreover, he wrote a paper to a *Festschrift* for Max Born saying that Bohm's model led to the unacceptable consequence that particles in stationary states, such as an electron in a hydrogen atom, were at rest. Einstein may have used the opportunity to distance himself from the widespread opinion that he was stubbornly attached to determinism. "For the presentation volume to be dedicated to you, I have written a little nursery song about physics, which has startled Bohm and de Broglie a little. It is meant to demonstrate the indispensability of your statistical interpretation of quantum mechanics, which Schrödinger, too, has recently tried to avoid. [. . .] This may well have been so contrived by that same 'non-dice-playing God' who has caused so much bitter resentment against me, not

---

[45] Bohm et al. (1955) and Bohm and Schiller (1955). On Tiomno, see Freire Jr. (1999, p. 95). Mario Bunge to the author, 1 Nov 1996, and 12 Feb 1997.

[46] Wang (1999, p. 279) and Schatzman (1953).

only amongst the quantum theoreticians but also among the faithful of the Church of the Atheists." Einstein, however, was kind enough to let Bohm read this paper before its publication and accepted Bohm's request to publish his reply in the same volume. Bohm showed that an adequate use of his model, including changes in the system due to measurements, could save it.[47]

Bohm's main hope for an ally among the foreign visitors he met in Brazil was Richard Feynman, who had been his colleague at Berkeley and spent his sabbatical year in 1951 at the Centro Brasileiro de Pesquisas Fisicas (CBPF) in Rio de Janeiro. Bohm liked Feynman's initial reaction: "At the scientific conference in Belo Horizonte, I gave a talk on the quantum theory, which was well received. Feynman was convinced that it is a logical possibility, and that it may lead to something new." Thus to Hanna Loewy:

> Right now, I am in Rio giving a talk on the quantum theory. About the only person here who really understands is Feynman, and I am gradually winning him over. He already concedes that it is a logical possibility. Also, I am trying to get him out of his depressing trap down long and dreary calculations on a theory [procedures of renormalization in Quantum Field Theory] that is known to be of no use. Instead maybe he can be gotten interested in speculation about new ideas, as he used to do, before Bethe and the rest of the calculations got hold of him.

This letter is evidence of how disconnected Bohm was at the time with the main themes of research on the physics agenda as he was criticizing as "dreary" the kind of calculations which were exciting not only Feynman and Hans Bethe, but almost all physicists involved with quantum field theories. Bohm's hopes about Feynman were unfounded as "in his physics Feynman always stayed close to experiments and showed little interest in theories that could not be tested experimentally" (Schweber 2005). The only reference Feynman made to hidden variables as a result of his Brazilian sabbatical was a mention, as a possible avenue for the development of theoretical physics. Furthermore, it came out in a general paper published in a Brazilian science journal. That could scarcely nourish Bohm's hopes.[48]

2.4.3 The Old Guard

From the old guard of quantum theory, let us now look at the cases of Niels Bohr, Erwin Schrödinger, and John von Neumann. Bohm particularly looked for reactions from Bohr and von Neumann, which is no surprise given that their views were the targets of his hidden variable interpretation. Bohm received the first report of Bohr's views through the American theoretical physicist Arthur Wightman, who

[47] Einstein to Born, 12 May 1952 and 12 Oct 1953 (Einstein et al. 1971; Einstein 1953; Bohm 1953b). For Einstein's stances, see Paty (1993, 1995).

[48] On Feynman in Brazil, see Lopes (1990) and Mehra (1994, pp. 333–342). David Bohm to Hanna Loewy, [w/d], 4 Dec 1951, BP (C.38) (Feynman 1954). For the role played by Feynman, Bethe, and the renormalization calculations in physics at that time, see Schweber (1994).

was then in Copenhagen. As Bohm wrote to Melba Phillips: "the elder Bohr [Niels] didn't say much to Art[hur] Wightman, but told him he thought it 'very foolish.'" The distinction between the "two Bohrs" was particularly important as Bohm had met the younger, Aage Bohr, in the spring of 1948 while at Princeton,[49] and was pleased to discover that Aage Bohr was more sympathetic to the causal interpretation than his father, Niels Bohr. As Bohm reported to Wightman, "I am glad that Aage Bohr admits its logical consistency."[50] Indeed, the younger Bohr [Aage] was more receptive to Bohm's proposal—"it would be nice to meet some time and discuss things, also the epistemological problems"—while he respected the value of the complementarity view: "there it seems to me that the very fact that one can give a logically consistent non-deterministic description of natural phenomena is a very great lesson which gives one a much freer way of thinking about things." The conversation continued and Bohm explained to Aage Bohr the two assumptions he considered to be "unnecessarily dogmatic" in the principle of complementarity: (1) "that the quantum of energy will remain indivisible and unanalyzable at all levels . . .", and (2) "that the statistical laws of quantum mechanics are final, in the sense that no deeper causal laws will ever be found . . .". [51] As for the elder Bohr, there was never any sign of empathy towards the causal interpretation, even after they had the opportunity of having personal conversations, for Bohm visited Copenhagen twice, in 1957 and 1958. As Bohm recorded 5 years later, Niels Bohr had "expressed especially strong doubts that such a theory [causal interpretation] could treat all significant aspects of the problem of *indivisibility* of the quantum of action" (Bohm 1962, p. 363).

However, the main interest of Aage Bohr in the exchange with Bohm was not related to the epistemological issues in quantum mechanics, but to Bohm and David Pines' work on plasma, metals as electron gas, and collective variables. Aage Bohr was extending the collective variable approach to his own work on nuclear physics. He sent Bohm a preprint of a paper written with Ben Mottelson, and observed, "I would be also very interested in any comments from you on this, admittedly still rather primitive attempt of ours to develop a more comprehensive and self-consistent treatment of a many-body system such as the nucleus. In some ways, there are parallelities, I think, to your treatment of the electron gas, even though the forces and the geometry are quite different." Bohm, who was still in Brazil, was interested in Aage Bohr's work on nuclear physics, comparing it with results from the Van der Graaf accelerator being built in São Paulo. This would produce slow neutrons with very accurately determined energy.[52]

[49] David Bohm to Melba Phillips, n.d., BP (C.46–C.48). Letter from Aage Bohr to the author, 17 Oct 1997.

[50] David Bohm to Arthur Wightman, [1953], Niels Bohr Archive, Copenhagen.

[51] Aage Bohr to David Bohm, 3 Oct 1953; Bohm to Aage Bohr, 13 Oct 1953, emphasis in the originals, Aage Bohr Papers, Niels Bohr Archive, Copenhagen.

[52] Aage Bohr to David Bohm, 3 Oct 1953; Bohm to Aage Bohr, 24 Sep 1953, ibid.

Thus in the late 1950s when Bohm was already in Israel and Pines visited Copenhagen, Bohm wrote to Aage Bohr. "I would very much like to spend [the summer] in Copenhagen and to work with Pines on plasma theory, on which subject both of us have interesting new ideas."[53] Bohm visited Copenhagen between 08 August and 29 September 1957 and then from 07 July 1958 to 13 September 1958. The influence of Pines and Bohm's plasma work on nuclear physics in Copenhagen was acknowledged by Ben Mottelson, the American physicist living in Copenhagen who went on, with Aage Bohr and Leo Rainwater, to win the 1975 Physics Nobel Prize for "the discovery of the connection between collective motion and particle motion in atomic nuclei and the development of the theory of the structure of the atomic nucleus based on this connection." In the Nobel acceptance speech, Mottelson recalled that: "It was a fortunate circumstance for us that David Pines spent a period of several months in Copenhagen in the summer of 1957, during which he introduced us to the exciting new developments in the theory of superconductivity. Through the discussions with him, the relevance of these concepts to the problem of pair correlations in nuclei became apparent."[54]

As for von Neumann, Bohm considered his reaction a little better than Bohr's. Bohm reported that "von Neumann thinks my work correct, and even 'elegant,' but he expects difficulties in extending it to spin." Von Neumann probably interested himself in Bohm's work in the 1950s while revising the English translation of his *Mathematische Grundlagen der Quantenmechanik* (1932), in which his famous proof appeared. To his publisher, he explained the difficulties, "the text had to be extensively rewritten, because a literal translation from German to English is entirely out of question in the field of this book. The subject-matter is partly physical-mathematical, partly, however, a very involved conceptual critique of the logical foundations of various disciplines." In a recent analysis, the philosopher Michael Stöltzner suggested that "von Neumann could accept Bohm's proposal as an interesting model, but not as a promising interpretation."[55] As for Schrödinger, in spite of criticisms of the complementarity view, his insistence on the wave function ontology of the quantum world and absence of interest in the recovery of determinism hampered the dialogue with those, such as Bohm, Vigier, and de

---

[53] Bohm to Aage Bohr, 18 Dec 1956, ibid. Aage Bohr replied, "I hope very much you can manage to come here next summer, when we also expect Pines to be here. We should, of course, be very pleased if you would tell us a little about plasma theory." Aage Bohr to Bohm, 26 Jan 1957, ibid. For the next summer, Aage mentioned they wanted to hear Bohm on superconductivity, reflecting the interest arose by the work of Bardeen and colleagues, Aage Bohr to Bohm, 25 Oct 1957, ibid.

[54] Visitors records, Niels Bohr Archive. "The Nobel Prize in Physics 1975", http://www.nobelprize.org/nobel_prizes/physics/laureates/1975/. "Ben R. Mottelson - Nobel Lecture", http://www.nobelprize.org/nobel_prizes/physics/laureates/1975/mottelson-lecture.html, on page 240. Both information accessed on 11 Jan 2014 (Bohr et al. 1958).

[55] J. von Neumann's reaction is in David Bohm to Wolfgang Pauli, [Oct 1951], in Pauli and Meyenn (1996, pp. 389–394). John von Neumann to H. Cirker, [President of Dover Pub], 3 Oct 1949. John von Neumann Papers (Box 27, Folder 8), Library of Congress, Washington, DC (Von Neumann 1955; Stöltzner 1999).

Broglie, who worked with a world populated by particles in a deterministic framework.[56]

2.4.4 *Bohm's Proposal and Philosophers of Science*

Bohm's causal interpretation also contributed to enticing philosophers of science to enter the quantum debate. Indeed they were never absent, as in the early stages of the debate some philosophers, such as Karl Popper, Hans Reichenbach, Gaston Bachelard, Grete Hermann, and Alexandre Kojève had ventured into this field.[57] Now, with the reheated controversy, there was new fuel for the philosophy of science. However, while in the 1930s philosophers mostly produced works more of an epistemological nature, in the sense of providing a critical analysis of an existent scientific theory, now they divided themselves along the same lines as the physicists. Some were sympathetic towards Bohm's enterprise, as in the case of Paul Feyerabend, who praised Bohm's *Causality and Chance* as containing "an explicit refutation of the idea that complementarity, and complementarity alone, solves all the ontological and conceptual problems of microphysics." Others aligned with Bohr's point of view, notably Norwood Hanson, who maintained that "when an interpretation of a theory has been as successful as this one [Copenhagen interpretation] has been, there is little practical warrant for the 'alternative interpretations' which have, since Bohm, been receiving prominence." And yet, there were cases, such as Bachelard, who retired from the debate as it became heated because de Broglie reconverted to the deterministic description of the quantum phenomena. Since then, the debate on the foundations of quantum physics has been an attractive topic for philosophers of science and deserves further historical research.[58]

[56] Schrödinger (1953). On Schrödinger's philosophical views, see Michel Bitbol's comments in Schrödinger and Bitbol (1992, pp. 140–141) and Bitbol (1996a). In private, Schrödinger kept high his fight against the complementarity view, as in this letter to Max Born, on October, 10, 1960: "The impudence with which you assert time and again that the Copenhagen interpretation is practically universally accepted, assert it without reservations, even before an audience of the laity–who are completely at your mercy–it's at the limit of the estimable ... Have you no anxiety about the verdict of history?" (Moore 1989, p. 479).

[57] Popper and Bartley (1982), Reichenbach (1944), Bachelard (1934), Hermann et al. (1996), and Kojève and Auffret (1990).

[58] Feyerabend (1960) and Hanson (1959). On Bachelard, see Freire Jr. (2004a). An illustrative example of how attractive this topic may be is Mara Beller's criticism of Kuhn's paradigms (Beller 1999). In her view, the appearance of the notion of paradigm is related to the quantum controversy. I discussed these issues in Freire Jr. (2014c). Popper, who was interested in the foundations of quantum mechanics from the 1930s, only became an active protagonist in the quantum controversy in the early 1980s. See Freire Jr. (2004b) and Popper and Bartley (1982).

2.5 Waning Causality and Disenchantment with Communism (Late 1950s–Early 1960s)

In 1955 David Bohm left Brazil for Israel and in 1957 moved again, this time to the UK, first to Bristol and then to the Birkbeck College in London. Bohm's main motivation for leaving Brazil was the possibility of travelling abroad—Europe—in order to discuss and defend his causal interpretation for quantum mechanics. Life, however, brings unexpected turns and he went on to experience a major intellectual change from 1956 on. Politically he broke his ideological ties with Marxism, in philosophical terms he weakened his beliefs on the centrality of causality for science and society, and in the scientific arena he gave up the causal interpretation discouraged with its developments. All these changes were not unrelated, as we will argue. He began to look for new research directions but they only would coalesce in the late 1960s. The peregrination to Israel was tainted by the worsening of his situation with the US government related to the confiscation of his passport. In order to travel to Israel he applied for Brazilian citizenship, which led to the loss of his American citizenship. Only in the UK, in 1960, would he face the obstacles to get it back. This period of transition did not only bring unpleasant experiences. He made new and lasting acquaintances; got married to Sarah Woolfson, with whom he would spend the rest of his days, in Israel; and met Basil Hiley, who would become his enduring collaborator on the new perspectives of research, in Britain. In addition, he found two graduate students deeply interested in touching upon the foundations of quantum mechanics, which, as we already saw, had become the intellectual pet of David Bohm. They were Yakir Aharonov in Haifa, and Jeffrey Bub in London.

2.5.1 Break with Communism

As one would expect, Bohm was very sensitive to the reactions to his reinterpretation of quantum theory in terms of hidden variables. In particular, he paid attention to the way Marxists, physicists and philosophers, reacted to it, which is no surprise given Bohm's Marxist background. He made much of the French work, no doubt in part because of Vigier's Marxist engagement: "I have heard from someone that in a debate on causality given in Paris, when our friend Vigier got up to defend causality, he was strongly cheered by the audience (which contained a great many students). I would guess that many of the younger people in Europe recognize that the question of causality has important implications in politics, economy, sociology, etc." The connection appeared so obvious to Bohm that he complained when fellow travelers like the American physicist Philip Morrison did not support him. "This type of inconsistency in Phil [Morrison] disturbs me. He should be helping, instead of raising irrelevant obstacles." And he wondered why the causal interpretation had appeared in the West and not in the USSR and why

Soviet physicists did not join him. "I ask myself the question 'Why in 25 years didn't someone in USSR find a materialistic interpretation of quantum theory?' … But bad as conditions are in U.S., etc, the only people who have thus far had the idea are myself in U.S., and Vigier in France."[59] If it is hard for historians to chart the precise influence on Bohm's shift towards the causal interpretation, there is no doubt that the influence of Marxism was effective in supporting the causal interpretation, especially among the French team led by Vigier, and that such support was influential on Bohm himself, albeit weaker than Bohm had hoped for. The unfulfilled expectations were mainly related to the USSR, as evidenced in a letter he sent in 1955 to the American physicist Melba Phillips[60]:

> At times I feel discouraged about the state of the world. A thing that particularly strikes home to me is the report I got from Burhop (confirmed by others) on Russian physicists. Apparently, they are all busy on doing calculations on electrodynamics according to Feynman, Dyson, et al. Their orientation is determined strongly by the older men, such as Fock and Landau, who in addition to their training, are influenced by the fear of a sort of "Lysenko affair" in physics. The typical physicist appears to be uninterested in philosophical problems. He has not thought much about problems such as the re-interpretation of qu. mchs, but tends to like the word of the "big-shots" that ideas on this such as mine are "mechanistic". Actually, the standard procedure is just to label such a point of view, and then most people accept the label without even bothering to read about such questions. There are some philosophers in Moscow who criticized the usual interpretation, but they haven't had much influence on the physicists. All in all, the situation in Soviet physics doesn't look very different from that in Western physics. It is disappointing that a society that is oriented in a new direction is still unable to have any great influence on the way in which people work and think.

What Bohm did not realize was that part of the support for complementarity and resistance to the causal interpretation was also based in commitment to Marxism. This was the case of Rosenfeld, as we have already seen, and also of Vladimir Fock, who supported Bohr's views in the USSR basing his position on dialectical materialism. From 1957 on, after Stalin's death and the ideological thaw in the USSR, Fock would become an outspoken defender of Bohr's views. In addition, a number of Soviet physicists, such as Blokhintsev and Terletsky, while being critics of complementarity were not supporters of the causal interpretation either. Indeed, the former became a leader in the defense of the ensemble interpretation, which says quantum theory does not describe states of single systems but only an equally prepared ensemble of them.[61] The latter devoted his energies to attempts to include non-linearities in the standard quantum mechanics, an approach which resonated with de Broglie's proposal of a "double solution." Indeed, we may see in hindsight,

---

[59] David Bohm to Miriam Yevick, 5 Nov 1954, BP. David Bohm to Melba Phillips, n.d. BP. David Bohm to Miriam Yevick, 7 Jan 1952, BP.

[60] Bohm to Melba Phillips, 18 March 1955, BP (C49). Andrew Cross (1991) saw Bohm's work as just a reflection of the ideological Marxist climate of the time; thus he missed the fact that the quantum controversy continued even when that climate faded. For the critique of this position, see (Freire Jr. 1992).

[61] For a description of the ensemble interpretation, see Home and Whitaker (1992).

the relationship between Marxism and the spectrum of stances in the quantum controversy was not one-to-one. Instead, Marxism influenced both critics as well as defenders of complementarity. This multi-sided relationship should be no surprise as when speaking of Marxism in the twentieth century it is better to use the plural Marxisms than the singular Marxism.[62]

At any rate in 1955 Bohm could still think that Soviet and Marxist physicists should support his causal interpretation in a stronger manner. The vicissitudes of the times, however, would make such a matter meaningless for Bohm. By late 1956 or early 1957, a crisis point in his commitment to Marxism was reached, triggered by Khrushchev's report on Stalin's crimes and by the invasion of Hungary by Soviet troops. Bohm's break with Communism, while he was visiting Paris to work with Jean-Pierre Vigier and Louis de Broglie, was witnessed by the physicist Jan Meyer and is well-recorded in two long letters to Melba Phillips. How dramatic Bohm's involvement was with these critical events may be seen from the following fragments[63]:

> It is clear from the above that what is needed in the left-wing movement today is a certain measure of disengagement from Russia. Russia has made an enormous number of errors. ... This raises the question of the probable future of the C.P.'s [Communist Parties] throughout the world. ... As soon as a man opposed the direction of the C.P. he became a traitor guilty of the most heinous crimes. Confessions were manufactured and extorted on a large scale. The truth had nothing to do with the case; what was published was only what would be convenient for the interests of the gov't. This was a direct perversion of the principle that dialectical materialism should be scientific and objective. Perhaps some people said that false confessions served the interests of a "larger truth". Similarly, Humanité [the official newspaper of the French Communist Party] still publishes lies about Hungary; quite cynically since the truth is evident. It is clear also that the Russian gov't publishes whatever it thinks is convenient about world affairs. Perhaps they have already ceased to lie consciously, and they may be only deceiving themselves.

Thus by 1958 Bohm's relation to Marxism came to an end. It had lasted from the late 1930s, when he approached the US Communist Party at Berkeley, in the wake of the Great Depression and the rise of Nazism in Europe, to 1956–1957 following Khrushchev's report and the invasion of Hungary by the USSR. That history had cost him the right to live in his home country and would still cause a lasting battle to recover his American citizenship. He lived the main political passions of his times and was a man trapped in the Cold War storm. And yet, his history, including adhesion to and a later break with Communism, was not exceptional, indeed it was

---

[62] On Marxism and the controversy over the interpretation of quantum theory, see Freire Jr. (2011c). See also Graham (1987, pp. 320–353), on Fock and Blokhintsev; Kuzemsky (2008), on Blokhintsev; Pechenkin (2012), on the early ensemble interpretation in the USSR and in the US; Forstner (2008), on Bohm; Jacobsen (2007, 2012), on Rosenfeld; Kojevnikov (2011), on ensembles; Pechenkin (2013), on Mandelstam; Kojevnikov (2004), on Soviet physics, and Besson (2011), on Vigier.

[63] For an account of those events, see Gaddis (2005, pp. 83–194). Jan Meyer, conversation with Olival Freire, 30 January 1997; Bohm to Phillips, undated, BP (C49). This rupture is also noted by Kojevnikov (2002, p. 191) and Peat (1997, p. 178).

typical of the generation of intellectuals in the mid-twentieth century, around World War II.[64]

2.5.2 *Causality Relativized*

After Bohm's break with Communism he made few references to Marxist ideas. However, one of them is very meaningful for the philosophy of science as it concerns the role of determinism in society. It appears in a letter to the American artist Charles Biederman, with whom he exchanged a large correspondence, over 4,000 pages between March 1960 and April 1969, now being edited by the Finnish philosopher Paavo Pylkkänen. The reference came in the middle of a discussion about determinism, on which I will comment later: "For they [Marxists] felt that by studying the evolutionary process of the past, they could pick out the main direction in which history was moving. They became so attached to their theories that they were unable to review their own role objectively, or to admit new and unexpected developments not fitting into these theories." How much Marx's historical materialism depends on adopting determinism in history is debatable, however. For the purposes of our analysis, nonetheless, it is enough to consider that Bohm's rupture with Marxism may have destroyed his general belief in determinism as a feature of society and its history.[65]

The connection between the break with Marxism and abandonment of determinism in science, particularly in physics, and not only in society, in Bohm's path is a guess, albeit a plausible one. The best evidence of how and when Bohm shifted his focus away from the philosophical priority for causal laws in physics can also be found in the correspondence with Biederman. The intellectual turn was acutely noted by Pylkkänen, "here we have Bohm, who is internationally known as a defender of a deterministic interpretation of the quantum theory, and thus for many a defender of strict determinism in nature, arguing strongly for the objective existence of properties such as contingency, chance, determinism, etc. Of course, Bohm does this already in *Causality and Chance*, but here the point is made more vividly, given that Bohm is defending the role of indeterminism rather than questioning it, as he most famously did in his 1952 papers."[66]

From this extensive correspondence between Bohm and Biederman, I have selected fragments from a few letters to provide the reader with an idea of the issues at stake. In his very first letter, in 1960, Biederman was clear-cut in his

[64] Ory and Sirinelli (2004), Hobsbawm (2011), Chaps. 11 and 14, Caute (1967).

[65] Bohm to Biederman, 2 February 1961, (Bohm et al. 1999, p. 95). As the historian Eric Hobsbawm remarked, at least two features of Marxism should not be abandoned unless one gives up historical materialism as a way to change the world: (a) the triumph of socialism is the logical end of all historical evolution until the present, and (b) socialism marks the end of prehistory as it cannot and will not be an antagonistic society (Hobsbawm 1997, Chap. 11).

[66] Paavo Pylkkänen's statement is in the introduction of Bohm et al. (1999, p. xix).

defense of determinism: "To explain my interest in your book [*Causality and Chance*]. To put it briefly, the notion of indeterminism has always seemed contrary to experience, which, even after reading your very fine book, I cannot accept even as an eventually limiting case." And yet, "I sympathize with your belief that a deeper penetration will reveal a nature of causality. But there is the possibility that this will also dispel the basis for the present 'lawless' view of nature and, rather than make it a limited case, will dispense with it entirely." Bohm's answer to Biederman is that time implies a certain ambiguity. "Thus, there is some ambiguity in past and future. We experience this ambiguity in certain ways directly. For when we try to say 'now,' we find that by the time we have said it, the time that we meant is already past, and no longer 'now.'" He continues, citing an example closer to physics, "and if we try to do it with clocks, so as to be more precise, quantum theory implies that a similar ambiguity would arise because of the quantal structure of matter. In fact, there is no known way to make an unambiguous distinction between past and future." Thus, "it becomes impossible that the past shall completely determine the future, if only because there is no way to say unambiguously what the past really was until we know its future." As Biederman might have compared that letter with the book which was the catalyst of their correspondence, Bohm anticipated this, "as you may perhaps have noticed, my ideas on determinism and indeterminism have developed since I wrote *Causality and Chance*, although what I now think about these questions was, to a considerable extent, implicit in the point of view expressed in the book." His conclusion, in short, is that "neither determinism nor indeterminism (causality or chance) is absolute. Rather, each is just the opposite side of the whole picture," and that "in the question of determinism vs. indeterminism, there is as I have said, a necessary complementary relation of the two ideas."[67]

Bohm's reference to *Causality and Chance* deserves some attention. The philosophical convictions he held while writing this book weakened the prominence he attributed to causal laws in science, as he concluded that causal and probabilistic laws should be accorded the same philosophical status. Also noteworthy is the fact that these philosophical studies were motivated, at least partially, by his ideological commitment to Marxism. For our purposes, however, the most meaningful remark in his letter to Biederman was the comment that "my ideas on determinism and indeterminism have developed since I wrote *Causality and Chance*." This can be seen as a clue to the kind of change Bohm experienced after writing the book and before the first letter from Biederman. The book was finished in 1955 while he was in Brazil, then he left for Israel, visited Paris and Bristol, and eventually settled in London, and the single most relevant change he experienced during this time was his break with Marxism.

[67] Biederman to Bohm, 6 March 1960; Bohm to Biederman, 24 April 1960; both in Bohm et al. (1999, pp. 3–4 and 8–19).

2.5.3 *Abandonment of the Causal Interpretation*

Throughout the 1950s Bohm worked consistently on the development of the causal interpretation. Two directions of research were particularly prioritized. The first was to develop a relativistic generalization of the initial approach and was considered by him and his supporters to be the main goal of their work. "The day that we defeat the Dirac equation, we are going to have a special victory party, with a case of champagne," he confessed to a correspondent in the mid-1950s. Till today this remains an unreached goal, considered by many to be a shortcoming of the causal interpretation. The second direction was related to Bohm's promises that his approach, conveniently modified, could approach the domain of intra-nuclear particles, which he labeled in 1952 as the domain of 10^{-13} cm distances. Bohm joined a collaboration between the French team led by Vigier and Japanese physicists which included the Nobel Prize winner Hideki Yukawa as its most prominent name. They looked to classify the myriad of recently discovered intra-nuclear particles through representing them as extended bodies in space-time and relating the number of degrees of freedom from these models to their quantum numbers.[68] While this approach was neither a clear-cut extension of the 1952 model of electrons nor based on the requirement of causality, before the quark model, in the late 1950s, this was an exciting adventure in a new physical territory. And yet, if philosophically it was not entirely based on the causal interpretation it was not strange to it. Indeed its philosophical assumptions were realism and the primacy of descriptions in the arena of space-time instead of abstract mathematical spaces.

In the late 1950s however, Bohm's research departed from that of his collaborators like Vigier and de Broglie. While they persevered in their research into the causal interpretation, Bohm gave it up. A number of factors may have played a role in his decision, including discouragement over the limited response to these ideas and, as he would acknowledge later, "because I did not see clearly, at the time, how to proceed further, my interests began to turn in other directions" (Bohm 1987, p. 40).[69] An inspection of the list of Bohm's publications related to the foundations of quantum mechanics suggests the late 1950s and the early 1960s as the time when this abandonment occurred. Indeed, while in the 1950s he wrote an average of 1.6 papers per year on these topics, in the 1960s and 1970s this figure drops to half, reversing in the 1980s to increase to 2.2 papers per year. Closer inspection reveals however, that most of the papers from the 1960s were related to the reaction to external challenges such as the appearance of John Bell's paper, a subject we will deal with in Chap. 7, or new perspectives he was adopting (Freire Jr. 1999, pp. 167–170). As a matter of fact from 1960 on Bohm gradually began to search for a new

---

[68] Bohm et al. (1960a, b); a review of the state of the art of this research is Broglie et al. (1963). Virgile Besson is studying the French side of the mentioned collaboration while Pablo Ruiz de Olano is studying the Japanese side.

[69] A balance of how far Bohm went with hidden variable theory is provided by Bohm (1962). See particularly pp. 359–363 for his evaluation of the criticisms it suffered.

approach to the interpretation of quantum mechanics. The new approach would take more than 10 years to mature. Indeed, it was only around 1970 that the first papers appeared suggesting "a new mode of description in physics" and taking "quantum theory as an indication of a new order in physics." We return to these new perspectives later.

2.5.4 Citizenship Lost, Dignity Preserved

Let us go back to early 1952 in Brazil, after the American officials confiscated Bohm's passport. Under pressure to travel abroad to discuss his causal interpretation with wider audiences he early began to consider applying for Brazilian citizenship. It would be a difficult choice, as he wrote to Hanna Loewy: "also, if I want, I can apply for citizenship. This would have some advantages: as with it, I could travel. But the disadvantage is that I could not return to the U.S., at least for a long time. For according to the McCarran act they can exclude any non-citizen from the U.S., who, in their opinion, was ever connected with Communism. So it's a tough decision, isn't it?" In the middle of 1954, anxious because of the tension of the political times (a year before the Rosenberg couple, accused of espionage, had been executed in the U.S., and in August 1954 the Brazilian President Getúlio Vargas had committed suicide in the middle of a serious political crisis), and having received a job offer from Israel sent by Nathan Rosen, Bohm decided to apply for Brazilian citizenship.[70] Helped by Brazilian scientists and politicians the whole process was quick. He applied for citizenship on 15 September 1954, received the presidential decree on 22 November 1954, and took the oath on 20 December 1954. In early 1955 he left the country for Israel. Getting Brazilian citizenship, however, was a fateful decision for Bohm as it led to the loss of his American citizenship. Indeed, in accord with the oath, he gave up his former citizenship. As early as April 1955, Marc Severe, an official from the American consulate, required the Brazilian Police Department to give the US government information about the Brazilian

[70] Bohm tried to convince Einstein to support his move to Israel, but Einstein was reluctant, writing, "to go there with the intention to leave on the first occasion would be regrettable." Einstein to Bohm, 22 Jan 1954. Bohm, however, was decided to go: "I have decided to go to Israel. This decision was precipitated by the receipt of an offer of a job in Haifa from Rosen [...] I have cited you as a possible recommendation, so you may be receiving a letter from them soon." He also promised to stay there for years to Einstein ("... do not plan to leave unless after several years of effort"), a promise he would not keep. In addition, Bohm was considering the possibility of getting a passport without losing American citizenship—"I am informed that the Israeli Embassy in Brazil may issue a passport for me to go to Israel, if the Technion request it."—which did not materialize, Bohm to Einstein, 3 Feb 1954. Then Einstein changed his views and supported Bohm's plans. Einstein to Bohm, 10 Feb 1954; and Einstein to Nathan Rosen, 11 March 1954. The Albert Einstein Archives, The Hebrew University of Jerusalem. My thanks to Michel Paty and Amit Hagar for providing me with copies of these letters.

nationality of Bohm. Brazilian authorities took time replying, but eventually they confirmed that Bohm had been granted Brazilian citizenship.[71]

Bohm lost his American citizenship on 5 December 1956, but only in 1960, already in London, did he try to recover it or even to get a visa, so that he could accept the position that Brandeis University had offered him. His attempts were unsuccessful. He tried again in 1965–1967, with the support of Stirling Colgate, President of the New Mexico Institute of Mining and Technology, in Socorro. Colgate became engaged in the fight as a result of the job offer he had made to Bohm. Again he did not succeed. The backgrounds of these attempts is revealing of the enduring constraints of the Cold War era. In 1960 in London, he was asked by the American Consul about his previous relationship with the Communist Party. Thus, Bohm made a notarized statement on 23 March 1960 about his former links with the Communist Party, and about his current withdrawal from Communist views. Although he had made a notarized statement, Bohm did not intend to make it public. However, this was exactly what the American officials expected from him. Indeed, it would be necessary to demonstrate an active attitude against Marxism, i.e. to make public pronouncements against Communism.[72]

At that stage, Bohm faced a dilemma: either to keep his dignity and not recover his American citizenship or recover it, even if it meant losing his dignity. Bohm decided not to pay the price required by the American authorities. His decision is well documented in a letter to Aage Bohr, "It seems that while they are satisfied that I am not a Communist, the McCarran act requires that I prove 'active anti-Communism', e.g. by writing political articles; and this I am not prepared to do." Later, in 1966, Bohm stuck to his decision, as one can see from the letter to Ross Lomanitz, who had been instrumental in recommending him to Colgate, "My principal objection to [publishing something of an 'anticommunist nature'] is that it is not really compatible with dignity. [. . .] I feel it wrong to say it [his criticisms to Communism] in order to regain American citizenship. For then, I am saying something not mainly because I think it is true, but rather, for some ulterior purpose. It's rather like writing a scientific article in order to impress one's superior, so as to get a better job." It is worth noting that Stirling Colgate understood and supported Bohm's attitude, writing to the US State Department, "He could apply for a visa as an immigrant, and I believe this would require a full demonstration of active opposition to communism with a question on his mind, I am sure, of just how

[71] Bohm to Hanna Loewy [Beginning of 1952]. BP (C.40). According to physicist José Leite Lopes [Interview with A.M.R. Andrade, 18 March 2003], Brazilian physicists had asked João Alberto Lins de Barros, a very influential politician and supporter of Brazilian physics, to accelerate Bohm's Brazilian citizenship application. File 40.135/54. Archives of the "Instituto de Identificação Ricardo Gumbleton Daunt", SSP—Polícia Civil, São Paulo.

[72] For the date of the "Certificate of Loss of Nationality", see Stirling Colgate folder in BP (C.8). "I would like very much to get the question of my US citizenship settled again". Bohm to Stirling Colgate, 28 April 1965, BP (C.8). I am thankful to Basil Hiley for his kindness in sending me a copy of the notarized documents.

active is active. This question relates, of course, to a sense of personal dignity among his friends and peers."[73]

Let us now break the chronology to report the outcome of Bohm's citizenship affair. In the twilight of the Cold War, Bohm eventually won the right to recover his American citizenship after living more than 30 years as a Brazilian citizen. He used his letters to Einstein written from Brazil, in which it was clear that Bohm did not intend to give up American nationality, and that he had applied for Brazilian citizenship only in order to get a passport. He succeeded in the legal process in 1986: "Dear Dr. Bohm. I am pleased to inform you that the Department of State has today notified the Embassy that your citizenship case has been reconsidered. It has now been determined that your naturalization which took place in Brazil, in November 1954, was an involuntary act. Consequently your loss of United States citizenship has been overturned; and the Certificate of Loss of Nationality that was initially prepared has been vacated." The victory came too late as he had no income to live in the US as a retiree. In Cold War times, keeping dignity came at a high price.[74]

2.5.5 *New Acquaintances: Students and Collaborators*

During this period of transition Bohm also had pleasant professional experiences meeting people who would collaborate in the new directions he would undertake. At Technion in Israel he met two new students, Yakir Aharonov and Gideon Carmi.[75] Aharonov analyzed the role of electromagnetic potentials in quantum theory and suggested a new effect, now known as the Aharonov-Bohm effect. Aharonov and Bohm illustrated this effect arguing that when an electron beam is

[73] Bohm to Aage Bohr, November 17, 1960, Aage Bohr Papers, Niels Bohr Archive, Copenhagen. The distinction between declaring not to be Communist and expressing active anti-Communism was not understood by Bohm's biographer David Peat (1997, pp. 254–255). Peat also asked "Why did he place his rejection of Communism at the end of the Second World War when in fact his letters from Brazil are staunchly pro-Communist?" I think Peat was not very sensitive to the carefully diplomatic manner in which Bohm wrote, in the statement previously cited: "Gradually however, and especially after the war was over, I began to see that…" He was simply avoiding any great disparity between that statement and what he had declared before the HUAC, in 1949–1950. Bohm to Ross Lomanitz, 21 Nov 1996, BP (C.42), underlined in the original. Stirling Colgate to George Owen (Deputy Director Visa Office—US State Dept), 4 Nov 4, 1966, BP (C.8).

[74] Bohm's lawyer, Edward S. Gudeon, based his petition on the decision of the Supreme Court, in 1967, in the case Afroyim v Rusk, which stated that an American citizen could only lose his citizenship if required by himself. Edward Gudeon to Ehud Benamy, 11 Feb 1986, BP [Probably C.8]. Richard Haegele—American Consul in London—to David Bohm, 11 Feb 1986, BP [Probably C.8]. "I cannot see how I could settle there permanently, because my pension could not be adequate for this". Bohm to Hanna Loewy, 3 March 1986, BP (C.41).

[75] David Bohm, interviewed by Maurice Wilkins, sessions 4 and 7, 25 Sept 1986 and 30 Jan 1987, Niels Bohr Library and Archives, American Institute of Physics, College Park, MD.

split around a region where an electromagnetic field is confined, the beam passing through a field-free region may undergo a physical change. They then argued that this was a quantum effect related to the vector potential, which is classically considered to be without physical meaning. This paper stirred up a flood of experiments and theoretical explanations and is by far the most influential paper authored by David Bohm, amounting to 3,500 citations as of May 2012 (Aharonov and Bohm 1959; Peshkin and Tonomura 1989).[76] It brought wide recognition to both, which included the 1998 Wolf Prize to Aharonov. However, the Aharonov-Bohm effect did not appear at Technion. Aharonov had followed Bohm to Bristol, where he got his PhD. Bristol was then a thrilling center for physics under the leadership of Maurice Pryce, the head of the Department. "Pryce appointed David Bohm (1917–1994), who arrived in 1957 with his student Yakir Aharonov (b. 1932). Their discovery [. . .] was central to the formulation of modern gauge theories of fundamental interactions." These are the recollections of Michael Berry and Brian Pollard (2008).

In London Bohm met Jeffrey Bub, who began to work with him as a graduate student on problems related to the foundations of quantum physics. Bub came from Cape Town, where he had become interested in foundational issues in quantum mechanics through the mathematician and mystic Michael Whiteman (Bub 1997, p. xi). He went to London to study under Karl Popper but at the time Popper was in the US. Bub was advised by G.J. Whitrow to work either with Bohm or Rosenfeld if he wanted to work on foundations of quantum mechanics. Bub chose Bohm because his scholarship funds were insufficient to support a move to Denmark, and he thought the language would present a problem. The research directions Bub would have followed, had he chosen Rosenfeld, we can only wonder. He began to work under Bohm in early 1963, however, Bohm was no longer interested in hidden variables. According to Bub's recollections,[77]

> At the time Bohm was no longer interested in hidden variables. He was trying to develop a general framework for physics based on a discrete space-time structure for events and held a weekly seminar where he discussed ideas on algebraic topology using Hodge's book on harmonic analysis. It was rather too abstract for me. We graduate students tried to make sense of Bohm's ideas with Hiley, but it seemed that every few days ideas he had talked about earlier were scrapped for new ideas, so it was rather frustrating.

In hindsight we can see that Bub was experiencing the attempts Bohm was making to develop new perspectives for his research. Eventually Bub found his own way through the reading of a paper by Margenau on the measurement problem; subsequently Bohm suggested he "read a paper by Wiener and Siegel, 'The differential space theory of quantum systems,' and consider treating the collapse problem in the framework of a hidden variables theory." More particularly, "Bohm's thought was that one should be able to exploit the Wiener-Siegel

---

[76] For the debate on the theoretical interpretation of the Aharonov-Bohm effect, see Lyre (2009).

[77] Talk with Jeffrey Bub, 22 May 2002, American Institute of Physics, College Park, MD. E-mails from Bub to the author, 29 May 2014.

'differential space' approach to quantum mechanics to construct an explicit nonlinear dynamical 'collapse' theory for quantum measurement processes" (Bub 1997, p. xii). Thus, with Bub as a student, Bohm came back to the hidden variable approach while in a different manner from that in the early 1950s. Bub coped with the suggestion and a thesis and papers resulted (Bohm and Bub 1966a, b).[78]

Bub was probably one of the first students to get a PhD in physics working on foundations of quantum mechanics. After a string of positions he eventually became a Distinguished Professor at University of Maryland. After working with Bohm, Bub's interests moved to quantum logic. Bub ultimately evolved for a kind of reconciliation between the two themes he had worked through his life: hidden variables and quantum logic (Bub 1997, p. xiii). In 1998 he won the prestigious Lakatos Award with the book *Interpreting the Quantum World* (Bub 1997) where this reconciliation is presented. Thus Bub's story is a success story of somebody who began and endured in the field of foundations of quantum mechanics. However, the very fact that most of his academic career was developed in philosophy departments, a standard followed by many quantum foundationalists till today, is reminiscent of the adversities such physics researchers have found among their fellow physicists. When he began his doctoral dissertation, Bohm had warned him that with such a subject he would not get a position in a physics department.[79] Bohm was premonitory.

2.6 New Perspectives: Wholeness and Implicate Order

Looking for new perspectives to understand quantum mechanics, Bohm drew heavily on analogies and images to convey the content of his new ideas on order, the most well-known being the image of a drop of ink falling into a rotating cylinder full of glycerin. When the cylinder rotates in one direction the ink disappears in the glycerin, which Bohm referred to as the implicate order. When it rotates in the opposite direction, the drop reappears, namely the explicate order. Bohm would associate the explicate order with classical or macroscopic phenomena and the implicate order with quantum phenomena. For Bohm, the usual interpretation of quantum mechanics was not the final word in quantum physics, and he went on to associate the implicate order with a physical theory yet to be worked out that has standard quantum mechanics as a limiting case.[80]

Bohm's ideas of implicate and explicate order resulted from diverse influences and inspirations. As he recalled, there was his search for new ideas and his enduring reflection about what was common to his previous approach and standard quantum

[78] Margenau's paper was Margenau (1963) and the papers by Norbert Wiener and Armand Siegel were Wiener and Siegel (1953, 1955) and Siegel and Wiener (1956).

[79] Jeffrey Bub, talk with the author, 3 April 2014.

[80] Bohm et al. (1970), Bohm (1971, 1973), Bohm (1981).

mechanics (a task that was eased by John Bell's 1965 work pointing to non-locality as the irreducible quantum feature, as we will see in Chap. 7). In addition, there were the insights from a TV program in which he saw the demonstration with ink and glycerin and the fruitful interaction with mathematicians and mathematical physicists. The question remains of how much Bohm was influenced in the early 1960s by his dialogues with the influential Indian writer Jiddu Krishnamurti, with whom Bohm kept a longstanding interaction (Peat 1997, Chap. 11). Bohm once acknowledged some influence from Krishnamurti's psychological ideas on the non-separability between observer and observed, which reinforced his ideas on the analogous problems in quantum measurement. Later, however, he did not mention this influence again. Basil Hiley, Bohm's longstanding collaborator to whom we will refer later, thinks that these dialogues were not influential in Bohm's physics; rather, they played a role in Bohm's reflections about society, thoughts, and creativity. A reflection on the relationship between observer and observed had been an essential feature of Bohm's early reflections on the foundations of quantum mechanics, see for instance how he treated measurement both in his 1951 book and 1952 causal interpretation. Thus, it seems that the influence of these dialogues on his physics, if any, was superseded by his enduring reflection on measurement in quantum physics.[81]

Implicate and explicate order would have remained mere philosophical or scientific intuitions if it had not been for the mathematical elaboration they later received. To accomplish this Bohm did not work alone. He counted on the collaboration of Basil Hiley, who was born in Burma, then part of the British Raj. He came to England when India gained independence. Hiley did his degree and doctoral studies at King's College working with the theory of condensed matter, but he was interested in abstract mathematics and foundational physics. He attended a lecture by Bohm at the end of his degree and was spellbound. Professional interaction with Bohm, however, came later, after Hiley was hired by Birkbeck College in 1961. Bohm was also there and he became Bohm's assistant. At the beginning of their collaboration there was no connection with Bohm's previous work on the causal interpretation. "When I started with Bohm we did not mention or discuss his '52 Hidden Variable approach at all" and "for about the first 10 years we didn't discuss the Hidden Variable Theory hardly at all," Hiley stated. Furthermore, according to Hiley's recollections, he "was brought up in an atmosphere where it was generally agreed that there was something basically wrong with the '52 paper of Bohm." Instead of hidden variable models, Hiley engaged with new mathematical objects with Bohm and the mathematician Roger Penrose, in a seminar they informally ran on Thursday afternoons.[82]

---

[81] Bohm (1982, 1987). Basil Hiley 2008, American Institute of Physics, ibid.

[82] Basil Hiley interviewed by Olival Freire, 11 Jan 2008, Niels Bohr Library and Archives, American Institute of Physics, College Park, MD. See also Basil Hiley interviewed by Alexei Kojevnikov, 05 Dec 2000, Niels Bohr Library and Archives, American Institute of Physics, College Park, MD.

Bohm and Hiley's strategy was to analyze the algebraic structures behind quantum mechanics' mathematical formalism and subsequently look for more general algebras which could be reduced to the quantum algebras as special cases. This strategy was informed by the fact that they did not want to take any kind of space-time geometry as assumptions in their reasoning. Instead they tried to develop algebraic structures from which space-time could emerge. Here the algebraic primary structure would be the implicate order and the emerging space-time geometry would be the explicate order. With the benefit of hindsight, we can identify Hiley's unique contribution in this sense. Indeed Hiley was, and still is, the mathematical mind behind the research program related to the idea of order. A number of different factors also contributed to the development of this mathematical approach, such as new and mathematically talented students including Fabio Frescura, interactions with the mathematician Roger Penrose at Birkbeck College, and inspiration from the Brazilian physicist Mario Schönberg's early works on algebras and geometry. Highly sophisticated from the mathematical point of view, such an approach has, however, suffered from little contact with experimental results, which could help to inform the mathematical choices to be made.[83]

2.6.1 Returning to the Quantum Potential

In the late 1970s a new stage in Bohm's quest for a new approach to quantum mechanics began, albeit strongly overlapping the previous one. To a certain extent it meant a return to Bohm's 1952 ideas. This return, almost 30 years later, is vividly described by Basil Hiley[84]:

> We had a couple of research students working for us, Chris Dewdney and Chris Philippidis. They came to me one day with Bohm's 1952 paper in their hand. And, they said, "Why don't you and David Bohm talk about this stuff?" And I then started saying, "Oh, because it's all wrong." And then they started asking me some questions about it and I had to admit that I had not read the paper properly. Actually I had not read the paper at all apart from the introduction! And when I took it and, so, you know, I was now faced with embarrassment that our research students [Laugh] were putting me in, in a difficult position, and so I went back home and I spent the weekend working through it. As I read it, I thought, "What on earth is wrong with this? It seems perfectly all right. Whether that's the way nature behaves is another matter." But as far as the logic, the mathematics, and the arguments were concerned, it was sound. I went back again to see the two again, I said, "Okay, let's now work out what the trajectories are, work out what the quantum potential looks like in various situations.

The students and the surprised Hiley went on to calculate the trajectories allowed by Bohm's quantum potential using the recently-arrived desktop computer

[83] See Bohm and Hiley (1981), Frescura and Hiley (1980a, b). Reference to Schönberg is in Frescura and Hiley (1980b).

[84] Basil Hiley, ibid.

resources to plot these trajectories, creating images of quantum phenomena (Philippidis et al. 1979). Motivated by students and collaborators, Bohm returned to his 1952 approach, but now he had a new problem: how to interpret such an approach and its deterministic trajectories shaped by the nonlocal physical interactions resulting from the quantum potential. Here there is a crucial point to consider while charting Bohm's thoughts on quantum mechanics. While he and his colleagues kept the mathematics and the model used in the 1952 paper, they changed many of their philosophical and conceptual assumptions. The quantum potential was no longer considered a new physical potential. Instead it was interpreted as an indication of a new order, in particular a kind of "active information." Emphasis was no longer put on the causality embedded in such an approach. According to Bohm and Hiley (1993), in the book synthesizing their ideas on quantum physics, *The Undivided Universe*, after considering terms such as "causal" and "hidden variable" interpretations "too restrictive" and stating that "nor is this sort of theory necessarily causal," they concluded that "the question of determinism is therefore a secondary one, while the primary question is whether we can have an adequate conception of the reality of a quantum system, be this causal or be it stochastic or be it of any other nature." Their main philosophical stance was to look for an ontological view of quantum phenomena, while the main scientific challenge remained how to tie such a requirement to the mathematical work related to the idea of an "implicate order." This challenge has survived Bohm and is a task on which Hiley remains focused.[85]

It is time now to ask about the share of continuity and the share of change in Bohm's enduring research on the foundations of the quantum theory. Continuity was related to the philosophical commitment to the quest for an ontology, an explanation of the kind of world described by quantum physics. From the *Quantum Theory* 1951 textbook to the 1993 *The Undivided Universe*, there was a permanent commitment to a kind of scientific realism. The changes were also formidable. Determinism, the leitmotif of the causal interpretation, was abandoned. The style of scientific research also changed along the way, with the building of physical models being replaced by a more abstract research on the algebras underlying the mathematical structure of quantum physics. Influences from Marxism were replaced by Eastern thinking. As influential as Bohm's thoughts on quantum physics may be, it has been hard to identify which part or stage of his thinking is being considered when his ideas are invoked by his current readers. An early example of this was Fritjof Capra and his bestseller *The Tao of Physics*, in which Bohm's ideas on order in quantum theory were presented while Bohm's previous ideas on a causal interpretation of the same theory were ignored. Bohm did not help his readers to make sense of the evolution of his thoughts and in the most widely influential of his books, *Wholeness and the implicate order*, he conflated different stages of his interpretation of quantum mechanics. Even in a paper showing the connections

[85] Philippidis et al. (1979) and Bohm and Hiley (1993, p. 2). For Hiley's recent work, see Hiley and Callaghan (2012).

between two of his most important approaches to quantum mechanics, when "asked to explain how [his] ideas of hidden variables tie up with those on the implicate order," he emphasized the continuity more than his change of emphasis.[86]

2.7 On the Legacy of a Notable Quantum Dissident[87]

Recognition was erratic in David Bohm's case. In the late 1940s he was considered among the most promising young American theoretical physicists. During the 1950s, his work on the causal interpretation was poorly received, casting doubts on that promise. Few were those who, like the Scottish engineer Lancelot L. Whyte, considered the causal interpretation anything but fleeting. Whyte considered Bohm's work comparable to Kepler's in mechanics, which was a compliment for a physicist. Later, in the 1980s, to some extent reflecting the physicists' changed mood about research on the foundations of quantum physics, his whole work fared better. A sign of the late prestige accorded to Bohm and to the field in which he mostly worked was the volume in honor of the centenary edition of *Physical Review*, the most influential American physics journal. It included commentaries and reprints from the most important papers ever published in this periodical. In the section on "Quantum Mechanics", edited by Sheldon Goldstein and Joel Lebowitz, all the papers, including Bohm's 1952 paper on the causal interpretation, concern the foundations of quantum mechanics and a photo of Bohm opens the section. The Festschrift honoring his 70th birthday had already brought tributes from scientists such as Ilya Prigogine, Maurice Wilkins, and Richard Feynman, all Nobel Prize laureates at the time the book appeared, Anthony Leggett, who would go on to win the 2003 Physics Nobel Prize, John Bell, Roger Penrose, David Pines, Bernard d'Espagnat, and Jean-Pierre Vigier, in addition to a number of Bohm's collaborators. The ultimate accolade was to be elected Fellow of the Royal Society in 1990. After his death, his prestige continued to grow, as remarked by his long-standing friend, the American physicist Melba Phillips (1907–2004), "it is too bad, very sad indeed, that he did not live to see how his reputation has shot up recently. His interpretation of quantum mechanics is becoming respected not only by philosophers of science but also by 'straight' physicists."[88]

---

[86] Capra (1991), Bohm (1981) and Bohm (1987).

[87] While the use of the term dissident for Bohm and his works in quantum mechanics is almost self-evident, I use the term for a wide description of those physicists who contributed to develop the research on the foundations of quantum mechanics after 1950. I postpone a justification for my use of the term to Chap. 9. The first to use a similar term in this context, as far as I know, was Karl Popper (Popper and Bartley 1982, p. 100): "Unlike the orthodoxy, the dissenters are far from united. Not two of them agree (except perhaps Bohm and de Broglie)."

[88] Stroke (1995). Lancelot Whyte to Léon Rosenfeld, 8 Apr 1958, RP (Bohm et al. 1987). Melba Phillips to David Peat, 17 Oct 1994, A22, BP.

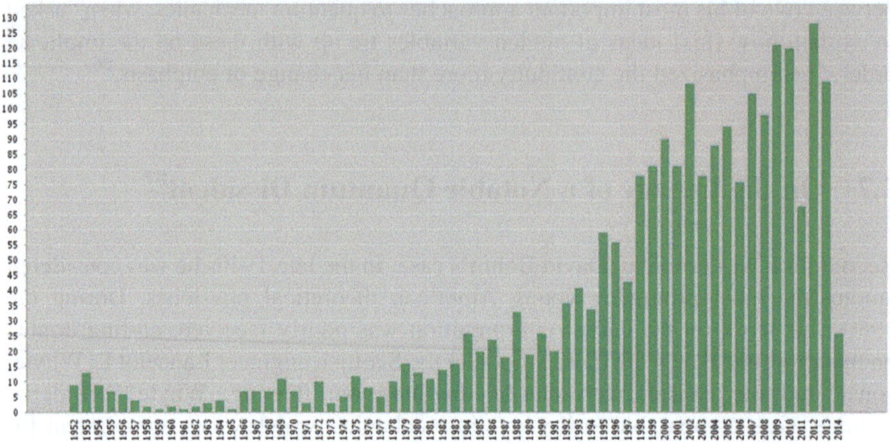

Picture 2.3 Citations of Bohm's causal interpretation papers from 1952 to 2014—Source of the data: Web of Science

With the benefit of hindsight, how can Bohm's legacy be assessed in the first decades of the twenty-first century? The question requires a multifaceted answer. First, there were achievements not directly related to his research on the foundations of quantum theory. This was the case of his work on collective variables, in plasma and metals, conducted with Eugene Gross and David Pines, a work which had begun before his shift towards the causal interpretation. And then, there was the Aharonov-Bohm effect, published in 1959, which became a landmark when one speaks about quantum effects without classical equivalent. These achievements are beyond debate; they are considered feats in the history of physics in the twentieth century.[89] In addition, if one takes scientometric data, the number of citations, into consideration, the aforementioned are by far among the most influential contributions by David Bohm. Second, the research lines on the interpretational issues he worked on have survived him and are fields of live research with their value to the future development of physics still subject to controversial assessments. They may be grouped into three different strands. The first continues work on Bohm's original 1952 proposal, not only trying to extend the first physical models but also keeping Bohm's early philosophical commitments to determinism and realism. This is, for instance, the path chosen by Peter Holland (1993). More recently, this trend has been renewed by Antony Valentini. He has worked with deterministic hidden-variables theories in the direction of relaxing the equality between distribution of hidden variables and probability distribution from standard quantum theory. As for him quantum physics may be a mere case of an effective theory of an equilibrium state, we should look for discrepancies between hidden variables and quantum

[89] Basil Hiley cited these achievements and Bohm's contributions to our understanding of quantum non-locality when asked for the background for Bohm's nomination for the Nobel Prize. B. Hiley to Sessler, 9 Jan 1989, A172, BP.

theory predictions in situations of nonequilibrium. Still, for him, we should look for this in astrophysical and cosmological tests (Valentini 2007, 2010). The second strand concerns Bohmian mechanics, a name coined by Detlef Dürr, Sheldon Goldstein, and Nino Zanghi. They construed Bohm's proposal in a very clean and elegant way. In his original paper Bohm had worked out analogies between Schrödinger's equation and classical Hamilton-Jacobi equations, which led to an emphasis on the role of the non-classical potential that Bohm christened the "quantum potential." Dürr and colleagues, however, adopted just two premises: the state which describes quantum systems evolves according to Schrödinger's equation and particles move, that is, they have a speed in the configuration space. Thus for them, "Bohmian mechanics is a version of quantum mechanics for non relativistic particles in which the word 'particle' is to be understood literally: In Bohmian mechanics quantum particles have positions, always, and follow trajectories. These trajectories differ, however, from the classical Newtonian trajectories." With this approach, without referring to the quantum potential and the difficult problem of its physical interpretation, they derived the same results one gets both with standard quantum mechanics and with Bohm's original approach for nonrelativistic phenomena. This approach has been useful for discussing quantum chaos, and for this reason it has received widespread acceptance, well beyond physicists just interested in the foundations of quantum mechanics. One should note that when these physicists define what they understand to be a Bohmian theory, the preference for determinism disappears and they consider that "a Bohmian theory should be based upon a clear ontology," meaning by ontology "what the theory is fundamentally about." While for non-relativistic physics they have adopted a particle ontology, they admit that they "have no idea what the appropriate ontology for relativistic physics actually is." This way, the commitment to a quantum ontology comes before an engagement with a causal pattern for physical theories, a position analogous to what was adopted by David Bohm and Basil Hiley since the 1960s.[90]

The third strand of Bohm's scientific legacy is represented by Basil Hiley, who continues to work on research that he and Bohm had been carrying out before Bohm's death. This research tries to connect the insights of implicate order and active information with the quest for algebraic structures able to underpin space-time geometry and standard quantum mechanics. This program has inherited from the causal interpretation the major challenge of obtaining a fully relativistic treatment in order to match the level attained by standard quantum mechanics with the Dirac equation.

Rather than one specific and lasting contribution, I think he should be acknowledged for his attitude to the importance of the research on the foundations of this theory. His late recognition was not independent of this role. The point is that the most influential single theoretical result in the foundations of quantum theory after WWII was Bell's theorem, which jointly with its experimental tests led to the

[90] Dürr et al. (1992, 1996, 2009).

recognition of entanglement as a physical property with far-reaching implications both for science and technology. However, John Bell's work has a close historical connection with Bohm's work on a hidden variable interpretation. Max Jammer wrote that "it was due to Bohm that many physicists and philosophers of science [...] examined more closely the logic of von Neumann's argument and that finally, in 1964, J. S. Bell clarified completely the nature of von Neumann's unnecessarily restrictive assumptions with the removal of which his proof breaks down." According to the recollections of Bell himself, "Smitten by Bohm's papers," he attempted to determine what was wrong with von Neumann's proof, since it did not allow for hidden variables in quantum mechanics. Here is not the place to chart the origins of Bell's theorem, which will be done in Chap. 7. For our purposes, suffice to say Bell was directly motivated by the very existence of Bohm's proposal and by its reception among physicists. His statements—"In 1952 I saw the impossible done," and "Bohm's 1952 papers on quantum mechanics were for me a revelation"—hide more truth than is usually recognized, "the impossible done" referring to the appearance of the causal interpretation which was considered by prevailing wisdom an impossible feat.[91]

2.7.1 Historiography on Bohm's Interpretation

The initial poor reception of Bohm's causal interpretation has attracted the attention of commentators. Some of them have looked to the political climate of the Cold War and Bohm's exile to explain this. "The political atmosphere in the U.S. at that time did not help rational debate and in consequence there was little discussion and the interpretation was generally ignored for reasons that had more to do with politics than science," stated Bohm's assistant, Basil Hiley. F. David Peat, a science writer and former Bohm collaborator, also advanced the political explanation for the unfavorable reaction to Bohm's work, but limited its force to the Princeton physics community. The historians Russel Olwell and Shawn Mullet blamed Bohm's Brazilian exile for the poor response to his causal interpretation theory. Others, such as James Cushing, underestimated the number of physicists who analyzed Bohm's papers, writing "[Bohm's proposal] was basically ignored, rather than either studied or rebutted." Our analysis, however, suggests otherwise, more related to the practice of physics as a cultural field. As pointed out by Max Jammer and Mara Beller, the dominance of the Copenhagen school in the early 1950s was very effective. The main critics of Bohm's ideas were Europeans, aligned with Bohr's complementarity, and not influenced by McCarthyism. Some of them were even Marxists. The record of debates about Bohm's papers and about his activities

[91] Jammer (1988, p. 694), Bernstein (1991, pp. 65–68) and Bell (1982, 1987). For the history of Bell's theorem and its experiments, see Chap. 7.

in Brazil and Israel should not lead us to underestimate these debates. In addition to the dominance of complementarity, other factors were also influential.

We have seen that Bohm and his collaborators searched in vain for predictions not foreseen by the usual quantum mechanics and also failed to find a satisfactory relativistic generalization of their approach.[92] Indeed, as most results of the causal interpretation were to replicate results already obtained with standard quantum physics, the idea grew that the controversy over the interpretation of quantum physics was a matter of philosophical taste, without implications for the workings of physics. Even physicists who were not open critics of the causal interpretation concluded this. We have seen in Chap. 1 the case of A. Messiah's influential textbook. He stated that the controversy "belongs to the philosophy of science rather than to the domain of physical science proper" (Messiah 1961, p. 48). A similar example is Fritz Bopp's statement: "what we have done today was predicting the possible development of physics—we were not doing physics but metaphysics" (in Körner 1957, p. 51). It was not by chance that in the 1950s the only conference dedicated to the subject was organized by philosophers rather than by physicists (Körner 1957). The idea of a philosophical controversy survived in the common discourse on the subject even when the context changed, as was the case when Max Jammer entitled his 1974 book *The Philosophy of Quantum Mechanics*. One should acknowledge, however, that at least in the special context of the young French Marxist physicists around Jean-Pierre Vigier, the philosophical bias of the dispute may have been considered more appealing than simply a diverting factor. And yet, the absence of new results reinforced the derogatory label of "philosophical" applied by the opponents of the causal interpretation, further discouraging young physicists from working on a subject that ultimately was more a question of philosophy than of physics.[93]

The ensemble of these reasons explains why Bohm's ideas challenged the dominance of the complementarity view among physicists, but did not weaken it enough to create a favorable space for the immediate development of alternative interpretations.

---

[92] Hiley (1997, p. 113), Peat (1997, p. 133) and Olwell (1999, p. 750). Shawn Mullet, "Political science: The red scare as the hidden variable in the Bohmian interpretation of quantum theory" (Senior thesis HIS679, University of Texas at Austin, unpub. paper, 1999). Mullet, after contact with sources from Bohm's stay in Brazil, has changed his views; cf. Shawn Mullet, "Creativity and the mainstream: David Bohm's migration to Brazil and the hidden variables interpretation," unpublished paper, Workshop on "Migrant scientists in the twentieth century," Milan, 2003. Cushing (1994, p. 144), Jammer (1974) and Beller (1999).

[93] However, Messiah did not please the hard core of the supporters of the Copenhagen interpretation. Rosenfeld wrote to him praising the book, but in disagreement with his diagnosis of the controversy. For Rosenfeld, "Ce n'est pas en effet d'expérience, mais bien de simple logique qu'il s'agit ici." Léon Rosenfeld to Albert Messiah, 16 Jan 1959, RP. About Bopp, by the way, he was then working on another alternative interpretation, the so-called "stochastic interpretation."

References

Aharonov, Y., Bohm, D.: Significance of electromagnetic potentials in the quantum theory. Phys. Rev. **115**(3), 485–491 (1959)

Anderson, P.: Considerations on Western Marxism. NLB, London (1976)

Andrade, A.M.R.: Físicos, mésons e política: a dinâmica da ciência na sociedade. HUCITEC, São Paulo (1999)

Bacciagaluppi, G., Valentini, A.: Quantum Theory at the Crossroads: Reconsidering the 1927 Solvay Conference. Cambridge UniversityPress, Cambridge (2009)

Bachelard, G.: Le Nouvel Esprit Scientifique. Félix Alcan, Paris (1934)

Bell, J.S.: On the impossible pilot wave. Found. Phys. **12**(10), 989–999 (1982)

Bell, J.S.: Beables for quantum field theory. In: Hiley, B.J., Peat, F.D. (eds.) Quantum Implications: Essays in Honour of David Bohm, pp. 227–234. Routledge & Kegan, London (1987)

Bell, J.S.: Speakable and Unspeakable in Quantum Mechanics: Collected Papers on Quantum Philosophy. With an Introduction by Alain Aspect. Cambridge University Press, Cambridge (2004)

Beller, M.: Quantum Dialogue—The Making of a Revolution. The University of Chicago Press, Chicago (1999)

Bernstein, J.: Quantum Profiles. Princeton University Press, Princeton (1991)

Berry, M., Pollard, B.: The Physical Tourist Physics in Bristol. Phys. Perspect. **10**, 468–480 (2008)

Besson, V.: Les premiers travaux de Jean-Pierre Vigier sur la théorie des quanta: une rencontre entre science et marxisme (1951-1954). Master dissertation, Université Claude Bernard Lyon 1 (2011)

Bitbol, M.: Schrödinger's Philosophy of Quantum Mechanics. Kluwer, Dordrecht (1996)

Blokhintsev, D.I.: Critique de la conception idéaliste de la théorie quantique. Questions scientifiques—Physique, pp. 95–129. Les éditions de la nouvelle critique, Paris (1952).

Bohm, D.: Quantum Theory. Prentice-Hall, New York (1951)

Bohm, D.: Reply to a criticism of a causal re-interpretation of the quantum theory. Phys. Rev. **87** (2), 389–390 (1952a)

Bohm, D.: A suggested interpretation of the quantum theory in terms of hidden variables—I & II. Phys. Rev. **85**(2), 166–179 and 180–193 (1952b)

Bohm, D.: Proof that probability density approaches (Psi)2 in causal interpretation of the quantum theory. Phys. Rev. **89**(2), 458–466 (1953a)

Bohm, D: A Discussion of Certain Remarks by Einstein on Born's Probability Interpretation of the [psi]—Function. Scientific Papers presented to Max Born. Edinburgh, Oliver and Boyd, pp. 13–19 (1953b)

Bohm, D.: Hidden variables in the quantum theory. In: Bates, D.R. (ed.) Quantum Theory—III—Radiation and High Energy Physics, pp. 345–387. Academic, New York (1962)

Bohm, D.: Quantum theory as an indication of a new order in physics. Part A. The development of new order as shown through the history of physics. Found. Phys. **1**(4), 359–381 (1971)

Bohm, D.: Quantum theory as an indication of a new order in physics. Part B. Implicate and explicate order in physical law. Found. Phys. **3**(2), 139–168 (1973)

Bohm, D.: Wholeness and the Implicate Order. Routledge & Kegan Paul, London (1981)

Bohm, D.: Interview. New Sci. **96**(1331), 361–365 (1982)

Bohm, D.: Hidden variables and the implicate order. In: Hiley, B., Peat, D. (eds.) Quantum Implications: Essays in Honour of David Bohm, pp. 33–45. Routledge, London (1987)

Bohm, D., Bub, J.: A proposed solution of measurement problem in quantum mechanics by a hidden variable theory. Rev. Mod. Phys. **38**(3), 453–469 (1966a)

Bohm, D., Bub, J.: A refutation of proof by Jauch and Piron that hidden variables can be excluded in quantum mechanics. Rev. Mod. Phys. **38**(3), 470–475 (1966b)

Bohm, D., Gross, E.P.: Theory of plasma oscillations. A. Origin of medium-like behavior. Phys. Rev. **75**(12), 1851–1864 (1949a)

Bohm, D., Gross, E.P.: Theory of plasma oscillations. B. Excitation and damping of oscillations. Phys. Rev. **75**(12), 1864–1876 (1949b)
Bohm, D., Hiley, B.J.: On a quantum algebraic approach to a generalized phase-space. Found. Phys. **11**(3–4), 179–203 (1981)
Bohm, D., Hiley, B.J.: The Undivided Universe : An Ontological Interpretation of Quantum Theory. Routledge, London (1993)
Bohm, D., Pines, D.: A collective description of electron interactions. 1. Magnetic interactions. Phys. Rev. **82**(5), 625–634 (1951)
Bohm, D., Pines, D.: A collective description of electron interactions. 3. Coulomb interactions in a degenerate electron gas. Phys. Rev. **92**(3), 609–625 (1953)
Bohm, D., Schiller, R.: A causal interpretation of the Pauli equation (B). Nuovo Cimento Suppl **1** (1), 67–91 (1955)
Bohm, D., Vigier, J.P.: Model of the causal interpretation of quantum theory in terms of a fluid with irregular fluctuations. Phys. Rev. **96**(1), 208–216 (1954)
Bohm, D., Vigier, J.P.: Relativistic hydrodynamics of rotating fluid masses. Phys. Rev. **109**(6), 1882–1891 (1958)
Bohm, D., Schiller, R., Tiomno, J.: A causal interpretation of the Pauli equation (A). Nuovo Cimento Suppl **1**, 48–66 (1955)
Bohm, D., Hillion, P., Takabayasi, T., Vigier, J.P.: Relativistic rotators and bilocal theory. Prog. Theor. Phys. **23**(3), 496–511 (1960a)
Bohm, D., Hillion, P., Vigier, J.P.: Internal quantum states of hyperspherical (nakano) relativistic rotators. Prog. Theor. Phys. **24**(4), 761–782 (1960b)
Bohm, D., Hiley, B.J., Stuart, A.E.G.: On a new mode of description in physics. Int. J. Theor. Phys. **3**(3), 171–183 (1970)
Bohm, D., Hiley, B.J., Peat, F.D.: Quantum Implications: Essays in Honour of David Bohm. Routledge & Kegan Paul, New York (1987)
Bohm, D., Biederman, C.J., Pylkkänen, P.: Bohm-Biederman Correspondence. Routledge, London (1999)
Bohr, N.: Discussion with Einstein on epistemological problems in atomic physics. In: Schilpp, P.A. (ed.) Albert Einstein—Philosopher-Scientist, pp. 199–242. The Library of the Living Philosophers, Evanston (1949)
Bohr, A., Mottelson, B.R., Pines, D.: Possible analogy between the excitation spectra of nuclei and those of the superconducting metallic state. Phys. Rev. **110**(4), 936–938 (1958)
Broglie, L.: Nouvelles Perspectives en Microphysique. Albin Michel, Paris (1956)
Broglie, L.: La théorie de la mesure en mécanique ondulatoire interprétation usuelle et interprétation causale. Gauthier-Villars, Paris (1957)
Broglie, L.D., Vigier, J.P., Bohm, D., Takabayasi, T.: Rotator model of elementary particles considered as relativistic extended structures in minkowski space. Phys. Rev. **129**(1), 438–450 (1963)
Brownell, G. L.: Physics in South America. Physics Today July, pp. 5–12 (1952)
Bub, J.: Interpreting the Quantum World. Cambridge University Press, Cambridge (1997)
Capra, F.: The Tao of Physics : An Exploration of the Parallels Between Modern Physics and Eastern Mysticism. Shambhala Publications, Boston, MA (1991)
Carson, C.: Heisenberg in the Atomic Age: Science and the Public Sphere. German Historical Institute, Washington, DC (2010)
Caute, D.: Le Communisme et les intellectuels français, 1914-1966. Gallimard, Paris (1967)
Cross, A.: The Crisis in Physics: Dialectical Materialism and Quantum Theory. Soc. Stud. Sci. **21**, 735–759 (1991)
Cushing, J.: Quantum Mechanics—Historical Contingency and the Copenhagen Hegemony. The University of Chicago Press, Chicago (1994)
Dürr, D., Goldstein, S., Zanghi, N.: Quantum chaos, classical randomness, and bohmian mechanics. J. Stat. Phys. **68**(1–2), 259–270 (1992)

Dürr, D., Goldstein, S., Zanghi, N.: Bohmian mechanics at the foundation of quantum mechanics. In: Cushing, J.T., Fine, A., Goldstein, S. (eds.) Bohmian Mechanics and Quantum Theory: An Appraisal, pp. 21–44. Kluwer, Dordrecht (1996)

Dürr, D., Goldstein, S., Tumulka, R., Zanghi, N.: Bohmian mechanics. In: Greenberger, D., Hentsche, K., Weinert, F. (eds.) Compendium of Quantum Physics—Concepts, Experiments, History and Philosophy, pp. 47–55. Springer, Berlin (2009)

Einstein, A.: Remarks to the essays appearing in this collective volume. In: Schilpp, P.A., Einstein, A. (eds.) Albert Einstein—Philosopher-Scientist, pp. 665–668. The Library of the Living Philosophers, Evanston, IL (1949)

Einstein, A.: Elementare Überlegungen zur Interpretation der Grundlagen der Quanten-Mechanik. Scientific Papers presented to Max Born, pp. 33–40. Oliver and Boyd, Edinburgh . French translation in Albert Einstein, Oeuvres Choisies, ed. by F. Balibar, B. Jech, and O. Darrigol, Paris: Editions du Seuil-CNRS, 1989, pp. 1251–1256 (1953)

Einstein, A., Born, M., Born, H.: The Born-Einstein Letters: Correspondence Between Albert Einstein and Max and Hedwig Born from 1916-1955, with Commentaries by Max Born. Macmillan, London (1971)

Epstein, S.T.: The causal interpretation of quantum mechanics. Phys. Rev. **89**(1), 319 (1953a)

Epstein, S.T.: The causal interpretation of quantum mechanics. Phys. Rev. **91**(4), 985 (1953b)

Feyerabend, P.: Professor Bohm's philosophy of nature. Br. J. Philos. Sci. **10**(40), 321–338 (1960)

Feynman, R.: The present situation in fundamental theoretical physics. An. Acad. Bras. Cienc. **26** (1), 51–60 (1954)

Fock, V.A.: On the interpretation of quantum mechanics. Czechoslov. J. Phys. **7**, 643–656 (1957)

Forstner, C.: The early history of David Bohm's quantum mechanics through the perspective of Ludwik Fleck's thought-collectives. Minerva **46**(2), 215–229 (2008)

Freire Jr., O.: The crisis in physics—comment. Soc. Stud. Sci. **22**(4), 739–742 (1992)

Freire Jr., O.: David Bohm e a controvérsia dos quanta. Centro de Lógica, Epistemologia e História da Ciência, Campinas [Brazil] (1999)

Freire Jr., O.: Gaston Bachelard et Louis de Broglie, ont-ils toujours été en synthonie? Cahiers Gaston Bachelard **6**, 160–166 (2004a)

Freire Jr., O.: Popper, probabilidade e teoria quântica. Episteme [Porto Alegre] **18**, 103–127 (2004b)

Freire Jr., O.: Science and exile: David Bohm, the cold war, and a new interpretation of quantum mechanics. Hist. Stud. Phys. Biol. Sci. **36**(1), 1–34 (2005)

Freire Jr., O.: Causality in physics and in the history of physics: a comparison of Bohm's and Forman's papers. In: Carson, C., Kojevnikov, A., Trischler, H. (eds.) Weimar Culture and Quantum Mechanics: Selected Papers by Paul Forman and Contemporary Perspectives on the Forman Thesis, pp. 397–411. Imperial College & World Scientific, London (2011a)

Freire Jr., O.: Continuity and change: charting David Bohm's evolving ideas on quantum mechanics. In: Krause, D., Videira, A. (eds.) Brazilian Studies in Philosophy and History of Science, pp. 291–299. Springer, Heidelberg (2011b)

Freire Jr., O.: On the connections between the dialectical materialism and the controversy on the quanta. Jahrbuch Für Europäihce Wissenschaftskultur **6**, 195–210 (2011c)

Freire Jr., O.: On the influence of science milestones on the history and philosophy of science. In: Blum, A., Gavroglu, K., Renn, J. (eds.) Towards a History of the History of Science: 50 Years Since "Structure". Edition Open Access, Berlin (2014c)

Freire Jr., O., Lehner, C.: 'Dialectical materialism and modern physics', an unpublished text by Max Born. Notes Rec. R. Soc. **64**(2), 155–162 (2010)

Freistadt, H.: The crisis in physics. Sci. Soc. **17**, 211–237 (1953)

Freistadt, H.: Connection between recent theories of Bohm, de Broglie, Dirac, and Schrodinger. Phys. Rev. **98**(4), 1176 (1955)

Freistadt, H.: The causal formulation of quantum mechanics of particles: The theory of de Broglie, Bohm and Takabayasi. Nuovo Cimento Suppl **5**, 1–70 (1957)

Frescura, F.A.M., Hiley, B.J.: The implicate order, algebras, and the spinor. Found. Phys. **10**(1–2), 7–31 (1980a)
Frescura, F.A.M., Hiley, B.J.: The algebraization of quantum-mechanics and the implicate order. Found. Phys. **10**(9–10), 705–722 (1980b)
Gaddis, J.L.: The Cold War: A New History. Penguin, New York (2005)
Graham, L.R.: Science and Philosophy in the Soviet Union. Knopf, New York (1972)
Graham, L.R.: Science, Philosophy, and Human Behavior in the Soviet Union. Columbia University Press, New York (1987)
Halpern, O.: A proposed re-interpretation of quantum mechanics. Phys. Rev. **87**(2), 389 (1952)
Hanson, N.R.: Copenhagen interpretation of quantum theory. Am. J. Phys. **27**(1), 1–15 (1959)
Harding, S.G. (ed.): Can Theories be Refuted? Essays on the Duhem-Quine Thesis. Synthese Library v 81. D. Reidel Publishing Co., Dordrecht (1976)
Heisenberg, W.: Physics and Philosophy; the Revolution in Modern Science. Harper, New York (1958)
Hermann, G., Soler, L., Schnell, A.: Les fondements philosophiques de la mécanique quantique introd., présentation, postf. critique par Lena Soler trad. par Alexandre Schnell en collab. avec Lena Soler préf. de Bernard d'Espagnat. J. Vrin, Paris (1996)
Hiley, B.J.: David Joseph Bohm. Biogr. Mem. Fellows R. Soc. **43**, 106–131 (1997)
Hiley, B.J., Callaghan, R.E.: Clifford algebras and the Dirac-Bohm quantum Hamilton-Jacobi equation. Found. Phys. **42**, 192–208 (2012)
Hobsbawm, E.J.: On history. Weidenfeld & Nicolson, London (1997)
Hobsbawm, E.J.: How to Change the World : Marx and Marxism, 1840-2011. Little, Brown, London (2011)
Hoffmann, D.: Robert Havemann: antifascist, communist, dissident. In: Macrakis, K., Hoffmann, D. (eds.) Science Under Socialism : East Germany in Comparative Perspective, pp. 269–285. Harvard University Press, Cambridge, MA (1999)
Holland, P.R.: The Quantum Theory of Motion: An Account of the de Broglie-Bohm Causal Interpretation of Quantum Mechanics. Cambridge University Press, Cambridge, England (1993)
Home, D., Whitaker, M.A.B.: Ensemble interpretations of quantum-mechanics—a modern perspective. Phys. Rep. **210**(4), 223–317 (1992)
Institut International de Physique Solvay: Electrons et photons—Rapports et discussions du Cinquième Conseil de Physique tenu à Bruxelles du 24 au 29 octobre 1927. Gauthier-Villars, Paris (1928) [English translation in Bacciagaluppi and Valentini 2009]
Jacobsen, A.: Léon Rosenfeld's Marxist defense of complementarity. Hist. Stud. Phys. Biol. Sci. **37**(Suppl), 3–34 (2007)
Jacobsen, A.: Léon Rosenfeld—physics, philosophy, and politics in the twentieth century. World Scientific, Singapore (2012)
Jammer, M.: The philosophy of quantum mechanics—the interpretations of quantum mechanics in historical perspective. Wiley, New York (1974)
Jammer, M.: David Bohm and his work on the occasion of his 70th-birthday. Found. Phys. **18**(7), 691–699 (1988)
Kaiser, D.: The atomic secret in red hands? American suspicions of theoretical physicists during the early Cold War. Representations **90**, 28–60 (2005)
Kaiser, D.: How the hippies saved physics: science, counterculture, and the quantum revival. W. W Norton, New York (2012)
Keller, J.B.: Bohm's Interpretation of the Quantum Theory in Terms of Hidden Variables. Phys. Rev. **89**(5), 1040–1041 (1953)
Kojève, A., Auffret, D.: L' Idée du déterminisme dans la physique classique et dans la physique moderne. Présentation de Dominique Auffret. Librairie générale française, Paris (1990)
Kojevnikov, A.: David Bohm and collective movement. Hist. Stud. Phys. Biol. Sci. **33**, 161–192 (2002)

Kojevnikov, A.: Stalin's Great Science: The Times and Adventures of Soviet Physicists. Imperial College Press, London (2004)
Kojevnikov, A.: Probability, marxism, and quantum ensembles. Yearb. Eur. Cult. Sci. **6**, 211–235 (2011)
Körner, S.: Observation and Interpretation; A Symposium of Philosophers and Physicists. Butterworths, London (1957)
Kuzemsky, A.L.: Works by D. I. Blokhintsev and the development of quantum physics. Phys. Part. Nucl. **39**(2), 137–172 (2008)
Lenin, V.I.: Materialism and Empirio-Criticism; Critical Comments on a Reactionary Philosophy. Foreign Languages Publishing House, Moscow (1947)
Lopes, J.L.: Richard Feynman in Brazil: personal recollections. Quipu **7**, 383–397 (1990)
Lyre, H.: Aharonov-Bohm effect. In: Greenberger, D., Hentschel, K., Weinert, F. (eds.) Compendium of quantum physics: concepts, experiments, history and philosophy, pp. 1–3. Springer, Berlin (2009)
Margenau, H.: Measurement and quantum states—I & II. Philos. Sci. **30**, 1–16 (1963). 138–157
Mehra, J.: The Beat of a Different Drum : The Life and Science of Richard Feynman. Clarendon, Oxford (1994)
Messiah, A.: Quantum Mechanics. North Holland, Amsterdam (1961)
Moore, W.J.: Schrödinger : Life and Thought. Cambridge University Press, Cambridge [England] (1989)
Mullet, S.K.: Bohm, David Joseph. New Dictionary of Scientific Biography, pp. 321–326. N. Koertge, New York, Thomson - Gale. I (2008b)
Mullet, S.K.: Little Man: Four Junior Physicists and the Red Scare Experience. PhD Dissertation, Harvard University (2008a)
New research techniques in physics (1954). Rio de Janeiro and São Paulo, July 15–29, 1952
Olwell, R.: Physical Isolation and Marginalization in Physics - David Bohm's Cold War Exile. ISIS **90**, 738–756 (1999)
Ory, P., Sirinelli, J.-F.: Les Intellectuels en France de l'affaire Dreyfus à nos jours. Perrin, Paris (2004)
Osnaghi, S., Freitas, F., Freire Jr., O.: The origin of the Everettian heresy. Stud. Hist. Philos. Mod. Phys. **40**(2), 97–123 (2009)
Paty, M.: Sur les 'variables cachées' de la mécanique quantique—Albert Einstein, David Bohm et Louis de Broglie. La pensée **292**, 93–116 (1993)
Paty, M.: The nature of Einsteins objections to the copenhagen interpretation of quantum-mechanics. Found. Phys. **25**(1), 183–204 (1995)
Pauli, W.: Remarques sur le problème des paramètres cachés dans la mécanique quantique et sur la théorie de l'onde pilote. In: George, A. (ed.) Louis de Broglie – Physicien et Penseur, pp. 33–42. Editions Albin Michel, Paris (1953)
Pauli, W., Meyenn, K.V.: Wissenschaftlicher Briefwechsel mit Bohr, Einstein, Heisenberg u. a. Band IV Teil I 1950-1952. Springer, Berlin (1996)
Pauli, W., Meyenn, K.V.: Wissenschaftlicher Briefwechsel mit Bohr, Einstein, Heisenberg u. a. Band IV Teil II 1953-1954. Springer, Berlin (1999)
Peat, F.D.: Infinite Potential : The Life and Times of David Bohm. Addison Wesley, Reading, MA (1997)
Pechenkin, A.: The early statistical interpretations of quantum mechanics in the USA and USSR. Stud. Hist. Philos. Mod. Phys. **43**(1), 25–34 (2012)
Pechenkin, A.: Leonid Isaakovich Mandelstam: Research, Teaching, Life. Springer, New York (2013)
Peshkin, M., Tonomura, A.: The Aharonov-Bohm Effect. Springer, Berlin (1989)
Philippidis, C., Dewdney, C., Hiley, B.J.: Quantum interference and the quantum potential. Nuovo Cimento della Societa Italiana di Fisica B—Gen. Phys. Relat. Astron. Math. Phys. Methods **52** (1), 15–28 (1979)
Pinault, M.: Frédéric Joliot-Curie. O. Jacob, Paris (2000)

Pines, D.: A collective description of electron interactions. 4. Electron interaction in metals. Phys. Rev. **92**(3), 626–636 (1953)
Pines, D., Bohm, D.: A collective description of electron interactions. 2. Collective vs individual particle aspects of the interactions. Phys. Rev. **85**(2), 338–353 (1952)
Popper, K.R., Bartley, W.W.: Quantum Theory and the Schism in Physics. Rowan and Littlefield, Totowa, NJ (1982)
Rattner, H.: Tradição e mudança (a comunidade judaica em São Paulo). Atica, São Paulo (1977)
Reichenbach, H.: Philosophic Foundations of Quantum Mechanics. University of California Press, Berkeley, CA (1944)
Rodrigues, L.M.: O PCB: Os dirigentes e a organização. História geral da civilização brasileira, tomo III, vol. 3: O Brasil republicano—sociedade e política (1930-1964). B. Fausto. Rio de Janeiro, Bertrand Brasil, pp. 361–443 (1996)
Rosenfeld, L.: (1953) L'évidence de la complementarité. Louis de Broglie—physicien et penseur. A. George. Paris, Editions Albin Michel, pp. 43–65 [A slightly modified English version of this paper is Strife about complementarity, Science progress. **163**, 1393–1410 (1953), reprinted in Robert Cohen and John Stachel (eds.). Selected papers of Léon Rosenfeld (Dordrecht, D. Reidel, 1979)
Rosenfeld, L.: A filosofia da física atômica. Ciência e cultura **6**(2), 67–72 (1954)
Rosenfeld, L.: Physics and metaphysics. Nature **181**, 658 (1958)
Rosenfeld, L.: Heisenberg, physics and philosophy. Nature **186**, 830–831 (1960)
Rosenfeld, L.: Berkeley redivivus. Nature **228**, 479 (1970)
Rosenfeld, L.: Classical Statistical Mechanics. Livraria da Física & CBPF, São Paulo (2005)
Saidel, R.G., Plonski, G.A.: Shaping modern science and technology in Brazil. The contribution of refugees from national socialism after 1933. Leo Baeck Inst. Year Book **39**, 257–270 (1994)
Schatzman, E.: Physique quantique et realité. La pensée **42–43**, 107–122 (1953)
Schilpp, P.A., Einstein, A.: Albert Einstein, Philosopher-Scientist. Library of Living Philosophers, Evanston, IL (1949)
Schönberg, M.: On the hydrodynamical model of the quantum mechanics. Nuovo Cimento **XII**(1), 103–133 (1954)
Schönberg, M.: Quantum theory and geometry. In: Kockel, B., Macke, W., Papapetrou, A. (eds.) Max-Planck-Festschrift 1958, pp. 321–338. Deutscher Verlag der Wissenschaften, Berlin (1959) [Reprinted in M. Schönberg, Obra Científica de Mario Schönberg, São Paulo: EDUSP, 2013]
Schönberg, M., Hamburger, A.I.: Obra Científica de Mario Schönberg—Volume 1—1936–1948. EDUSP, São Paulo (2009)
Schönberg, M., Hamburger, A.I.: Obra Científica de Mario Schönberg—Volume 2—1949–1987. EDUSP, São Paulo (2013)
Schrecker, E.: No Ivory Tower : McCarthyism and the Universities. Oxford University Press, New York (1986)
Schrödinger, E.: The meaning of wave mechanics. In: George, A. (ed.) Louis de Broglie—physicien et penseur, pp. 16–32. Albin Michel, Paris (1953)
Schrödinger, E., Bitbol, M.: Physique quantique et représentation du monde introd. et notes par Michel Bitbol. Ed. du Seuil, Paris (1992)
Schweber, S.S.: QED and the Men Who Made It : Dyson, Feynman, Schwinger, and Tomonaga. Princeton University Press, Princeton, NJ (1994)
Schweber, S.S.: Feynman, Richard. In: Heilbron, J.L. (ed.) The Oxford Guide to the History of Physics and Astronomy, pp. 118–120. Oxford University Press, New York (2005)
Siegel, A., Wiener, N.: Theory of measurement in differential-space quantum theory. Phys. Rev. **101**(1), 429–432 (1956)
Stöltzner, M.: What John von Neumann thought of the Bohm interpretation. In: Greenberger, D., Reiter, W.L., Zeilinger, A. (eds.) Epistemological and Experimental Perspectives on Quantum Mechanics, pp. 257–262. Springer, Dordrecht (1999)

Stroke, H.H.: The Physical Review—The First Hundred Years: A Selection of Seminal Papers and Commentaries. American Institute of Physics, New York (1995)

Takabayasi, T.: On the formulation of quantum mechanics associated with classical pictures. Prog. Theor. Phys. **8**(2), 143–182 (1952)

Takabayasi, T.: Remarks on the formulation of quantum mechanics with classical pictures and on relations between linear scalar fields and hydrodynamical fields. Prog. Theor. Phys. **9**(3), 187–222 (1953)

Terletsky, I.P.: Problèmes du développement de la théorie quantique. Questions scientifiques—Physique, pp. 131–146. Les éditions de la nouvelle critique, Paris (1952)

Valentini, A.: Astrophysical and cosmological tests of quantum theory. J. Phys. A Math. Theor. **40** (12), 3285–3303 (2007)

Valentini, A.: Inflationary cosmology as a probe of primordial quantum mechanics. Phys. Rev. D **82**(6) (2010)

Vals, A.V.: Louis de Broglie et la diffusion de la mécanique quantique en France (1925-1960), PhD dissertation, Université Claude Bernard Lyon 1 (2012)

Vieira, C.L., Videira, A.A.P.: Carried by history: Cesar Lattes, nuclear emulsions, and the discovery of the Pi-meson. Phys. Perspect. **16**(1), 3–36 (2014)

Von Neumann, J.: Mathematical Foundations of Quantum Mechanics. Princeton University Press, Princeton, NJ (1955)

Wang, J.: American Science in an Age of Anxiety: Scientists, Anticommunism, and the Cold War. University of North Carolina Press, Chapel Hill, NC (1999)

Whyte, L.L.: The Scope of Quantum-Mechanics—Discussion. Br. J. Philos. Sci. **9**(34), 133–134 (1958)

Wiener, N., Siegel, A.: A new form for the statistical postulate of quantum mechanics. Phys. Rev. **91**(6), 1551–1560 (1953)

Wiener, N., Siegel, A.: The differential-space theory of quantum systems. Il Nuovo Cimento **2** (4 Suppl), 982–1003 (1955)

Chapter 3
The Origin of the Everettian Heresy

Abstract In 1956, Hugh Everett, then a PhD student at Princeton, proposed his "relative state" formulation of quantum mechanics. John Wheeler, who was Everett's advisor, recognized the originality and importance of such a proposal, but he denied that its non-conventional approach to measurement questioned the orthodox view. Indeed, Wheeler made serious efforts to obtain the blessing of Niels Bohr for Everett's ideas. These efforts gave rise to a lively debate with the Copenhagen group, the existence and content of which have been only recently disclosed by the discovery of unpublished documents. The analysis of such documents opens a window on the conceptual background of Everett's proposal, and illuminates at the same time some crucial aspects of the Copenhagen view of the measurement problem. Also, it provides an original insight into the interplay between philosophical and social factors which underlay the postwar controversies on the interpretation of quantum mechanics.

3.1 Introduction

The "relative state" formulation of quantum mechanics, put forward by Hugh Everett III in his doctoral dissertation (Everett 1957a), has become popular as one of the most heterodox interpretations of quantum mechanics. This is due, in the first place, to its non-conventional treatment of the measuring process. Remarkably, however, John A. Wheeler, who was Everett's advisor at Princeton University and a dedicated Bohrian, thought that Everett's proposal was not meant to *question* the orthodox approach to the measurement problem.[1] Indeed, Wheeler made serious efforts to obtain Bohr's blessing for Everett's ideas. In 1956, when he left Princeton to spend one semester in Leiden, he sent a draft of Everett's dissertation to Bohr and

---

This chapter is a roughly reproduction of Stefano Osnaghi, Fábio Freitas, and Olival Freire Jr, "The Origin of the Everettian Heresy," *Studies in History and Philosophy of Modern Physics*, 40, 97–123, 2009. Some references were updated and references and footnotes were adjusted and minor changes were made in the text in order to adjust to the book's style. Spelling was kept as in the original paper. We are grateful to Elsevier for allowing its reproduction. Credits and acknowledgments are recorded in the original paper.

[1] John A. Wheeler to Alexander Stern, 25 May 1956, *WP* (Series 5—Relativity notebook 4, p. 92).

O. Freire Junior, *The Quantum Dissidents*, DOI 10.1007/978-3-662-44662-1_3

went personally to Copenhagen in order to discuss it with him and his collaborators. The debate went on in the following months, culminating in a visit paid by Everett to Bohr in 1959, 2 years after the publication of the dissertation. Notwithstanding Wheeler's reiterated efforts, however, the Copenhagen group remained not only unsympathetic to Everett's ideas, but also reluctant to attach any relevance to them.

The existence of this early debate on Everett's ideas has remained unknown until recently,[2] and its content has not been exhaustively analysed so far. More generally, in spite of the increasing attention that the relative state formulation is receiving from physicists and philosophers,[3] the context of its birth and that of its early reception have not been thoroughly investigated.[4] The purpose of the present paper is to fill this lacuna. We will analyse Everett's first manuscripts, as well as the criticisms raised in Copenhagen and the way Everett replied to them. This analysis is not meant to solve the problems that beset Everett's programme, nor to provide grounds for one particular interpretation of his ideas over the others. Nevertheless, it can contribute to the clarification of some controversial passages in his published papers,[5] and help to appraise the overall coherence of his project.

There is, however, another reason for which the reconstruction of the early debate on Everett's dissertation is valuable, namely that such a reconstruction sheds light on the role that Bohr played in the controversies over the foundations of quantum theory in the 1950s. Two issues are involved here.

The first is Bohr's approach to the measurement problem. This is a rather controversial (and poorly documented) topic,[6] on which the documentary material that we have uncovered provides interesting insights. We will examine in particular some letters in which Bohr's collaborators spell out their view of the problem and contrast it with the approaches inspired by von Neumann's theory of measurement. These letters, together with the replies of Everett and Wheeler, document the misunderstandings that hindered the comprehension of Bohr's ideas and made their epistemological and methodological implications so difficult to grasp for those who did not belong to the inner circle of his collaborators. It is quite revealing that even someone like Wheeler, who had worked with Bohr and considered

[2] See Freire Jr. (2004), Freire Jr. (2005), and Byrne (2007). See also Freitas (2007). After the appearance of our paper Osnaghi et al. (2009), other works have dealt with the context of this debate in Copenhagen over Everett's thesis, see Byrne (2010) and Everett et al. (2012). For our review on the latter, see Freire Jr (2014).

[3] See Barrett (1999), Butterfield (2002), and references therein. See also Ben-Dov (1990) and Lehner (1997).

[4] Cassinello (1994) contains some historical remarks concerning the origin of Everett's thesis. E. B. Shikhovtsev, Biographical sketch of Hugh Everett, III, 2003 Niels Bohr Library, American Institute of Physics, College Park, MD, Unpublished paper (http://www.hep.upenn.edu/~max/everett/everettbio.pdf) provides more information. Both papers, however, overlook the discussions which took place with the Copenhagen group.

[5] See Barrett (1999, Chap. 3).

[6] See Teller (1981) and Murdoch (1987).

himself an orthodox Bohrian, seemed not to be aware of the chasm that separated the epistemological presuppositions of Bohr's and Everett's programmes.

This brings us to another important issue involved in our analysis, namely the historiographical problem of elucidating the rise and fall of what Jammer has called the "monocracy of the Copenhagen school" (Jammer 1974, p. 250). The story of Everett's dissertation can be regarded as a paradigmatic example of how strong the influence of Bohr was, even in the American context of the 1950s. However, as we will see, the very factors which ensured the supremacy of the so-called Copenhagen interpretation harboured the premises of its eventual decline. As a fine-grained analysis will reveal, such premises were already apparent in the Everett episode.

Section 3.2 outlines briefly the historical context in which Everett's proposal was conceived, focusing in particular on the attitude of the physics community towards Bohr's ideas in the 1950s. Section 3.3 describes the genesis of Everett's dissertation, whose content is discussed in Sects. 3.4 and 3.5. Special attention will be paid to the conceptual background of Everett's ideas and to their relationship to other research programmes that were developed in the same period. Section 3.6 provides a historical reconstruction of the various stages of the debate that opposed Wheeler and Everett to the Copenhagen group. The conceptual and philosophical content of the debate is analysed in Sect. 3.7. In Sect. 3.8, after relating the epilogue of the thesis affair, we focus on the early reception of Everett's ideas. In order to elucidate the psychological, social and cultural factors which influenced the discussion in the 1950s, it will also prove enlightening to take into account the subsequent evolution of Wheeler's and Everett's ideas and careers. Section 3.9 summarises our conclusions.

3.2 Historical Background: The Twilight of the "Copenhagen Monocracy"

In this section we outline the context in which the relative state formulation appeared. We focus in particular on Niels Bohr and the so-called "Copenhagen school", whose important (and complex) role within such a context needs to be spelled out before addressing the Everett affair itself.

3.2.1 General Attitude Towards the Foundational Issues in the US

In the US, which after the Second World War became the central stage of research in physics in the West, the discussions about the interpretation of quantum mechanics had never been very popular.[7] A common academic policy was to gather theoreticians and experimentalists together in order to favour experiments and concrete applications, rather than abstract speculations (Schweber 1986). This practical attitude was further increased by the impressive development of physics between the 1930s and the 1950s, driven on the one hand by the need to apply the new quantum theory to a wide range of atomic and subatomic phenomena, and on the other hand by the pursuit of military goals. As pointed out by Kaiser (2002, pp. 154–156), "the pedagogical requirements entailed by the sudden exponential growth in graduate student numbers during the cold war reinforced a particular instrumentalist approach to physics." In this context, "epistemological musings or the striving for ultimate theoretical foundations—never a strong interest among American physicists even before the war—fell beyond the pale for the postwar generation and their advisors." A few textbooks, like for example David Bohm's *Quantum theory* (Bohm 1951), discussed some issues of interpretation. However, as a rule, the textbooks in use in the 1950s (in America as well as elsewhere) did not reflect much concern at all about the interpretation of the theory (Mehra and Rechenberg 2001, p. 1194).

A consequence of this attitude was that little attention was paid to Bohr's complementarity, which, according to Heilbron (2001), was perceived as an eminently philosophical approach, an especially obscure one indeed.[8] Kragh (1999, p. 211) has observed that "the uncertainty principle was eagerly taken up by several American physicists [. . .], but they showed almost no interest in Bohrian complementarity." According to him: "Most textbook authors, even if sympathetic to Bohr's ideas, found it difficult to include and justify a section on complementarity. Among 43 textbooks on quantum mechanics published between 1928 and 1937, 40 included a treatment of the uncertainty principle; only eight of them mentioned the complementarity principle."

Bohr's epistemological reflections were circulated in papers presented at conferences and published in scientific journals and anthologies. Such publications were unlikely to have any direct influence on the background of young physicists, which depended mainly on textbooks.[9] In a referee's report of 1957, Léon Rosenfeld, who was one of Bohr's closest collaborators since the 1930s, complained about this state of affairs: "There is not a single textbook of quantum mechanics in any language in which the principles of this fundamental discipline are adequately treated, with proper consideration of the role of measurements to

[7] Referring to the attitude of American physicists towards the early debate on the foundations of quantum mechanics, Cartwright (1987) has observed that "Americans in general had little anxiety about the metaphysical implications of the quantum theory; and their attitude was entirely rational given the operationalist-pragmatist-style philosophy that a good many of them shared." According to Kragh (1999, p. 211), the "interest in foundational problems among the Americans [. . .] went in different directions and was on a less grand scale than in Denmark and Germany." See also Sopka (1980, pp. 3.67–3.69) and Assmus (1992).

[8] Chevalley (1997, pp. 598–600) and Chevalley (1999).

[9] See Kuhn (1970).

define the use of classical concepts in the quantal description."[10] In a letter to Bohr of the same year, Rosenfeld remarked: "There is great interest in the topic among chemists and biologists, but there is no book that one can refer them to and that could protect them from the confusion created by Bohm, Landé, and other dilettantes." And he concluded: "I will now do my bit here in Manchester by giving a lecture for chemists and biologists; but nothing can replace the book that *you* must write."[11] As is well known, Bohr did not comply.

Even the circumstances that counterbalanced the scarce propensity of American physicists towards foundational issues ran against the general endorsement of Bohr's views. For example, a number of distinguished scholars who had taken part in the early debate on the significance of quantum mechanics, such as von Neumann, Wigner and Einstein, moved subsequently to the US. But none of them were particularly well disposed towards complementarity. Furthermore, in the 1950s, the circumscribed but increasing interest in cosmology and general relativity boosted a highly speculative field of research, in which American theorists were faced with the fundamental problem of reconciling quantum mechanics with gravitation. However, the approach based on complementarity was generally considered to be unsuited to deal with such a problem.[12]

3.2.2 Bohr and the Quantum Orthodoxy

The existence of an "orthodox view" of quantum mechanics was generally taken for granted since the 1930s. However, the meaning of such a label was far from being univocally determined.[13] Several factors contributed to keeping its definition vague, and by the same token to reinforcing the impression that an orthodox view did indeed exist. The very term "Copenhagen interpretation", introduced in the late 1950s to denote the orthodox view,[14] was in the first place intended to underpin the myth of a monolithic "Copenhagen school" acting as the guardian of the quantum

[10] In 1957, Rosenfeld was requested to give an opinion about the possible translation of Louis de Broglie's *La théorie de la mesure en mécanique ondulatoire* into English. The quotation is from the (negative) referee's report he wrote on that occasion (Léon Rosenfeld. Report on: Louis de Broglie, *La théorie de la mesure en mécanique ondulatoire* (Paris: Gauthier-Villars), 1957, *RP*, Niels Bohr Archive, Copenhagen, Unpublished paper).

[11] Léon Rosenfeld to Niels Bohr, 14 Jan 1957, *BSC* (reel 31).

[12] This is quite apparent from the 1957 papers of Everett and Wheeler (see Sect. 3.4.2). This point was explicitly discussed by DeWitt in a lecture of 1967 (DeWitt 1968).

[13] See Scheibe (1973, p. 9), Beller (1999b, pp. 187–188), Camilleri (2009).

[14] The term was probably introduced by Heisenberg in his contribution to the volume celebrating Bohr's 70th birthday (Pauli 1955). The usage of such a label was criticised by Rosenfeld, because it implicitly allowed the existence of other interpretations (Freire Jr. 2005, p. 28). Howard (2004) suggested that Heisenberg had in fact personal reasons—namely, the wish to break his isolation after WWII—for assimilating his own position to that of Bohr, whose ideas on complementarity he actually never endorsed.

orthodoxy. Such a myth was to some extent constructed retrospectively to serve the purposes of the parties involved in the controversies of the 1950s, a period marked by hidden variables and Marxist materialism.[15]

Faye (2002) has argued that the label "Copenhagen interpretation" was used "by people opposing Bohr's idea of complementarity, to identify what they saw as the common features behind the Bohr–Heisenberg interpretation as it emerged in the late 1920s." It was generally assumed that these "common features" were conveyed by the "standard" *formulation* of quantum mechanics, whose most popular and mathematically sound version was provided by John von Neumann's *Mathematische Grundlagen der Quantenmechanik*. This book, published in 1932, had several reprints and translations (the English version appeared in 1955). It provided an axiomatic theory in which some aspects of the presentation given by Dirac in his *The Principles of Quantum Mechanics* (1930) received a more rigorous formulation. Thus, for example, the so-called postulate of projection formalized Dirac's idea that, when a system is measured, it "jumps" into an eigenstate of the measured observable.[16]

Von Neumann's formalism can be interpreted in different ways and it is not a priori incompatible with Bohr's view. Yet, von Neumann's presentation may appear "misleading in several respects" when regarded from a Bohrian standpoint. Thus, for example, in the abovementioned report, Rosenfeld observed that "v. Neumann's book 'Foundations of Quantum Mechanics' [. . .], though excellent in other respects, ha[d] contributed by its unhappy presentation of the question of measurement in quantum theory to create unnecessary confusion and raise spurious problems."[17] (Rosenfeld, 1957, *op. cit.*) Indeed, as Kragh puts it, "the 'measurement problem' was not the same for Bohr and von Neumann."[18] The reason why von Neumann's formulation was nonetheless routinely associated with the

[15] See Pauli (1955), Freire Jr. (2005, p. 28), Howard (2004).

[16] Dirac (1958, p. 36). For a discussion see Barrett (1999, pp. 22–37).

[17] Rosenfeld's *Report* contains further considerations about the treatment of measurement in the textbooks of quantum mechanics: "The nearest to a really good treatment is found in Landau and Lifschitz's outstanding treatise: but it is too short and not explicit enough to be a real help to the student. The only books which are purposely devoted to an exposition of the principles are v. Neumann's aforementioned treatise and a little book by Heisenberg: the first is (as stated above) misleading in several respects, the second is too sketchy and on the subject of measurements it even contains serious errors (however surprising this may appear, the author being one of the founders of the theory). As to Bohr's authoritative article, it is in fact only accessible to fully trained specialists and too difficult to serve as an introduction into this question." (Rosenfeld, 1957, *op. cit.*)

[18] "Bohr tended to see it as a problem of generalizing the classical framework in order to avoid contradictions between two mutually incompatible classical concepts, both necessary in the description of experiments. His solution was complementarity." In contrast, "to von Neumann, [. . .] the problem of measurement meant the mathematical problem of proving that the formalism gave the same predictions for different locations of the 'cut' between observer and object" (Kragh 1999, p. 214). In the 1960s this difference in the approach to measurement gave rise to what has been called the "Princeton school". This term refers in particular to Eugene Wigner's view of measurement; see Home and Whitaker (1992) and Freire Jr. (2007).

"Copenhagen interpretation" is that what people meant by such a term had in most cases little to do with Bohr's complementarity.[19] This is not too surprising, since even within the "Copenhagen scholars", there existed divergent interpretations of Bohr's approach.[20] We are therefore faced with two questions. First, why was the existence of a standard view of quantum mechanics taken for granted? And second, why was such a view so often associated with Bohr?

As for the first question, it must be observed that, in spite of the existence of important differences, both the intellectual backgrounds and the scientific views of people like Bohr, Pauli, Heisenberg, Born, and Jordan, who had been working together on the collective construction of quantum mechanics,[21] had several points in common. All of them endorsed both indeterminism and the assumption of the corpuscular and discrete nature of atomic phenomena. They also firmly believed in the completeness of quantum theory and were prepared to dispense with the isomorphism between the symbolic structures of physics and the pictorial representation of microscopic objects. To them, the main issue raised by quantum mechanics was not one of interpretation, but rather one of epistemology (Heilbron 2001): how must our view of physical knowledge be amended in order to accommodate the implications of the discovery of the quantum of action? In this sense, they were attached to the revolutionary character of quantum mechanics,[22] and were unsympathetic to any attempt to restore such classical ideals like causality and visualizability in microphysics.

As for the second question, the reason why the standard view of quantum mechanics was commonly attributed to Bohr (and indeed termed the *Copenhagen* interpretation) is undoubtedly related to Bohr's intellectual charisma and to his role in the construction of quantum mechanics.[23] Bohr's personal influence upon his colleagues is legendary and has been exhaustively analysed by Chevalley (1997). Beller (1999b, pp. 254–257) has described Bohr as a "charismatic leader". "As the founder of the philosophy of complementarity, Bohr was declared by his followers to be not merely a great philosopher, but a person of exceptional—perhaps superhuman—wisdom, both in science and in life." Thus, for example, in a recollection

---

[19] "The Copenhagen interpretation [. . .] is a mixed bag, consisting of the errors and misunderstandings and superficialities of many people. [. . .] Hence, putting your hand into this bag you may come up with almost anything you want". Paul Feyerabend, letter to Imre Lakatos, 28 Jan 1968, in Lakatos et al. (1999, p. 127). Feyerabend is here defending Bohr's original view against Popper's criticisms, and arguing that Popper mispresented Bohr, just as "almost all physicists" did.

[20] See Howard (2004), Camilleri (2009), Jacobsen (2007).

[21] See Rozental (1967), Heilbron (2001).

[22] In a conversation with Everett, which occurred in the 1970s, Charles Misner, who had been Everett's roommate at Princeton and a student of Wheeler's, recalled that, as an undergraduate, he was "taught by people who had learned quantum mechanics in the 1930s." He remarked that "to them, quantum mechanics was really a big philosophical change, and they were shocked by the whole ideas," whereas he and Everett "[. . .] felt that well, you know, every new course in physics you get some new kind of nonsense which seems to make sense a little bit later [. . .]." (Hugh Everett interviewed by Charles Misner, May 1977, p. 9, *EP*.)

[23] See e.g. Bohr (1949).

of the 1980s, Wheeler, compared Bohr's wisdom with that of Confucius and Buddha, Jesus and Pericles, Erasmus and Lincoln (Wheeler 1985, p. 226). Besides setting the agenda for the development and comprehension of quantum mechanics, Bohr and the Institute of Theoretical Physics of Copenhagen, which he had founded in 1921, provided guidance for a whole generation of physicists, including Heisenberg, Pauli, Dirac, Landau, Weisskopf, Wheeler and many others (Rozental 1967; Bohr et al. 1985). As emphasised by Beller, all those who visited the Institute were deeply impressed by the experience. However, "while in matters of complementarity philosophy not directly relevant to research, physicists were willing to repeat 'Bohr's Sunday word of worship', in physics proper they maintained a fruitful balance between humble reverence and free creativity" (Beller 1999b, p. 257)—a balance similar to that which characterized Wheeler's attitude in the Everett affair.

During the 1920s and 1930s, the ideas which were to be identified with the "orthodox view" of quantum mechanics became quite popular. The positivist flavour of the approach developed by Heisenberg, Jordan, Born and Pauli was not only in tune with the cultural climate of continental Europe between the two wars,[24] but was also well suited to cope with the change of paradigm that atomic phenomena seemed to demand. Bohr's arguments were generally taken as a warrant that such an approach was free from inconsistency and could be accommodated in a coherent conceptual framework, although the acknowledgement of Bohr's authority implied neither the conscious adhesion to, nor the clear understanding of, his philosophy (Heilbron 2001). Did this state of affairs give rise to "a somewhat dictatorial imposition of what was called 'the Copenhagen dogma' or 'orthodox view' upon the younger generation of physicists" (Jammer 1974, p. 250) ? To be sure, the defence of the orthodox ideas by a group of physicists whose outstanding prestige was unanimously acknowledged was not always carried out according to the polite rules of an open and rational discussion.[25] However, it is likely that both

[24] See e.g. Jammer (1966, Sect. 3.4.2), Forman (1971), Brush (1980).

[25] This observation does not apply solely to the old guard of the Copenhagen school. "Some of the most vitriolic comments directed at people who questioned the Copenhagen Doctrine were given by Rosenfeld. He's written some papers that have taken the young people who were wanting to probe a little more deeply to task". (Bryce S. DeWitt & Cecile M. DeWitt-Morette interviewed by Kenneth W. Ford, 28 Feb 1995, p. 18, *AIP*.) Rosenfeld's attitude is apparent from his letters, some of which are quoted in the remainder of this paper. In 1972, he wrote for example to Frederik Belinfante: "Not only [. . .] is it futile to speak of two Copenhagen schools; but it is even wrong to speak of one Copenhagen school; there has never been any such thing and I hope there will never be. The only distinction is between physicists who understand quantum mechanics and those who do not." Léon Rosenfeld to Frederik J. Belinfante, 22 Jun 1972, *RP*. Feyerabend argued that the vagueness of the principles defining the Copenhagen interpretation allowed its defendants "to take care of objections by *development* rather than by *reformulation*", a procedure which—he added—"serves to create the impression that the correct answer has been there all the time and that it was overlooked by the critic." Hence, according to Feyerabend, the attitude of Bohr and his followers "has very often been one of people who have the task to clear up the misunderstandings of opponents rather than to admit their own mistakes" (Feyerabend 1964, p. 193, quoted in Home and Whitaker 1992, pp. 258–259). Beller (1999a, p. 191) has described the dialectical strategy of the Copenhagen scholars as "the rhetoric of finality and inevitability", arguing that they

the existence of an "orthodox view" and the unsharpness of its definition met the needs of the majority of the physics community, which was not concerned with the foundations of quantum mechanics in so far as the theory could be efficiently used to perform calculations and experiments. Not only did vagueness act as a protective belt which prevented the users of the theory from being faced too crudely with the alleged flaws in its foundations, but it also made possible the identification between the orthodox view and Bohr's, thereby allowing them to rely on Bohr's undisputed authority when adopting such an uncritical attitude.[26] As regards the dissenters, the possibility of contrasting their original proposals with a dominant view could offer both psychological and rhetorical advantages. Generally, by the label "orthodox" (or the equivalent "official", "usual", etc.), the dissenters meant the instrumentalist attitude that rejected any attempt to provide a coherent pictorial model of the world allegedly underlying the quantum phenomena. This was of course a dramatic simplification of Bohr's stance. But identifying it with the "orthodox view" allowed the dissenters to avoid coming to grips with the more sophisticated (and, to many, obscure) aspects of Bohr's doctrine.

3.2.3 The Revival of Dissidence and the Measurement Problem

Notwithstanding some disagreements about the philosophical interpretation of complementarity (Camilleri 2009), between the 1930s and the end of the 1940s the "monocracy of the Copenhagen school in the philosophy of quantum mechanics" remained "almost unchallenged" (Jammer 1974, p. 250). Einstein, who was one of the earliest and most influential critics, did not renew his attacks after the discussions on the EPR paper in the mid-1930s. Schrödinger dismissed his "wave interpretation" of 1926, and his analysis focused on the epistemic interpretation of the state vector which he regarded as the "official" one. Even de Broglie repudiated his pilot-wave theory and joined the orthodox camp (Jammer 1974, pp. 113–114). In the early 1950s, however, the situation began to change. "The appearance in 1949 of the often quoted Einstein volume edited by Schilpp and Einstein (1949) which contained Bohr's debate with Einstein, Einstein's self-written 'obituary' and

---

"advocated their philosophy of physics not as a possible interpretation but as the only feasible one." This attitude was often pointed out by those who, like Einstein, were dissatisfied with the Bohr–Heisenberg "religion" Albert Einstein to Erwin Schrödinger, 31 May 1928, *apud* Murdoch (1987, p. 101); see also Heilbron (2001, pp. 222–223). Thus for example, in a paper that appeared in *Physics Today* in 1954, Henry Margenau (1954, p. 9) observed that Bohr's complementarity "relieved its advocates of the need to bridge a chasm in understanding by declaring that chasm to be unbridgeable and perennial; it legislated a difficulty into a norm."

[26] In one of his Dublin seminars (1949–1955), Schrödinger remarked: "Philosophical considerations about quantum mechanics have gone out of fashion. There is a widespread belief that they have become gratuitous, that everything is all right in this respect for we have been given the marvellously soothing word of *complementarity* [. . .]" (Apud Bitbol 1996a, pp. 212–213).

his candid 'reply to criticisms' and which was widely read by philosophizing physicists contributed considerably to the creation of a more critical atmosphere toward the complementarity philosophy."[27] In the same period, in his Dublin seminars, Schrödinger presented his critical reflections on the orthodox view, which were subsequently developed in a series of papers that appeared in the 1950s. In these papers Schrödinger sharpened his criticisms and sketched a sophisticated philosophical framework (differing substantially from that of 1926) for his wave interpretation.[28]

In contradistinction to the previous decades, a number of physicists belonging to the new generation, who—to paraphrase John Bell—had not sat at the feet of Bohr, were sympathetic to such criticisms (Bell 2004, p. 271). The social and cultural context of fundamental research had undergone deep changes following the WWII. On one hand, in the West, the intellectual environment resulting from the Americanization of research was not very favourable to the understanding of Bohr's ideas, although for the reasons highlighted in Sect. 3.2.1, this did not immediately produce a hostile attitude.[29] On the other hand, in the Soviet Union, such ideas, which had been previously tolerated, were accused of promoting idealist trends in science and were almost banished (Graham 1988). The repercussions of the Soviet polemics were enhanced by the context of the Cold War. Marxist physicists in the West were stimulated to take sides with the critics of Bohr's views. Some of them endorsed either the "stochastic" or the "statistical" interpretations, which seemed to fit the materialist framework better than complementarity.[30] However, the main challenge to the orthodox view came from David Bohm, a brilliant young physicist and American Marxist. This challenge was analysed in Chapter 2. In 1952 he proposed a hidden variable theory in which particles had well-defined (though not entirely determinable) trajectories. Such a theory challenged a famous no-go theorem stated by von Neumann (which was supposed to rule out hidden variables) and called into question the need to resort to complementarity when dealing with atomic phenomena. Bohm's theory was generally regarded with scepticism. Yet it gathered some important supporters, including Jean-Pierre Vigier, Mario Bunge, and Hans Freistadt. De Broglie himself, stimulated by Bohm's work, resumed his pilot-wave programme with renewed enthusiasm.[31]

---

[27] Jammer (1974, p. 250). Einstein's late objections against the "orthodox view" are discussed in Howard (1985). See also Paty (1995).

[28] See Bitbol (1996a).

[29] As late as in 1970, DeWitt (1970, p. 159), in introducing what he called the "'conventional' or 'Copenhagen' interpretation", observed: "If a poll were conducted among physicists, the majority would profess membership in the conventionalist camp, just as most Americans would claim to believe in the Bill of Rights, whether they had ever read it or not."

[30] See Jammer (1974, Chaps. 9 and 10). There were, however, important exceptions, like for example Rosenfeld and the Soviet physicist Vladimir Fock. About Marxism and quantum mechanics, see Freire Jr. (2011).

[31] For an elementary account of Bohm's theory, see Barrett (1999). The role played by Bohm's Marxist ideas in his search for a new interpretation of quantum mechanics is discussed in Forstner (2008). For an analysis of the reception of Bohm's proposal, see Freire Jr. (2005). A survey of the "causal interpretations" proposed in the early 1950's can be found in Scheibe (1973, p. 2). See also Jammer (1974, pp. 287–288).

In 1957, some of these alternative views on quantum mechanics were debated at an international conference held in Bristol. Besides Bohm, Rosenfeld and other distinguished physicists, a number of philosophers—such as Adolf Grünbaum, Norwood Hanson, and Paul Feyerabend—attended the meeting and took part in the discussions.[32] Though such discussions were probably not given much importance in Copenhagen,[33] the fact that three of the founding fathers of quantum mechanics, all of which Nobel Prize winners, had resumed their earlier criticisms could not go unnoticed.[34] As pointed out by Camilleri (2009), "in the context of the emergence of a new threat from Bohm, de Broglie and Vigier, as well as Soviet physicists such as Blokhintsev and Alexandrov, the different schools of thought [which had been involved in the previous decades in the dispute on the true meaning of complementarity] closed ranks in identifying themselves with Bohr—the canonical author—whose writings were taken as a direct expression of the 'authentic' Copenhagen interpretation." Indeed, Pauli, Heisenberg, Born and Rosenfeld all wrote papers to rebut the objections of Schrödinger and other dissenters. Bohm's work, in particular, was virulently criticised.[35]

The controversies in the first half of the 1950s revolved mainly around the possibility of providing a "causal interpretation" of quantum mechanics—possibly "completing" it with "hidden parameters". In the second half of that decade, however, the problematic aspects of measurement in quantum physics started to receive increasing attention. An important part of Heisenberg's contribution to the volume celebrating Bohr's 70th birthday, in which the author presented the Copenhagen interpretation and replied to recent criticisms, was dedicated to spelling out what Heisenberg considered to be the orthodox approach to measurement. Heisenberg quoted in particular an assertion by Lajos Janossy to the effect that, since the "reduction of wave-packets" cannot be deduced from Schrödinger's equation, there must be "an inconsistency in the 'orthodox' interpretation."[36]

The doubts raised by the "reduction of wave-packets" were certainly not new (they went back to the Fifth Solvay conference of 1927 and had been discussed for example at an international conference held in Warsaw in 1938, which both von

[32] See Körner (1957). Karl Popper, who was not able to attend, sent a written report.

[33] Rosenfeld advised Bohr not "to waste his time in reading [the proceedings of the conference]", but rather suggested that Petersen might look through them and tell him "about the worse nonsense" he would find there. (Léon Rosenfeld to Niels Bohr, 21 Oct 1957, *BSC*, reel 31.)

[34] "This comedy of errors [the attempt to develop a "theory of measurement" based on the "causal interpretation" of quantum mechanics] would have passed unnoticed, as the minor incident in the course of scientific progress which it actually is, if it had not found powerful support in the person of L. de Broglie, who is now backing it with all his authority." (Rosenfeld 1957, *op. cit.*)

[35] See e.g. George (1953), Born (1953), Pauli (1955). As regards the criticisms addressed to Bohm, see Chap. 2.

[36] Heisenberg (1955, p. 23). Such statements are not unusual in the literature of the 1950s. Schrödinger, for example, repeatedly criticised the collapse of the wave function (Bitbol 1996a, p. 111): see for instance Schrödinger (1953, pp. 18–20). See also Margenau (1958), in which the objections of de Broglie are discussed (pp. 31–32). (Margenau's own criticisms went back to the 1930s.)

Neumann and Bohr attended). In the 1930s and 1940s, there had been some sporadic contributions intended to clarify the puzzling aspects of von Neumann's postulate of projection. These contributions included a couple of works which attributed a crucial role to mental faculties such as volition and consciousness in the measuring process.[37] Far from committing himself to such approaches, Bohr put much emphasis in his writings on the fact that the physical account of measurement by no means required a *conscious* observer.[38] While it is likely that such emphasis reflected the worry that his view could be confused with what the Soviets regarded as "idealistic vagaries",[39] there is no doubt that it also expressed a deep conviction of his. The role played by the observer in the epistemological framework of complementarity was not to be understood in terms of idealistic doctrines, but rather in connection to a *pragmatic* analysis of the conditions under which one can acquire objective knowledge.[40] However, for many scholars, denying the subjectivist character of Bohr's approach amounted to dismissing at once his pragmatic analysis. Along these lines, Bohr's *functional* distinction between object system and measuring instrument was presented as a crude *physical* assumption according to which macroscopic systems behave classically. In other words, according to this reading, Bohr's approach just split the physical world into a quantum microcosm and a classical macrocosm.[41]

In the second half of the 1950s there was a rise of studies on the measurement problem,[42] from which emerged in particular the "thermodynamic approach"

[37] The first was a little book by Fritz London and Edmond Bauer (1939), and the second was a paper by Carl Friedrich von Weizsäcker (who was a close collaborator and former student of Heisenberg). See Jammer (1974, pp. 482–489).

[38] Thus, for example, in a paper of 1958, Bohr (1963, p. 3) stressed that the description of atomic phenomena has "a perfectly objective character, in the sense that no explicit reference is made to any individual observer." It is worth noting that, in 1957, Fock, who had been a prominent and tenacious advocate of complementarity in the Soviet Union, visited Copenhagen and had a few conversations on the philosophical significance of quantum mechanics with Bohr. According to the Soviet commentators, Bohr's efforts to avoid any "subjectivist" ambiguity in his late writings were an outgrowth of such conversations (Graham 1988, pp. 311–313).

[39] See Graham (1988). Heisenberg's epistemic interpretation of the wave function was often considered to imply a "subjectivist" view, see Stapp (1994), Howard (2004). Since Heisenberg was considered to be a member of the "Copenhagen school", the charge of subjectivism was sometimes extended to Bohr; Howard (2004) discusses in particular the use of this rhetorical strategy in Popper's writings.

[40] These aspects are discussed in Sect. 3.7.

[41] See e.g. Bell (2004, pp. 188–189). A good example is provided by the celebrated course of theoretical physics of the Soviets Lev Landau and Evgenij Lifshitz (whose first edition in English, supervised by John Bell, appeared in 1958). Their account of measurement, which was traditionally considered to be quite close to Bohr's (Bell said that it was perhaps "the nearest to Bohr that we have"; *Ibid*, p. 217), postulated—in Bell's words—that macroscopic systems "spontaneously" jump into a definite macroscopic configuration which, in the case of a "classical" apparatus, corresponds to an eigenstate of the "reading" (i.e. a so-called "pointer state").

[42] See Margenau (1963) and references therein.

developed by Günther Ludwig.[43] By treating macroscopic measuring apparatus as thermodynamic systems, such a programme purported to explain, within the framework of ordinary quantum mechanics, the fact that measurements have *definite* outcomes. After Bohr's death, those of his disciples who were committed to materialism, like Rosenfeld, saw in such a programme the possibility of providing a rigorous physical foundation for Bohr's approach, thereby dispelling the misunderstandings surrounding the alleged subjectivism of the Copenhagen view. Thus, when Wigner (1963) took up the banners of the approach which attributed a role to the observer's mind, claiming that it fitted the orthodox view of Heisenberg and von Neumann, Rosenfeld reacted by strongly supporting the theory of measurement that Adriana Daneri, Angelo Loinger and Giovanni Maria Prosperi (1962) had proposed in the framework of the thermodynamic approach.[44]

3.3 The Genesis of Everett's Thesis

3.3.1 Everett at Princeton

Everett enrolled himself at Princeton University in 1953, after obtaining a bachelor's degree in chemical engineering at the Catholic University of America in Washington, where he had shown exceptional mathematical ability.[45] In his 1st year Everett took the course of Quantum Mechanics with Robert Dicke.[46] In May 1955 he passed the general exams and undertook his doctoral research on the "Correlation Interpretation of Quantum Mechanics" under the supervision of Wheeler.

Wheeler was a prominent figure at Princeton. He had given important contributions to nuclear physics and had served in the Manhattan project. When he met Everett, at some moment between 1954 and 1955, he was just beginning to get involved in the research in cosmology. Wheeler had been acquainted with Bohr since the mid-1930s, when he had spent some time at the Institute of Theoretical Physics of Copenhagen with a Rockfeller post-doctoral fellowship (Wheeler 1985, p. 125). In 1939, Bohr visited Princeton bringing the news of the first observations of nuclear fission, and they started a collaboration that led to the theory of fission based on the liquid drop model. They remained friends until Bohr's death.[47] In an address delivered at Princeton University in 1955, Wheeler described Bohr's

[43] See Jammer (1974, pp. 488–490).

[44] See Rosenfeld (1965) and the discussion of Sect. 3.7.3. For a detailed analysis of the dispute between Rosenfeld and Wigner, which went on till the early 1970s, see Chaps. 4 and 5.

[45] For Everett's biography, see Byrne (2010). The information about Everett's curriculum is taken from the Princeton alumni file, *GAR*.

[46] From Dicke's textbook (Dicke and Wittke 1960) we can conjecture that the course paid little attention to interpretive issues.

[47] In 1957, Bohr earned the Atoms for Peace Award. In reply to Wheeler's congratulations, Bohr wrote to him: "In these weeks I have with gratitude dwelt with many memories and not least with our cooperation through the years and your faithful friendship." (Niels Bohr to John A. Wheeler,

complementarity as "the most revolutionary philosophical conception of our day."[48] Therefore his decision of discussing Everett's ideas with Bohr in person shows to what extent he must have been impressed by them. Indeed, Wheeler's letters prove that he held Everett in high esteem.[49]

With regard to the origin of Everett's ideas on quantum mechanics, our main source is an interview recorded at a party in 1977 (*op. cit.*). The interview is in fact an informal discussion with Charles Misner, who had done his PhD in cosmology under Wheeler in the same years as Everett. According to Everett's and Misner's recollection, the choice of the topic of Everett's thesis was influenced by the discussions which they both had with Bohr's assistant Aage Petersen, who was then visiting Princeton.[50] In the interview,[51] Everett remarks that Petersen was the only one who "took seriously" the issues relating to the foundations of quantum mechanics, and in his letters to Petersen he repeatedly expresses the desire of renewing their "always enjoyable arguments."[52]

In one of his papers, Everett quotes an address delivered by Einstein (who had been working at the Institute for Advanced Studies of Princeton since 1933) in the spring of 1954.[53] On that occasion, according to Everett, Einstein had colourfully expressed his discomfort with the idea that simple acts of observation can bring about drastic changes in the universe.[54] This is a good example of the kind of atmosphere that Everett could respire at Princeton, even though the emphasis put by

---

12 Apr 1957, BSC, reel 33). Bohr received the Award at a ceremony which was attended by President Eisenhower and for which Wheeler delivered an address.

[48] Wheeler (1956, p. 374); quoted in Jammer (1974, p. 74).

[49] Thus, for example, referring to the necessity to dispel the misunderstandings which could arise from Everett's work, Wheeler wrote to him: "This appallingly difficult job I feel you (among very few in this world) have the ability in thinking and writing to accomplish". And, alluding to Bohr, he added: "The combination of qualities, to accept corrections in a humble spirit, but to insist on the soundness of certain fundamental principles, is one that is rare but indispensable; and you have it. But it won't do much good unless you go and fight with the greatest fighter." (John A. Wheeler to Hugh Everett, 22 May 1956 [2nd letter], *ME*.)

[50] Everett interview, op. cit., p. 9. Petersen was educated at the University of Copenhagen and became Bohr's assistant in 1952. According to Everett, he spent 1 year in Princeton (Hugh Everett to Max Jammer, 19 Sep 1973, *ME*). This occurred probably in 1954–1955, because Petersen accompanied Bohr when Bohr visited Princeton in the autumn of 1954 (see Sect. 3.6). (Felicity Pors, priv. comm., 16 Oct 2007.)

[51] *Ibid,* p.10.

[52] Hugh Everett to Aage Petersen, 31 May 1957, WP (Series I—Box Di—Fermi Award #1—Folder Everett). See also Hugh Everett to Aage Petersen [draft], summer of 1956, *ME*.

[53] Everett (1973, p. 116). Wheeler (1979b, p. 184) recalled: "We persuaded him [Einstein] to give a seminar to a restricted group. In it the quantum was a central topic."

[54] According to Everett's recollection, Einstein said that he "could not believe that a mouse could bring about drastic changes in the universe simply by looking at it". However, the quotation might have been reported to Everett by others, since in his 1977 interview (*op. cit.*, p. 4) he did not remember having attended the seminar.

Misner on Einstein's seminar in the interview suggests that such occasions were in fact rare.[55] Princeton hosted some of the most distinguished experts of the foundations of quantum mechanics: John von Neumann, whose textbook was the main reference of Everett's work (see Sect. 3.4.1), was at the Institute for Advanced Studies; and Eugene Wigner was Everett's professor of Methods of Mathematical Physics at Princeton University.[56] Also, it was at Princeton that, a few years earlier, David Bohm had worked out his hidden variable theory. Everett did not meet Bohm personally, since Bohm had to leave Princeton in 1951, as a consequence of McCarthyism (Olwell 1999; Freire Jr. 2005). However, Everett's manuscripts show that he was acquainted with Bohm's work on hidden variables. Moreover, Bohm's textbook of quantum mechanics (which presented the standard formulation, but also discussed some issues of interpretation such as the measurement problem and the EPR paradox) seems to have been one of Everett's main sources for the study of the Copenhagen views on measurement (see Sect. 3.4.2).

It is reasonable to think that, in this context, a critical attitude towards the orthodox view of quantum mechanics might emerge occasionally in discussions and seminars, and that non-conventional ideas circulated more freely in Princeton than elsewhere. The very fact that Wheeler accepted the supervision of a PhD research like Everett's shows that he had an open-minded attitude with regard to such issues.[57] Indeed, 15 years earlier Wheeler had been the supervisor of Richard Feynman, who, in his PhD thesis had set the basis of the path-integral formulation of quantum mechanics.[58] Even though Everett denied having received any external input for undertaking his work,[59] in the interview he and Misner allude to the

[55] Everett interview, *op. cit.*, p.4. Wheeler (1979b) reported a few occasions when he and Einstein discussed issues of fundamental physics. In May 1953, for example, Einstein invited Wheeler and his students to his home for tea and answered questions about his view of quantum mechanics.

[56] Von Neumann and Wigner were not directly involved in the public debate on the interpretation of quantum mechanics in the 1950s. However, von Neumann's persistent concern with the epistemological issues raised by quantum mechanics is borne out by the efforts he devoted to the revision of the English translation of his book (Freire Jr. 2005, p. 27). See also Rédei and Stöltzner (2001), and, with regard to von Neumann's opinion on Bohm's proposal, (Stöltzner 1999). As for Wigner, his dissatisfaction with Bohr's complementarity predated his involvement in the debates of the 1960s (Freire Jr. 2007; Camilleri 2009). Interestingly, in the notes taken by Wheeler in Copenhagen in 1956 (John A. Wheeler, *Notes taken in Copenhagen*, 3 May 1956, *ME*), Aage Petersen refers to von Neumann's theory of measurement as "von N[eumann] + Wig[ner]" "stuff".

[57] In the interview with Everett (*op. cit.*, p. 5), Misner says: "You probably already had these quantum mechanical ideas and just needed someone to talk to about them and he [Wheeler] was obviously the kind of person who. . ."

[58] Feynman might have read some version of Everett's dissertation (or might have been informed about it by Wheeler), since at the beginning of 1957 he already knew the general lines of Everett's work (see Sect. 3.6).

[59] As we will see in Sects. 3.4 and 3.5, two important sources of inspiration for Everett's work were the hidden variable theories on the one hand, and Schrödinger's "wave interpretation" on the other. Schrödinger was sent a pre-print of Everett's paper in 1957, but, in so far as we know, he did not reply.

influence that Wheeler's characteristic approach to theoretical physics might have exerted on the development of the relative state formulation. Misner says: "He [Wheeler] was preaching this idea that you ought to just look at the equations and if there were the fundamentals of physics [. . .] you followed their conclusions and gave them a serious hearing. He was doing that on these solutions of Einstein's equations like Wormholes and Geons". And Everett replies: "I've got to admit that that is right, and might very well have been totally instrumental in what happened."[60]

Picture 3.1 Bohr converses with group at Princeton. (L-R): Charles W. Misner, Hale F. Trotter of Kingston, Niels Bohr, Hugh Everett, and David K. Harrison of Belmont, MA. Photograph by Alan Richards, courtesy AIP Emilio Segre Visual Archives

The analysis of Everett's early writings does not indicate that his search for an original approach to quantum mechanics was inspired by issues of cosmology. Yet, there is little doubt that Wheeler's interest in Everett's ideas was enhanced by his recent involvement in that area of research. This is mostly apparent from the final version of the dissertation, in the drafting of which Wheeler took an important part. Therefore, if Everett's ideas received some attention when they were first put forward, this might be partly due to the circumstance that, at the time, Princeton was in the small minority of places in the US at which physicists were interested in general relativity and cosmology. (As we will see, Bryce DeWitt, who was a friend of Wheeler's and the head of the cosmology group of the University of North

[60] Everett interview, *op. cit.*, pp. 9–10.

Carolina at Chapel Hill, was to play a crucial role in the diffusion of Everett's ideas.)

3.3.2 The Steps Towards the Dissertation

Everett's dissertation *On the Foundations of Quantum Mechanics* (Everett 1957a) was submitted in March 1957. Except for the abstract[61] and a few minor stylistic alterations, the dissertation is identical to the paper published in July 1957 in the *Reviews of Modern Physics,* with the title *"Relative State" Formulation of Quantum Mechanics* (1957b). It is a rather small manuscript (36 pages), which was written in the winter of 1956–1957. In an introductory note, Everett mentions "an earlier less condensed draft of the present work, dated January 1956", which he says "was circulated to several physicists" (Everett 1957a, p. 1). There is good evidence that the longer draft "circulated to several physicists", whose title was *Wave Mechanics Without Probability,*[62] was very similar, if not identical, to a paper of 137 pages published many years later (in 1973) as *The Theory of the Universal Wave Function.*[63] Henceforth we will refer to this paper as the "long thesis".[64]

The documentary material that will be discussed in the following sections indicates that the manuscript that was read in Copenhagen (*Wave Mechanics Without Probability*) was the *second* version of the thesis.[65] This does not

---

[61] See Barrett (1999, p. 65).

[62] See Alexander Stern to John A. Wheeler, 20 May 1956, *ME*; Wheeler, *Notes*, 1956, *op. cit*; Hip J. Groenewold to Hugh Everett and John A. Wheeler, 11 Apr 1957, *ME.*

[63] The paper was published in a collective volume edited by DeWitt and Graham (DeWitt et al. 1973). There is a letter from Everett to Jean-Marc Lévy-Leblond (15 Nov 1977, *EP*) which seems to support the hypothesis that the title of the original manuscript was indeed changed in the process of publication.

[64] A copy of the long thesis was sent to Copenhagen in April 1956, and a second one seems to have followed a few weeks later (Everett to Petersen [draft], 1956, *op. cit.*).We were unable to locate either. However, a draft of the long thesis is deposited in the *EP* archive (Hugh Everett, 1956, Wave Mechanics Without Probability, EP (Box 1, Series II, Folder 1), Niels Bohr Library, American Institute of Physics, College Park, MD, Unpublished paper; hereafter Everett, 1956). It contains some handwritten corrections which were incorporated in the paper published in 1973. The *EP* manuscript lacks the cover (hence we can only guess its title). However, a cover with the title *Wave Mechanics Without Probability*, which might have belonged to the *EP* manuscript, was unearthed by Peter Byrne among the papers in possession of Everett's son (Everett 2012 [1955a]). If the *EP* manuscript is the one that Everett sent to DeWitt in 1971 (after removing the cover, in which there appeared a title that Wheeler found inappropriate; John A.Wheeler to Niels Bohr, 24 Apr 1956, *BSC*, reel 34; also in *WP*, Series I, Box Boh-Bu, Folder Bohr, N. #2), this would explain why the title of the version published in DeWitt, Everett et al. (1973) differed from the original. Almost all of the unpublished documents concerning the origin and reception of Everett's thesis are now published in Everett et al. (2012).

[65] See John A. Wheeler to Hugh Everett, 22 May 1956 [1st letter], *WP* (Box Di-Fermi #2, Folder Everett);Wheeler to Bohr, 24 Apr 1956, *op. cit.*; John A.Wheeler to Allen Shenstone, 28 May

necessarily imply that a *first* structured version, differing substantially from the long thesis, actually existed.[66] Nevertheless, the hypothesis that the bulk of the long thesis had already been worked out early in 1955 is supported by the analysis of both the original manuscript and a few unpublished papers.[67] Besides a small paper entitled *Objective vs Subjective probability* (Everett 2012 [1955a]),[68] which outlines the "Wigner's friend"-type argument that forms the core of Everett's critique of the standard formulation in the long thesis,[69] the archives contain two manuscripts which were probably written in the summer of 1955 (see Sect. 3.6). One of them, *Quantitative Measure of Correlation* (Everett 2012 [1955b]), summarises the mathematical results of the second chapter of the long thesis (on correlation theory).[70] The other (Everett 2012 [1955c]) is a short paper (9 pages) whose title *Probability in Wave Mechanics* suggests a close relationship with the second version of the thesis. Indeed, this paper expounds all the relevant results concerning the interpretation of quantum mechanics that one finds in the long thesis.[71] Even though the presentation is made in a non-technical language devoid of formulas, it seems unlikely that Everett had reached all his conclusions without relying on a

---

1956, *WP* (Box Di-Fermi #2, Folder Everett); Groenewold to Everett & Wheeeler, *op. cit.*; Aage Petersen to Hugh Everett, 28 May 1956, *ME*.

[66] The archives contain no document that may correspond to such a first version. However, the recent discovery of some folders containing Everett's personal papers (Byrne 2007; Everett et al. 2012) may hopefully provide further insight into the very first steps of Everett's doctoral research.

[67] For example, as pointed out by DeWitt (DeWitt interview, *op. cit.*, p. 6), the first draft of the last chapter of the long thesis was probably written prior to Einstein's death (April 1955), since Einstein is referred to as if he were still alive (Everett 1973, p. 112). Admittedly, the long thesis contains references to three books published in 1955, one of which (von Neumann's *Mathematical foundations*) is also extensively quoted. Yet in the original manuscript of the long thesis deposited in the *EP* archive [(Everett 1956), "Wave mechanics without probability" [?]. EP (Box1, Series II, Folder 1)], the quotations from von Neumann's book appear to have been added later. Moreover, the reference to a paper that appeared in an issue of the *Supplemento al Nuovo Cimento* printed on 22 November 1955 lacks the volume and page number (they were added in the version published in 1973), which suggests that Everett read the pre-print.

[68] Everett (1955a), "Objective vs subjective probability." EP (Box 1, Folder 6). Printed in Everett (2012 [1955a]).

[69] The opening sentence of this manuscript ("Since the root of the controversy over the interpretation of the formalism of quantum mechanics lies in the interpretation of the probabilities given by the formalism, we must devote some time to discussing these interpretations"; Everett, 1955a) suggests that it was— or was intended to be—part of a larger work. Indeed, the structure of the paper resembles that of the introduction of the long thesis, although the projection postulate is not given the same central place. Moreover, in this early manuscript, Everett's own proposal is not mentioned.

[70] Everett (1955b). "Quantitative measure of correlation." EP (Box 1, Folder 6).

[71] Everett (1955c). "Probability in wave mechanics." EP (Box 1, Folder 6). In particular, the "emergence" of objects from correlations is discussed by means of an example which is reproduced almost literally on p. 86 of the long thesis.

formal analysis. Therefore, by the summer of 1955, Everett had probably already outlined both the mathematical and the conceptual framework of his approach.[72] In the light of this reconstruction, one can understand why Everett, who we know had continued to work "madly" on the draft to be sent to Copenhagen until Wheeler's departure to Europe in April 1956,[73] in later recollections always stated that the long thesis had been written in 1955.[74]

Here is a tentative chronology of the thesis versions and of the related papers:

(1a) *Objective vs Subjective probability*, short manuscript (first half of 1955).
(1b) *Quantitative Measure of Correlation*, short manuscript (summer 1955).
(1c) *Probability in Wave Mechanics*, short manuscript (summer 1955).
(2) *Wave Mechanics Without Probability*, second version of the dissertation (the long thesis) (winter 1955–1956), published as *The Theory of the Universal Wave Function* (1973).
(3) On the Foundations of Quantum Mechanics, final dissertation (winter 1956–1957), published as "Relative State" Formulation of Quantum Mechanics (July 1957).

3.4 The Reasons for Everett's Discontent

3.4.1 Standard Formulation

Everett's proposal stems from his dissatisfaction with von Neumann's formulation of quantum mechanics—"the more common (at least in this country) form of quantum theory", as he says in a letter to Petersen.[75] Both of Everett's published papers contain explicit references to von Neumann's *Mathematical Foundations*, whose English translation appeared in 1955, exactly when Everett's ideas were taking shape.[76] A central assumption in Everett's understanding of the standard

[72] Interestingly, in (Everett, 1956, op. cit.), the chapter on Observation, which forms the core of Everett's proposal, appears to have been imported from an earlier (and arguably shorter) manuscript (witness the old numbering of pages which appears in the upper margin).

[73] Nancy Gore Everett, Diary, entry of 28 Mar 1956, *ME* (Peter Byrne, priv. comm.). Nancy was Everett's wife.

[74] Everett interview, *op. cit.*, p. 6; Hugh Everett to Jean-Marc Lévy-Leblond, 15 Nov 1977, *EP*; Everett to Raub, 1980, *op. cit.*; Hugh Everett to Bill Harvey, 20 Jun 1977, *EP*, Series I-8. According to the recollection of Everett's wife, who typed the manuscript (Everett interview, *op. cit.*, p. 6), the thesis was written in the winter of 1954–1955 (Nancy Gore Everett, *Calendar of events*, *EP*, Box 1, Folder 1). (But this information could simply be inaccurate: the manuscript that Nancy Everett had in mind might actually be the second version, which was written in the winter of 1955–1956.)

[75] Everett to Petersen, 1957, *op. cit.*

[76] Everett probably had a working knowledge of German, and might have read von Neumann's book in the original.

formulation is that the state vector mirrors the physical state of a system (i.e. its putative *objective properties*) (Everett 1973, p. 63). Based on this hypothesis, von Neumann's account of observation can be regarded as involving a *physical* process in which the state of the observed system undergoes in general an acausal transition (from a superposition of eigenstates of the measured observable to the specific eigenstate corresponding to the observed value).[77] Such a process, whose outcomes can be statistically predicted using the Born rule, is considered to be responsible for the probabilistic features of quantum phenomena.[78] Unlike other critics of the postulate of projection,[79] therefore, Everett does not regard the collapse of the wave function as a formal trick, which the epistemic construal of the state vector requires in view of the intrinsic indeterminism of the quantum phenomena. Rather, he believes that in the standard formulation, the collapse of the wave function is what *prescribes* the probabilities of the various possible outcomes (Everett 1957b, p. 142). According to him, therefore, the postulate of projection instantiates a particular interpretation of quantum indeterminism, namely that of "objective chance". Although there are no grounds for endorsing or rejecting such an interpretation a priori, Everett contends that the odd implications of the projection postulate compel us to look for an alternative in which the probabilistic features of quantum mechanics can be understood in terms of "subjective chance".[80]

What Everett finds disturbing in the projection postulate is, first of all, the artificial way in which it splits the dynamics of the theory. It appears to be "a 'magic' process in which something quite drastic [occurs] (collapse of the wave function), while in all other times systems [are] assumed to obey perfectly natural continuous laws'".[81] The *ad hoc* nature of the projection postulate is borne out by the fact that, being designed to account for *idealized* observations, it is unsuited to deal with realistic models of the measurement interaction (Everett 1973, pp. 100–103). More generally, if one tries to understand measurements as just a physical interaction occurring between measuring apparatus and systems, the theory "leaves entirely unknown" which interactions are to be regarded as measurements.[82] Everett illustrates the consequences of this situation by means of a

[77] This reading of von Neumann has been thoroughly criticised by Becker (2004). The way to understand the postulate of projection changes depending on one's interpretation of the state vector. The interpretation that Everett seems to take as the "conventional" one is not inconsistent with that which seems to underlie some statements made by "orthodox" scholars. See, for example, Dirac's assertion that "the theory describes the state of the world at any given moment by a wave function" (Institut International de Physique Solvay 1928; Bacciagaluppi and Valentini 2009). See Bitbol (2000, pp. 72–83) for a discussion.

[78] In a letter of 1973 to Max Jammer (*op. cit.*), Everett identifies the "probability interpretation of quantum mechanics" with the assertion that "somehow the measuring process [is] 'magic' and subject to a separate axiom governing the collapse of the wave function."

[79] See e.g. Bohm (1952), Margenau (1958), Schrödinger (1953), Schrödinger (1958).

[80] Everett (2012 [1955a]).

[81] Everett to Jammer, 1973, *op. cit.*

[82] Everett (1955a, p. 4), in Everett (2012 [1955a]).

Wigner's-friend-type argument (see Sect. 3.3.2), from which he infers that a consistent application of the projection postulate within the standard theory implies the commitment to solipsism, i.e. to the hypothesis that there is only one observer in the universe who is responsible for the "collapse" of the state of observed systems.[83] Everett sees basically two ways to avoid this conclusion. Either one denies that measurement interactions fall into the domain of applicability of microphysics, or one postulates that the quantum description is simply incomplete, and must be supplemented with hidden parameters that can also characterise measurements. Both these solutions are at variance with the idea that the state vectors provide a complete model of the world, an idea to which Everett is strongly committed.

3.4.2 Dualistic Approach

The first way to avoid the alleged paradoxes of the standard formulation is to assume that "not every physical system possesses a state function, i.e. that even in principle quantum mechanics cannot describe the process of measurement itself." Everett considers this option "somewhat repugnant, since it leads to an artificial dichotomy of the universe into ordinary phenomena, and measurements."[84] In the long thesis he gives a further reason for rejecting this view, namely that it "does violence to the so-called principle of psycho-parallelism" stated by von Neumann.[85]

[83] The argument, which came subsequently to be known as the "Wigner's friend" paradox, appeared in a paper of Wigner's dated 1961. Given the resemblance between Wigner's and Everett's formulation, one may wonder whether Wigner picked up the argument from Everett's thesis, which he might have read. (However, of course, the converse might also be true, i.e. Everett might have been inspired by discussions with Wigner.) In a paper of 1958, Schrödinger (1958, pp. 168–169) alludes to the same argument: "But jokes apart, I shall not waste the time by tritely ridiculing the attitude that the state-vector (or wave function) undergoes an abrupt change, when 'I' choose to inspect a registering tape. (Another person does not inspect it, hence for him no change occurs.) The orthodox school wards off such insulting smiles by calling us to order: would we at last take notice of the fact that according to them the wave function does not indicate the state of the physical object but its relation to the subject; this relation depends on the knowledge the subject has acquired, which may differ for different subjects, and so must the wave function." This ironical presentation of the problem suggests that, had Schrödinger read the pre-print of Everett's paper that he was sent by Wheeler, he would have found Everett's arguments quite naïve. Nevertheless, Schrödinger was opposed to the epistemic interpretation of the state vector and he believed, like Everett, that "the Kopenhagen epistemology [. . .] leads to the physics of solipsism." (*Ibid.*)

[84] Everett (1955a, p. 3). In Everett (2012 [1955a]).

[85] Everett (1973, p. 7). The principle was stated by Von Neumann (1955, p. 418) in the following terms: "[. . .] it must be possible so to describe the extra-physical process of the subjective perception as if it were in reality in the physical world—i.e. to assign its parts equivalent physical processes in the objective environment, in ordinary space."

In the introduction of the long thesis, Everett makes a distinction between this view and Bohr's. After outlining the former approach together with other possible solutions, he says: "We have omitted one of the foremost interpretations of quantum theory, namely the position of Niels Bohr" (Everett 1973, p. 8). He discusses the latter in the conclusion, but then one gets the impression that, in Everett's eyes, the Copenhagen interpretation (which is the label he uses to denote what he takes to be Bohr's approach[86]) is closely related to the dualistic view presented earlier.[87] The criticisms he addresses to the Copenhagen interpretation in the long thesis (Everett 1973, p. 111) are summarised and developed in a letter to Petersen of May 1957, in which he says that, while his paper of 1957 addresses "mostly" von Neumann's formulation, he finds Bohr's approach "even more unsatisfactory", although "on quite different grounds."[88] The main objections appearing in the letter of 1957 are similar to those raised in the long thesis of 1955–1956. (Incidentally, this shows that Everett had not changed his mind-notwithstanding the fact that, for reasons on which we will return, his criticisms do not appear in the final version of the dissertation.) What Everett finds "irritating" in the Copenhagen interpretation is on the one hand the "complete reliance on classical physics from the outset (which precludes even in principle any deduction at all of classical physics from quantum mechanics, as well as any adequate study of measurement processes)", and, on the other hand, the "strange duality of adhering to a 'reality' concept for macroscopic physics and denying the same for the microcosm."[89]

In the letter to Petersen, Everett develops his critique, pointing out other alleged deficiencies of the Copenhagen approach:

> You talk of the massiveness of macrosystems allowing one to neglect further quantum effects (in discussions of breaking the measuring chain), but never give any justification for this flatly asserted dogma. Is it an independent postulate? It most certainly does *not* follow from wave mechanics [. . .]. In fact, by the very formulation of your viewpoint you are

[86] Everett found the term "Copenhagen interpretation" in the above mentioned book edited by Pauli (1955), which is cited in the long thesis.

[87] The introduction and the conclusion of the long thesis were arguably written at different times. The first and third "interpretations" outlined in the conclusion are explicitly put into correspondence with the first and fourth "alternatives" appearing in the introduction (solipsism and hidden variables respectively). Everett avoids emphasising the correspondence between the second interpretation (Copenhagen) and the second alternative (dualistic view), but it is quite clear that he sees a link between them.

[88] Everett to Petersen, 1957, *op. cit.*

[89] *Ibid.* It is instructive to recall the discussion about the "relationship between Quantum and Classical concepts" which Everett found in Bohm's textbook. In his presentation of the orthodox view, Bohm said that "in order to obtain a means of interpreting the wave function, we must [. . .] *at the outset* postulate a classical level in terms of which the definite results of a measurement can be realized." He also asserted that "classical concepts cannot be regarded as limiting forms of quantum concepts", and that "without an appeal to a classical level, quantum theory would have no meaning" (Bohm 1951, pp. 624–626).

> totally incapable of any justification and *must* make it an independent postulate—that macrosystems are relatively immune to quantum effects.
>
> You vigorously state that when apparatus can be used as measuring apparatus then one cannot simultaneously give consideration to quantum effects—but proceed blithly to apply [the uncertainty relations] to such devices, tacitly admitting quantum effects.[90]

Furthermore, Everett claims that while the Copenhagen interpretation takes "the fundamental irreversibility of the measuring process" to be what "allows the destruction of phase relations and make possible the probability interpretation of quantum mechanics", "there is nowhere to be found any consistent explanation of this 'irreversibility'." And he concludes: "Another independent postulate?"

In the light of these criticisms one may find surprising Everett's assertion, stated elsewhere, that the Copenhagen interpretation is "undoubtedly safe from contradiction" (Everett 1973, p. 111). Indeed, Everett is prepared to concede that the Copenhagen interpretation avoids inconsistency, but he believes that this is achieved at the cost of endorsing a strongly dualistic approach. Such an approach is at odds with the task of providing a coherent and all-inclusive model of the world, which is, for Everett, the very goal of physics. Hence, the Copenhagen interpretation is to him "hopelessly incomplete."[91]

The final version of the dissertation, in which Everett criticises what he calls the "external observation formulation", contains a remark which can be interpreted as a further objection to the Copenhagen interpretation. As we will see, the label "external observation formulation" denotes a dualistic approach in which the state reduction is brought about by an "external" observer that cannot in principle be described by the formalism. Such a view is clearly reminiscent of the one that Everett associated with the Copenhagen interpretation, and this association is indeed made explicit by Wheeler (1957, p. 151). The question of whether the *pragmatic* aspects of Bohr's view, and in particular his *functional* distinction between measuring apparatus and object system, can really be expressed in the dualistic terms of the external observation formulation is postponed to Sect. 3.7. Certainly Wheeler and Everett thought that they could, and interpreted Bohr's remarks on the necessity to frame the quantum predictions in a well-defined experimental context as implying that von Neumann's measurement chain needed to be "cut" into two parts, one of which could not be described by quantum mechanics. This view, they argued, led to critical problems "in the case of a closed universe", since then "there is no place to stand outside the system to observe it. There is nothing outside it to produce transitions from one state to another" (Everett 1957b, p. 142). The external observation formulation appears thus unsuited to providing a description of the whole universe; and this, in turn, precludes any possibility of a synthesis with general relativity.[92]

---

[90] Everett to Petersen, 1957, *op. cit.*

[91] Hugh Everett to Bryce S. DeWitt, 31 May 1957, courtesy of Eugene Shikhovtsev.

[92] Interestingly, such an objection is not mentioned in Everett's letter to Petersen, though the letter was written *after* the paper. This suggests that this objection reflected in fact a concern of Wheeler's.

3.4.3 Hidden Variables

From the manuscript *Objective vs Subjective Probability*, it is clear that at first Everett regarded hidden variable theories as a promising approach to overcome the paradoxes of the standard formulation. In later writings, he still acknowledges their "great theoretical importance" and undisputable appeal, but he emphasises that they are unnecessarily "cumbersome and artificial" as compared to his own proposal.[93]

Bell has pointed out some structural analogies between Everett's and the hidden variable approaches.[94] Indeed the conceptions of physical theories which underlie the two approaches are closely related to each other, and so are the strategies adopted to fit the quantum indeterminism into them. Like the hidden variable theorists, Everett held that theories must supply an exhaustive model of the world, including observers and measurement interactions,[95] although, unlike them, he believed that the state vectors alone can provide such a model. Everett claimed that the indeterministic features of quantum phenomena only appear within *subjective* experience. According to him, this point of view was similar to that adopted by the advocates of hidden variables, for whom "the probabilities occurring in quantum mechanics are not objective" since "they correspond to our ignorance of some hidden parameters."[96] However, Everett's proposal did not stem from an aprioristic commitment to determinism.

> From my point of view there is no preference for deterministic or indeterministic theories. It is quite conceivable that an adequate *stochastic* interpretation could be developed [. . .] where the fundamental processes of nature are pictured as stochastic processes *whether or not* they are undergoing observation. I only object to mixed systems where the character changes with mystical acts of observation.[97]

In the long thesis, following Schrödinger (1952), Everett nonetheless criticised the stochastic interpretations because of their "desire to have a theory founded upon particles", while it seems "much easier to understand particle aspects from a wave picture [. . .] than it is to understand wave aspects [. . .] from a particle picture" (Everett 1973, p. 114).

More generally, Everett seemed to agree with Schrödinger that "the demand for a non-subjective description is inevitable, of course without prejudice whether it be deterministic or otherwise" (Schrödinger 1958, p. 162). If Everett is so concerned with probability (think of the titles of his earliest manuscripts) this is because, for him, probabilities arise within the conventional formulation as a consequence of

---

[93] Everett (1973, p. 113); Everett to DeWitt, 1957, *op. cit.*

[94] Bell (2004, pp. 93–99) made a comparison between Everett's approach and de Broglie's pilot wave theory. See also Barrett (1999, Chap. 5). This point is discussed by DeWitt in a letter sent to Wheeler and Everett in 1957. (Bryce S. DeWitt to John A. Wheeler & Hugh Everett, 7 May 1957, *WP*, Series I—Box Di—Fermi Award #1—Folder Everett).

[95] See e.g. Körner (1957, p. 61).

[96] Everett (1955a, p. 4), in Everett (2012 [1955a]).

[97] Everett to DeWitt, 1957, *op. cit.*

state reduction, and state reduction requires in turn the intervention of an external observer, thereby undermining the very possibility of an objective description.

3.5 Everett's Project

3.5.1 A Unitary Model of the World

Everett outlines his conception of theories in an appendix of the long thesis. The relationship between such a conception and his formulation of quantum mechanics is discussed in a letter to DeWitt, some passages of which are quoted in a note added in proof to the paper published in 1957.

> To me, any physical theory is a logical construct (model), consisting of symbols and rules for their manipulation, *some* of whose elements are associated with elements of the perceived world.[98]

The "perceived world" or "world of experience" is to be understood as "the sense perceptions of the individual, or the 'real world'—depending upon one's choice in epistemology." As to *his* choice, Everett is quite reticent. His theory deals ultimately with the content of the observers' memories. However, he proposes to identify the "subjective knowledge (i.e. perceptions)" of the observers with "some objective properties (states)" of theirs (Everett 1973, p. 63).

Remarkably, all throughout Everett's writings, the terms "real" and "reality" (as well as "actual", "branching process", "branches") appear systematically in quotes. Indeed, Everett emphasises that the meaning of terms such as "reality" ought to be understood on the basis of their usage in scientific *practice*.[99]

> When one is using a theory, one naturally pretends that the constructs of the theory are "real" or "exist". If the theory is highly successful (i.e. correctly predicts the sense perceptions of the user of the theory) then the confidence in the theory is built up and its constructs tend to be identified with "elements of the real physical world". This is however a purely psychological matter. No mental construct (and this goes for everyday, prescientific conceptions about the nature of things, objects, etc. as well as elements of formal theories) should ever be regarded as more "real" than any others. We simply have more *confidence* in some than others.[100]

In the long thesis, the point is illustrated by the following example:

> The constructs of classical physics are just as much fictions of our minds as those of any other theory we simply have a great deal more confidence in them. It must be deemed a mistake, therefore, to attribute any more "reality" here than elsewhere. (Everett 1973, p. 134)

---

[98] Everett to DeWitt, 1957, *op. cit.*

[99] See for example Everett (1973, p. 116).

[100] Everett to DeWitt, 1957, *op. cit.*

Everett's attitude shows some analogy with Schrödinger's "methodological realism" or "quasi-realism", in which any naïve metaphysical commitment is explicitly rejected.[101] Although Everett holds that the primary purpose of theoretical physics is to build useful models, he does not bother about their ontological status, since, he says, models "serve for a time and are replaced as they are outworn" (Everett 1973, p. 111). This attitude is also apparent in Everett's critique of Bohr's doctrine of classical concepts (which we shall discuss in detail in Sect. 3.7). Far from attributing any special status to classical concepts, Everett urged their replacement by quantum ones. This position was not based on ontological considerations. Rather, Everett thought that since *all* concepts serve to deal with a *reality-in-quotes*, there is no reason to stick to a particular set of concepts: our concepts can evolve just as our models of "reality" do.[102]

The "conceptual model of the universe" that Everett proposes "postulates only the existence of the universal wave function which obeys a linear wave equation" (Everett 1973, p. 117). In such a theory, "one can regard the state functions themselves as the fundamental entities, and one can even consider the state function of the whole universe."[103] In one of the manuscripts of 1955, Everett put it as follows: "The physical 'reality' is assumed to be the wave function of the whole universe itself."[104] In the long thesis, comparing his programme to the existing interpretations of quantum mechanics, Everett explicitly refers to a paper in which Schrödinger contrasts the continuous description provided by the wave function with the "quantum jumps" of the "current probability view".[105] Indeed, in Schrödinger's writings of this period, one can easily find passages which are amazingly in tune with Everett's views:

> [...] at the present stage and as long as the state vector plays the role it does it must be taken to represent 'the real world in space and time', it ought not to be sublimed into a probability function for the purpose of making forecasts [. . .] changing abruptly when somebody (who?) cares to inspect a photograph or a registering tape. (Schrödinger 1958, p. 169)

What Everett has in mind when he talks of "model" is an "objective description" of "reality". Such a description must leave no room for mental entities and processes which exceed the boundaries of quantum physics.[106] In accordance with von Neumann's principle of psycho physical parallelism, which Everett interprets as implying that an observer (including their perceptions) is completely characterised by her/his physical state, the observers and their mental states must be described by a state vector. The universal wave function includes therefore an exhaustive model of all existing observers and of their interactions with the

[101] See Bitbol (1998, pp. 182–184).

[102] For a comparison with the debate that Schrödinger had with Bohr on this issue, see Murdoch (1987, p. 101), Bitbol (1996a, pp. 22–23).

[103] *Ibid*, p.9.

[104] Everett (1955c, p. 9), in Everett (2012 [1955c]).

[105] The paper cited by Everett (1973, p. 115) is Schrödinger (1952).

[106] This is explicitly stated in a letter of 1980 (Everett to Raub, 1980, *op. cit*).

observed systems. This is perhaps why Everett contends that, unlike the conventional formulation, his theory "sets the framework for its interpretation".[107] In the methodological appendix of the long thesis, Everett says that each theory must contain an "interpretive part", i.e. "rules which put some of the elements of the formal part into correspondence with the perceived world" (Everett 1973, p. 133). Thus one might possibly argue that the universal wave function "sets the framework for its interpretation" because it is isomorphic to the "world" perceived by all observers (inasmuch as it mirrors the properties of the observers' brains which correspond to their "subjective perceptions").[108] From Everett's standpoint, the same cannot be said of the conventional formulation, in which "pure wave mechanics" must be supplemented with the postulate of projection if one wants to put the symbolism (which, in general, describes a system by means of a *superposition* of "absolute" states) into correspondence with the "perceived world" (in which the system is described by a *single* element of the superposition).[109]

Everett is committed to an ideal of unity, simplicity and completeness.[110] The structural features of his theory reflect this commitment. Firstly, there is no dualism in the dynamics: the projection postulate is relinquished and the universal wave function evolves according to a continuous and deterministic process. Secondly, this simplification is purportedly achieved without introducing supplementary "artificial" variables (see Sect. 3.4.3).

3.5.2 *Objective Description and Correlations*

While Everett's motives, goals and assumptions are similar to those of other critics of the conventional formulation of quantum mechanics, his strategy to make a descriptive interpretation of the theoretical symbolism viable is completely original. The cornerstone of this strategy is what Everett names "the fundamental principle of the relativity of states". Suppose that the universal wave function is expanded as a linear combination of the vectors of some basis. According to the principle of the relativity of states, if, in a given component of this expansion, a system is represented by the eigenvector of an observable A corresponding to the eigenvalue a_i, then the system can be said to *have* the property "$A = a_i$" (i.e. to be in the corresponding state), but this assertion is true only *relative* to the properties that

[107] Everett (1957b, p. 142). See also Wheeler (1957, p. 152).

[108] To be sure, this point of view is quite problematic. Its meaning and implications are analysed in the following subsections.

[109] This reasoning assumes that, in the conventional formulation, there is a straightforward link between state vectors and physical states. As we have seen, this assumption was part of Everett's reading of von Neumann's formulation.

[110] "We have a strong desire to construct a single all-embracing theory which would be applicable to the entire universe." (*Ibid*, p. 135).

the other systems "have" in the same component of the expansion (i.e. to their state in that component).

On one hand, in virtue of the principle of the relativity of states, the state vectors need no longer to undergo an abrupt, acausal change in order to provide a consistent *description* of the properties which measurements are supposed to reveal.

> From the viewpoint of the theory, all elements of a superposition (all "branches") are "actual", none any more "real" than another. It is completely unnecessary to suppose that after an observation somehow one element of the final superposition is selected to be awarded with a mysterious quality called "reality" and the others condemned to oblivion. We can be more charitable and allow the others to coexist—they won't cause any trouble anyway because all the separate elements of the superposition ("branches") individually obey the wave equation with complete indifference to the presence or absence ("actuality" or not) of the other elements.[111]

On the other hand, properties are now intrinsically "relative":

> All statements about a subsystem [. . .] become relative statements, i.e. statements about the subsystem relative to a prescribed state for the remainder. (Everett 1973, p. 118)

In this way Everett thinks that he has managed to construe the quantum theory as an "objective description", although of course the description is objective not in the sense that it captures *the* "actual value" of each observable, but because it provides a symbolic structure which connects *any* possible value of a given observable to a particular state of the whole universe (which includes a specific state of every conceivable observer).[112] What quantum mechanics describes are the *correlations* occurring in nature.[113]

Everett argues that, in this framework, even *objects* should be understood in terms of correlations, no matter whether their size is atomic or macroscopic:

> [If we] consider a large number of interacting particles [. . .], throughout the course of time the position amplitude of any single particle spreads out farther and farther, approaching uniformity over the whole universe, while at the same time, due to the interactions, strong correlations will be built up, so that we might say that the particles have coalesced to form a solid object. That is, even though the position amplitude of any single particle would be "smeared out" over a vast region, if we consider a "cross section" of the total wave function

[111] Everett to DeWitt, 1957, *op. cit.*

[112] In this case too, it is interesting to compare Everett's position to Schrödinger's. Commenting on our "yearning for a complete description of the material world in space and time", Schrödinger (1958, p. 169) remarked: "[. . .] It ought to be possible, so we believe, to form in our mind of the physical object an idea (Vorstellung) that contains in some way everything that *could be* observed in some way or other by any observer, and not only the record of what *has been* observed simultaneously in a particular case."

[113] Everett's mathematical work on correlations was probably undertaken independently of his reflection on quantum mechanics. Indeed, the chapter of the long thesis dedicated to correlation theory contains a lot of mathematical details that are not essential to the remainder. The chapter on correlation theory was not reproduced in the final dissertation. However, it gave rise to a paper (Everett 1955b), in Everett (2012 [1955b]), which remained unpublished (albeit Wheeler considered it "practically ready" for submission; John A. Wheeler to Hugh Everett, 21 Sep 1955, *EP* (Box 1, Folder 9)).

> for which one particle has a definite position, then we immediately find all the rest of the particles nearby, forming our solid object.[114]

As an example, Everett analyses the formation of a hydrogen atom in a box containing a proton and an electron. He concludes that:

> What we mean by the statement, "a hydrogen atom has formed in the box", is just that this correlation has taken place—a correlation which insures that the relative configuration for the electron, for a definite proton position, conforms to the customary ground state configuration. (Everett 1973, p. 86)

This example is also discussed in the manuscript *Probability in Wave Mechanics* of 1955, in which one finds the same emphasis on correlations, though the notion of "relative state" is not yet explicitly stated there. More generally, Everett claims that "all [physical] laws are correlation laws".[115] These passages help us to understand how Everett can claim that his "universal wave function model" is complete, notwithstanding the fact that it contains no information about *which* branch represents "actuality". Indeed, from Everett's point of view, such a question is not one that can or must be answered by physics, for the simple reason that it cannot be formulated in terms of correlations. In 1957, Everett wrote to Norbert Wiener:

> You also raise the question of what it means to say that a fact or a group of facts is actually realized. Now I realize that this question poses a serious difficulty for the conventional formulation of quantum mechanics, and was the main motives for my reformulation. The difficulty is removed in the new formulation, however, since it is quite unnecessary in this theory ever to say anything like "Case A is actually realized."[116]

Thus Everett can consistently hold that his model provides a complete description of "reality". There remains a crucial problem, however, to be solved "investigat [ing] the internal correlations in the universal wave function" (Everett 1973, p. 118), namely, how to put this description into correspondence with the correlations that *we* observe. As we will now see, for Everett even this problem can be settled without singling out a unique "actual" branch.

3.5.3 Subjective Experience and Probabilities

How does Everett's theory account for the "perceptions" of a typical observer engaged in experimental activity?

> For this purpose it is necessary to formulate abstract models for observers that can be treated within the theory itself as physical systems, to consider isolated systems containing such model observers in interaction with other subsystems, to deduce changes that occur in an observer as a consequence of interaction with surrounding subsystems, and to interpret the changes in the familiar language of experience. (Everett 1957b, p. 142)

---

[114] Everett (1955c, p. 6), in Everett (2012 [1955c]).

[115] See Everett (1973, pp. 118, 137).

[116] Hugh Everett to Norbert Wiener, 31 May 1957, *ME*.

More specifically, it must be shown that the memory contents of a typical observer described by Everett's theory are consistent with the qualitative and quantitative features that are commonly ascribed to the results of the observations carried out in atomic physics: the "appearance of the collapse" (i.e. the invariance of the result when a measurement is immediately repeated, and the consistency of the results recorded by different observers who measure the same observable) on the one hand, and the statistical distributions predicted by the Born rule for ensembles of measurements carried out on identical systems on the other.

In accordance with the principles of the relativity of states and psychophysical parallelism, these features of empirical data must in the first place be expressed in terms of correlations between memory states of the observer. For example, the repeatability requirement will be expressed by the following proposition (*R*): Consider an observer *O* who, after measuring some observable, has immediately repeated the measurement. If r_1 and r_2 are the values recorded by *O*'s memory as the results of the two observations, then $r_1 = r_2$. We note that the correlation between subsequent measurement outcomes has been reduced to "some present properties" of the observer's memory which can be identified "with features of the past experience". The idea behind this move is that

> in order to make deductions about the past experience of an observer it is sufficient to deduce the present contents of the memory as it appears within the mathematical model. (Everett 1973, p. 144)

Secondly, one must be able to deduce, from the model provided by the universal wave function at a given instant, that (*R*) has probability 1 of being true. Everett assumes that this second condition is fulfilled if the set of the branches in which the state of *O*'s memory contradicts (*R*) has vanishing measure in the Hilbert space. As for the measure to be used, Everett proposes, on the basis of a plausibility argument that he finds compelling, a function which is analogous to the probability function appearing in the Born rule. This choice enables Everett to claim that, in the case in which O has performed the same measurement upon an infinite collection of identical systems, the statistical results predicted by the conventional theory are recovered (since they correspond to the statistical distribution recorded by all memory sequences "except for a set [y] of measure zero"). Assuming that "the actions of the [observer] at a given instant can be regarded as a function of the memory contents only", this is supposed to demonstrate why we use standard quantum mechanics to predict experimental results (Everett 1973, pp. 148, 144).

We have so far considered the empirical domain of atomic physics. By the same type of argument, Everett also claims that "the classical appearance of the macroscopic world to us can be explained in the wave theory." In quantum mechanics, the general state of a system of macroscopic objects does not ascribe any nearly definite positions and momenta to the individual bodies. Yet, such a state can "at any instant be analyzed into a *superposition* of states each of which *does* represent the bodies with fairly well defined positions and momenta." Hence if one considers the result of an observation performed upon a system of macroscopic bodies in a general state, the observer

> will not see the objects as 'smeared out' over large regions of space [. . .] but will himself simply become correlated with the system—after the observation the composite system of objects + observer will be in a superposition of states, each element of which describes an observer who has perceived that the objects have nearly definite positions and momenta, and for whom the relative system state is a quasi-classical state [. . .], and furthermore to whom the system will appear to behave according to classical mechanics if his observation is continued. (Everett 1973, pp. 89–90)

Based on the foregoing arguments, Everett maintains that his theory can account for both classical determinism and quantum indeterminism in terms of "subjective experience". In particular, he believes that he has shown "how pure wave mechanics, without any initial probability assertions, can lead to these notions on a subjective level, as appearances to observers."[117] Hence, he claims that, whereas in the conventional formulation the "probabilistic features are postulated in advance instead of being derived from the theory itself", in the relative state formulation

> the statistical assertions of the usual interpretation do not have the status of independent hypothesis, but are deducible (in the present sense) from the pure wave mechanics that starts completely free of statistical postulates. (Everett 1957b, p. 149)

In the last two decades, several commentators, e.g. Barrett (1999) and Kent (1990), have pointed out that Everett's argument is wanting. There is perhaps no need of a statistical postulate in order to "interpret" each branch of the universal wave function *individually*, i.e. to state which occurrences in the "perceived world" that particular branch describes. Yet, the theory provides us with infinite branches, and *this is* the formal structure from which we have to extract empirical information. *Here* we need what Everett calls the "interpretive part" of the theory. As a matter of fact, Everett *does use* an interpretive rule in his deduction, which is similar to that of classical statistical mechanics, although logically weaker. Unlike the measure of the set of trajectories in the phase space of statistical mechanics, the measure of the set of branches is not straightforwardly interpreted as a statistical weight for empirical statements. Nevertheless, such an interpretation is indeed assumed in the limiting case: true statements are those which hold for all but a set of branches of measure zero. Everett himself asserts that "the situation here is fully analogous to that of classical statistical mechanics" and develops the analogy in detail. The very constraints from which Everett derives the mathematical function to be used as a measure in the Hilbert space reflect in his eyes "the only choice which makes possible any reasonable statistical deductions at all", just as "the choice of Lebesgue measure on the phase space can be justified by the fact that it is the only choice for which the 'conservation of probability' holds" (Everett 1973, pp. 147–149).[118] In his assessment of 1957, Wheeler makes a quick allusion to

---

[117] Everett (1973, p. 78); see also p. 142.

[118] For example, the additivity requirement, which plays a crucial role in the deduction, is so chosen as "to have a requirement analogous to the 'conservation of probability'." (*Ibid.*) In his letter to Max Jammer (*op. cit.*), Everett insisted that his "deduction of the probability interpretation" was "just as 'rigorous' as any of the deductions of classical statistical mechanics, since in

Laplace's universe. From a pencilled note in the margin of a letter, we learn that the analogy he saw between Everett's and Laplace's theories was in fact quite deep and general. "In Laplace description," he says, "we don't know what's going to happen tomorrow morning, but we have a scheme within which it fits." And he adds: "How to do the same in qm description of nature."[119]

It is unlikely that Everett would have endorsed a *postulate* stating the interpretive rule his argument seems to rest on. One often gets the impression that he believed that the rule simply followed from an adequate interpretation of branches. But the few passages that are explicitly intended to clarify the controversial aspects of such an interpretation, either in published papers or in private correspondence, can hardly be said to shed any light on the issue (Barrett 1999, pp. 86–90). In the last decades, the attempts to provide a consistent interpretation of branches have given rise to a growing family of disparate approaches, ranging from many-worlds and many-minds to consistent histories and relational interpretations. For almost all these approaches it is important to define the ontological status of branches—a problem that Everett systematically avoids, talking at most of a *language* difficulty in connection to the "splitting" of the observer state when a measurement is performed (Everett 1973, p. 68). In the light of Everett's pragmatic conception of reality, the question of whether his pictorial language is to be understood literally or metaphorically may appear immaterial. Yet, among the 1955 manuscripts, there is a paper (Everett 1955c, in Everett (2012 [1955c])) in which Everett seems to take rather seriously the "splitting" process and its possible effects as seen "from within". In that paper he says for example that, after a measurement, "the observer himself has split into a number of observers, each of which sees a definite result of the measurement."[120] Or that the price to be paid in order to have a complete theory "is the abandonment of the concept of the uniqueness of the observer, with its somewhat disconcerting philosophical implications."[121] He also draws a detailed analogy with the case of a splitting amoeba. On this passage, Wheeler, who read the manuscript, annotated: "This analogy seems to me quite capable of misleading readers in what is a very subtle point. Suggest omission." And elsewhere: "Split? Better words needed." While acknowledging the value of the paper, Wheeler wrote to Everett that it had to be reformulated in order to avoid "mystical misinterpretations

both areas the deductions can be shown to depend upon an 'a priori' choice of a measure on the space." And he continued: "What is unique about the choice of measure and why it is forced upon one is that in both cases it is the only measure that satisfies a law of conservation of probability through the equations of motion. Thus, logically, in both classical statistical mechanics and in quantum mechanics, the only possible statistical statements depend upon the existence of a unique measure which obeys this conservation principle."

[119] Stern to Wheeler (1956, *op. cit*).

[120] Everett (1955c, p. 5), in Everett (2012 [1955c]).

[121] *Ibid*, p. 8. In a note of 1956, Everett wrote: "Statistical ensemble of observers is, within the context of the theory, a *real,* in distinction to a *virtual*, ensemble!" (*Notes on Stern's letter*, 1956, *ME*).

by too many unskilled readers."[122] From these remarks, it would seem that Wheeler considered the references to branches and splitting as a matter of form, rather than one of substance. Certainly, however, he was aware that Everett's pictorial phrasing might not only be confusing, but might also conceal some real shortcoming. In replying to the claim of the Copenhagen group that there was no relationship at all between "Everett's system" and "physics as we do it", Wheeler said:

> No, because Everett traces out a correspondence between the 'correlations' in his model universe on the one hand, and the on the other hand what we observe when we go about making measurements. [. . .] Has the nature of the correspondence been made clear [. . .]? Far from it.[123]

3.6 Striving for Copenhagen's Imprimatur

At the beginning of the fall term of 1955, Everett submitted *Quantitative Measure of Correlation* and another paper (probably *Probability in Wave Mechanics*) to Wheeler. In his response, after approving the former, Wheeler observed: "As for the 2nd one, I am frankly bashful about showing it to Bohr in its present form, valuable and important as I consider it to be."[124] Remarkably, the reference to Bohr comes without any introductory comment. Since it must have been quite unusual for a Princeton student to have his drafts read by Bohr in person, this suggests that the possibility of sending the paper to Copenhagen had already been discussed. When exactly we do not know. In October 1954, Bohr had visited Princeton, and we know that he met Everett.[125] But it is unlikely that any serious discussion between them took place on that occasion. The project to get Bohr involved in the assessment of Everett's thesis could have originated from Wheeler. The aforementioned note shows that Wheeler was impressed by Everett's qualities and ideas since the beginning (see Sect. 3.3.2). Furthermore, as we shall see, although Wheeler endorsed Bohr's doctrine, he was puzzled by some aspects of it, and probably saw Everett's proposal as an opportunity to sound Bohr out about the necessity to "generalize" the orthodox view.

In 1956, Wheeler was invited by the university of Leiden to hold the Lorentz Chair for one semester. Before leaving in April, he received from Everett a bound copy of *Wave Mechanics Without Probability*, which he mailed to Copenhagen

---

[122] This remark is contained in a note that Wheeler sent to Everett in September 1955 (Wheeler to Everett, 1955, *op. cit.*). That Wheeler was indeed referring to *Probability in Wave Mechanics* is actually only a conjecture, though a plausible one.

[123] Wheeler to Stern, 1956, *op. cit.*

[124] Wheeler to Everett, 1955, *op. cit.*

[125] There is a photograph, which appeared in a local journal, portraying Bohr holding a discussion with a group of students, and Everett is among them. See Picture 3.1.

soon after his arrival in Leiden.[126] In the letter accompanying the manuscript, Wheeler appears quite cautious. "The title itself," he says, "[. . .] like so many ideas in it, need further analysis and rephrasing."[127] A few days later, Wheeler went to Copenhagen in order to discuss the draft with Bohr and Petersen. Shortly after returning to Leiden, he wrote to Everett:

> We had three long and strong discussions about it. [. . .] Stating conclusions briefly, your beautiful wave function formalism of course remains unshaken; but all of us feel that the real issue is the words that are to be attached to the quantities of the formalism. We feel that complete misinterpretation of what physics is about will result unless the words that go with the formalism are drastically revised.

Wheeler also added that Bohr had promised to write to him about Everett's work and that "he was arranging [. . .] for Stern to give [. . .] a seminar report on [Everett's] thesis, so it could be thoroughly reviewed before he wrote."[128] The same day, Wheeler forwarded to Everett the notes that he had taken during the discussion with Petersen (when he was with Bohr, he said, he wrote "almost never"[129]), together with a second letter in which he outlined his plan of action. In this letter, besides insisting on the necessity of removing any possible source of misunderstanding (though this was going to take "a lot of heavy *arguments* with a practical tough minded man like Bohr"), Wheeler tried to make clear what he considered to be the main issue at stake:

> I don't think, because I don't make out Bohr's case well, that it isn't strong or convincing: that the words you use in talking about things in your formalism have nothing to do with words + concepts of everyday physics; that one will give rise to a complete misunderstanding of what is going on to use the same words.[130]

After some time, Wheeler, who had not received any news from Bohr, wrote to Stern. Stern answered that he had indeed given a seminar on Everett's paper, and added that "Prof. Bohr was kind enough to make a few introductory remarks and open the discussion." The outcome of this discussion was a merciless criticism of Everett's "erudite, but inconclusive and indefinite paper."

> In my opinion, there are some notions of Everett's that seem to lack meaningful content, as, for example, his universal wave function. Moreover he employs the concept of observer to mean different things at different times [. . .].

[126] Wheeler to Everett, 1956 [I], *op. cit.*

[127] John A. Wheeler to Niels Bohr, 24 Apr 1956, *BSC* (reel 34). This letter contains a passage (in which Wheeler refers to the "second draft of the thesis of Everett") that seems to confirm that Bohr already knew about Everett, and that the first version of the thesis had already been mentioned to him. That Petersen was acquainted with Everett's former writings is suggested by a passage of a letter, in which, besides other things, he says: "I also had the opportunity to read the *new* draft of your thesis." (Aage Petersen to Hugh Everett, 28 May 1956, *ME*; our emphasis).

[128] Wheeler to Everett, 1956 [I], *op. cit.* Alexander Stern was an American researcher then at the Institute of Theoretical Physics of Copenhagen.

[129] Wheeler to Everett, 1956 [II], *op. cit.*

[130] *Ibid.*

> I do not follow him when he claims that, according to his theory, one can view the accepted probabilistic interpretation of quantum theory as representing subjective appearances of observers.
>
> But, to my mind, the basic shortcoming in his method of approach [. . .] is his lack of an adequate understanding of the measuring process.
>
> His claim that process I [the Schrödinger equation] and process II [the collapse of the wave function] are inconsistent when one treats the apparatus system and the atomic object system under observation as a single composite system and if one allows for more than one observer is, to my mind, not tenable.[131]

Wheeler's reply is a long and detailed defence of Everett's proposal, which aims to dispel the impression that Everett's purpose was to criticise the orthodox approach. In the preamble of his letter, Wheeler reassured Stern about his own intentions:

> I do not in any way question the self consistency and correctness of the present quantum mechanical formalism [. . .]. On the contrary, I have vigorously supported and expect to support in the future the current and inescapable approach to the measurement problem. To be sure, Everett may have felt some questions on this point in the past, but I do not.

About Everett, Wheeler observed that

> [. . .] this very fine and able and independently thinking young man has gradually come to accept the present approach to the measurement problem as correct and self consistent, despite a few traces that remain in the present thesis, draft of a past dubious attitude.[132]

(Of this alleged conversion there is no trace in Everett's writings; see Sect. 3.4.2.) Although Wheeler believed that "the concept of 'universal wave function'" was indeed "an illuminating and satisfactory way to present the content of quantum theory", he was prepared to "recognise that there are many places in Everett's presentation that are open to heavy objection, and still more that are subject to misinterpretation." He added that "to make the whole discussion consistent at every point" he would "make sure that Everett [had] the benefit of a number of weeks in Copenhagen." The importance that Wheeler attached to this plan is also apparent from his previous letters to Everett:

> I told Bohr I'd arrange to pay [. . .] half your minimum rate steamship fare New York to Copenhagen; I think there's an appreciable chance Bohr would take care of the other half, according to what he said. He would welcome very much a several weeks' visit from you to thrash this out. You ought not to go of course except when he signifies to you that you are picking a time when he can spend a lot of time with you. Unless and until you have fought out the issues of interpretation one by one with Bohr, I won't feel happy about the conclusions to be drawn from a piece of work as far reaching as yours. Please go (and see me too each way if you can!).

To this request, Wheeler added the following remark: "So in a way your thesis is all done; in another way, the hardest part of the work is just beginning". And he

---

[131] Stern to Wheeler 1956, *op. cit.*

[132] Wheeler to Stern, 1956, *op. cit.*

concluded: "How soon can you come?"[133] This letter was dictated by phone or telex in order to reach Everett as soon as possible, and, as previously mentioned, it was followed by another sent the same day. In the second letter, Wheeler reiterated his plea and argued that Everett's qualities would not have done much good unless he went and fought "with the greatest fighter" (in which case, he pledged to go to Copenhagen during part of Everett's time there "if that might help"). Wheeler also said that in his annual letter of assessment to the National Science Foundation Fellowship Board (which sponsored Everett's studies), he had urged the need for Everett to go to Copenhagen "with this sentence: 'I feel Everett's very original work is destined to become widely known, and it ought to have the bugs ironed out of it before it is published rather than after!'"[134] In the same period, Wheeler wrote to Bohr, arguing that Everett should "discuss the issue with [him] directly and arrive at a set of words to describe his formalism that would make sense and be free from misunderstandings for this purpose."[135] Wheeler's strategy is outlined in a letter to Allen Shenstone, the chairman of the Physics Department of Princeton University:

> After a first review in Copenhagen of Everett's Thesis in its present only partly satisfactory second draft, I have urged him to come and struggle it out in person with Bohr for a few weeks. I would like to see the thesis reach a form where it will be accepted for publication in the Danish Academy. I think his very original ideas are going to receive wide discussion. [. . .] Since the strongest present opposition to some parts of it comes from Bohr, I feel that acceptance in the Danish Academy would be the best public proof of having passed the necessary tests. Because of my feeling of the importance of this mutual agreement before publication, I am contributing $260 towards Everett's travel out of my very small Elementary Particle Research Fund.[136]

The project of having Everett's work published by the Royal Danish Academy of Sciences had already been mentioned to Everett in the two letters of May 22nd:

> I also feel that the Danish Academy and under Bohr's auspices is the best possible plan for you to publish your work: a full length presentation, going to a wide audience.[137]

When Everett got the news from Wheeler, he phoned him. Following their conversation, Wheeler cabled to Bohr:

> Everett now Princeton phone asking confer with you hopes fly almost immediately but must return in midjune you cable him if convenient my great hope thesis suitable Danish academy publication after revision have answered Stern regards.[138]

[133] Wheeler to Everett, 1956 [I], *op. cit.*

[134] Wheeler to Everett, 1956 [II], *op. cit.*

[135] John A. Wheeler to Niels Bohr, 24 May 1956, *BSC* (reel 33).

[136] John A. Wheeler to Allen G. Shenstone, 28 May 1956, *WP* (Series I—Box Di—Fermi Award #1—Folder Everett).

[137] Wheeler to Everett, 1956 [II], *op. cit.* In the other letter of the same day, he says "I would like to feel happier than I do with the final product; then I would like to see it published in the Danish Academy in full—that's the perfect place for it." (Wheeler to Everett, 1956 [I], *op. cit.*)

[138] John A. Wheeler to Niels Bohr, Cable, 26 May 1956, *BSC* (reel 33).

Shortly afterwards, Everett received a cable from Copenhagen, in which some reservations about this plan were expressed.[139] The cable was followed by a letter which Petersen wrote after consulting with Wheeler. In the letter, Petersen assured Everett that Bohr would have very much liked to discuss his ideas with him, but he added that a period of 2 or 3 months was in their opinion necessary to "come to the bottom of the problems."[140] Since he had in the meantime returned Everett's dissertation to Wheeler (with a note enclosed explaining that Bohr had been too busy to send comments "on the question discussed in the thesis", but hoped to write to him in more detail "about the status of observers in the complementary mode of description"[141]), Petersen requested Everett to send a new copy. He also suggested that "as a background of [his] criticism", Everett should give "a thorough treatment of the attitude behind the complementary mode of description" and state as clearly as possible "the points where he [thought] that this approach [was] insufficient."[142] In the middle of June, Everett was expected to start a job at the Weapon Systems Evaluation Group of the Pentagon in Arlington, which was incompatible with the conditions laid out by Bohr for the visit to Copenhagen. Even though Everett had not excluded the possibility of being allowed to leave in the fall, the project was eventually abandoned.[143]

Wheeler came back from Europe at the end of September 1956. By that time, Everett had passed his final examination and left Princeton for Arlington.[144] However, it took 6 more months for the thesis to be finally submitted (it was defended in April 1957 and graded "very good"[145]). Bohr and his collaborators (including Rosenfeld and Hip Groenewold,[146] who had not attended Stern's seminar, but had read the manuscript) were "warmly thanked" in a note "for the useful objections" (Everett 1957a, p.1). An obvious question is why the thesis, whose second version had been achieved in the 1st months of 1956, was submitted only 1 year later. We know from a letter of Wheeler's that, for administrative reasons related to military service, Everett wished to remain registered at Princeton

[139] Aage Petersen to Hugh Everett, 28 May 1956, *ME*.

[140] Petersen to Everett, 1956, *op. cit.*

[141] Aage Petersen to John A. Wheeler, 26 May 1956, *BSC* (reel 33).

[142] Petersen to Everett, 1956, *op. cit.* To this suggestion, Everett replied: "[. . .] while I am doing it you might do the same for my work." (Everett to Petersen [draft], 1956, *op. cit.*) Everett agreed to send a new copy of the thesis and remarked: "Judging from Stearn's [sic] letter to Wheeler, which was forwarded to me, there has not been a copy in Copenhagen long enough for anyone to have read it thoroughly, a situation which this copy may rectify. I believe that a number of misunderstandings will evaporate when it has been read more carefully (say 2 or 3 times)."

[143] Petersen to Everett, 1956, *op. cit.*; Everett to Petersen [draft], 1956, *op. cit.*

[144] Wheeler to Shenstone, 1956, op. cit.; Petersen to Everett, 1956, *op. cit.* Everett to Petersen [draft], 1956, *op. cit.*, Nancy Everett's calendar of events, *op. cit.*

[145] *GAR*.

[146] Groenewold had been at the University of Groningen since 1951 (he became professor in 1955). He had made his doctorate at the university of Utrecht under the supervision of Rosenfeld, with a dissertation entitled *On the Principles of Elementary Quantum Mechanics*.

University at least until 1956.[147] In the course of 1956, as we have seen, he moved to the Pentagon, where he was no longer in danger of being drafted, but probably had little time to devote to the thesis.[148] Besides these practical reasons, however, it is likely that the revision of the second version in the light of the objections raised in Copenhagen took a good deal of time. In his autobiography, Wheeler remembers that he worked with Everett "long hours at night in [his] office to revise the draft" (Wheeler and Ford 1998, p. 268). In an interview with Kenneth Ford, DeWitt reported Wheeler's recollection more colourfully, saying that Wheeler told him many years later that "he sat down beside Everett and told him precisely what to write."[149] Elsewhere, DeWitt expressed the belief that "Wheeler felt that the Uhrwerk [the long thesis] might offend his hero Bohr."[150] Wheeler explains in his autobiography that "his real intent was to make [Everett's] thesis more digestible to his other committee members" (Wheeler and Ford 1998, p. 268). Bohr and the debate with the Copenhagen group are not mentioned.[151] Yet, there is little doubt that the revision also aimed at making Everett's ideas "more digestible", or at least more comprehensible, to *Bohr*.

The external observation formulation, with which Everett contrasts his approach in the final version of his thesis, is associated, if only obliquely, with Bohr's view—which was not the case for the "conventional formulation" that Everett criticised in the long thesis. At the same time, the emphasis is no longer on the alleged shortcomings of the orthodox view, but on the limitations which seem to restrict its domain of applicability. In his assessment, Wheeler is careful to stress that Bohr's view provides a consistent interpretation of the conventional theory. He points out that the "'external observation' formulation of quantum mechanics has the great merit that it is dualistic" (Wheeler 1957, p. 151)—which is a remarkably gentle way of saying that it "splits the world in two" (Wheeler and Ford 1998, p. 269). We know that Everett regarded this "artificial dichotomy" as "a philosophic monstrosity" (see Sect. 3.5.2),[152] but Wheeler himself, in his autobiography, refers to it as "a difficulty that still deeply troubles me and many others" (Wheeler and Ford 1998, p. 269). Wheeler's cautiousness is jokingly pointed out by DeWitt in a letter of 1967:

---

[147] Wheeler to Everett, 1956, *op. cit.* In the interview with Misner (*op. cit.*, p. 6), Everett himself alludes to the risk of being enlisted in the army upon finishing his studies, and this circumstance is confirmed by DeWitt (Bryce S. DeWitt to Eugene Shikhovtsev, [w/d], courtesy of Eugene Shikhovtsev).

[148] Petersen to Everett, 1956, *op. cit.* Everett to Petersen [draft], 1956, *op. cit.*, Nancy Everett's calendar of events, *op. cit.*; DeWitt to Shikhovtsev, [w/d], *op. cit.*

[149] DeWitt interview, *op. cit.*, p. 6.

[150] DeWitt to Shikhovtsev, [w/d], *op. cit.*

[151] Nor are the discussions with Bohr mentioned in Wheeler's interviews deposited in the archives of the American Institute of Physics.

[152] Everett (1955a, p. 3), in Everett (2012 [1955a]); Everett to DeWitt, 1957, *op. cit.*

> [. . .] I can only say 'Good Old John!'. It always amused me to read your Assessment of Everett's theory [. . .] how highly you praised Bohr, when the whole purpose of the theory was to undermine the stand he had for so long taken!"[153]

In 1956, writing to Stern, Wheeler had been unequivocal: "Everett's thesis is not meant to question the present approach to the measurement problem, but to accept it and generalize it."[154] Indeed, in the introduction and in the conclusion of the paper of 1957, the relative state formulation is not presented as an alternative to the orthodox approach, but rather as a new theory which generalizes it.

> The aim is not to deny or contradict the conventional formulation of quantum theory, which has demonstrated its usefulness in an overwhelming variety of problems, but rather to supply a new more general and complete formulation, from which the conventional interpretation can be *deduced.* (Everett 1957b, p. 141)

Everett's dissertation was published in the *Reviews of Modern Physics*, within a collection of papers "prepared in connection with the Conference on the Role of Gravitation in Physics" held at Chapel Hill in January of 1957. Everett did not attend the conference. Yet, his ideas were mentioned in the discussions,[155] and his paper was submitted by Wheeler to DeWitt, who was the editor in charge for the section of the July issue of the Reviews containing conference papers.[156] The paper was published together with a "companion piece" written by Wheeler, since, notwithstanding the thorough revision, Wheeler was not yet completely satisfied and feared the possible misunderstandings (Wheeler and Ford 1998, p. 268). In his assessment, Wheeler discussed some aspects of Bohr's epistemological analysis explicitly, showing how they could be reformulated in the framework of Everett's theory. These were certainly not the optimum publishing conditions for Everett's work to receive the wider recognition that Wheeler had originally hoped for. Pre-prints were nonetheless sent to many distinguished physicists, including Schrödinger, van Hove, Oppenheimer, Dyson, Yang, Wiener, Wightman, Wigner, and Margenau, besides of course Bohr and his collaborators.[157]

The responses of DeWitt, Wiener and Margenau were quite favourable[158] Groenewold sent a long letter, in which he said that although he found the new draft much improved compared to that he had borrowed 1 year earlier in

---

[153] Bryce S. DeWitt to John A. Wheeler, 20 Apr 1967. *WP* (Series I—Box Co-De Folder DeWitt). DeWitt refers to Wheeler (1957) (see below).

[154] Wheeler to Stern, 1956, *op. cit.*

[155] For example, Feynman, who attended the conference, made some critical remarks on "the concept of a 'universal wave function'." (This fact was brought to our attention by H. Dieter Zeh, who saw the report of the proceedings of the conference deposited in the Wright Air Development Center, Ohio.). This report is now published in DeWitt-Morette and Rickles (2011).

[156] DeWitt to Shikhovtsev, [w/d], *op. cit.*

[157] John A. Wheeler, *Note*, 10 Mar 1957, *WP* (Series I—Box Di—Fermi Award #1—Folder Everett).

[158] DeWitt to Wheeler, 1957, *op. cit.*; Wiener to Wheeler and Everett, 1957, *op. cit.*; Henry Margenau to John A. Wheeler & Hugh Everett, 8 Apr 1957, *WP* (Series I—Box Di—Fermi Award #1—Folder Everett).

Copenhagen, "with regard to the fundamental physical and epistemological aspects" he "still profoundly disagree[d]."[159] Once more, Bohr answered that, although he had no time to write down his comments, he would have asked Petersen to report their discussions. His only remark was that the argumentation contained "some confusion as regards the observational problem."[160] Perhaps he had in mind this "confusion" when, 2 months later, he wrote to Wheeler that he was preparing a new collection of his papers on the epistemological problems in quantum physics (*Atomic Physics and Human Knowledge* was to appear the following year) and that he hoped that, "in spite of all present divergences", this would "help to appreciate the clarification of our position in this field of experience", which, according to his conviction, had been obtained.[161]

Petersen's letter followed indeed, as Bohr had promised. It rejected Everett's approach as a whole, defending the Copenhagen approach to measurement and pointing out Everett's alleged misunderstandings. In his answer, besides spelling out his criticisms of Bohr's approach (see Sect. 3.4.2), Everett mentioned the possibility of being "sent to Europe in the fall on business", in which case he "could probably take a few weeks off and come to Copenhagen."[162] But something hindered this second attempt. The meeting between Everett and Bohr that Wheeler had longed for eventually occurred 2 years later, in March 1959. During the 6 weeks he spent in Copenhagen, Everett met Bohr, but, according to the recollections of his wife, no real discussion on Everett's ideas took place.[163] In Everett's interview, the comments on his visit to Copenhagen are lost in background noise, and we are left with only a few fragments ("that was a hell. . .doomed from the beginning"), which are however quite telling.[164] A much more explicit account is contained in a letter written by Rosenfeld (who had moved to Copenhagen in 1958) many years later[165]:

> With regard to Everett neither I nor even Niels Bohr could have any patience with him, when he visited us in Copenhagen more than 12 years ago in order to sell the hopelessly

[159] Groenewold to Everett and Wheeler, 1957, *op. cit.*

[160] Bohr to Wheeler, 12 April 1957, *op. cit.*

[161] Niels Bohr to John A. Wheeler, 6 Aug 1957, *BSC* (reel 33).

[162] Everett to Petersen, 1957, *op. cit.*

[163] Nancy Everett recalled that "during our visit [. . .] Niels Bohr was in his 80s and not prone to serious discussion of any new (strange) upstart theory." Nancy Gore Everett to Frank Tipler, 10 Oct 1983, *EP* (Box 1, Folder 9). (Bohr was actually 73.) Wheeler gave a similar account in a letter to Max Jammer (19 Mar 1972, *WP*, Series I—Box I—Jason—Folder Jammer).

[164] Everett interview, *op. cit.*, p.8

[165] Rosenfeld to Belinfante, 22 Jun 1972, *op. cit.* In a letter of 1971, Rosenfeld congratulated John Bell for having succeeded in giving "an air of respectability" to "Everett's damned nonsense". (Léon Rosenfeld to John S. Bell, 30 Nov 1971. *RP*.) (Rosenfeld referred to a talk given by Bell at an international conference held at the Pennsylvania State University, in which Bell had presented Everett's theory as a "refurbishing of the idea of preestablished harmony".) Rosenfeld's words should of course be placed in the context of the 1970s (see Sect. 3.8). We are thankful to Anja Jacobsen for having brought the correspondence of Rosenfeld with Belinfante and Bell to our attention.

wrong ideas he had been encouraged, most unwisely, by Wheeler to develop. He was undescribably stupid and could not understand the simplest things in quantum mechanics.

3.7 The Issues at Stake in the Debate

The fact that Wheeler was persuaded that Everett's ideas might obtain Bohr's approval is puzzling. It shows that we should not confine an analysis of the discussions about Everett's proposal to overt disagreements. We must address in the first place the misunderstandings surrounding the Copenhagen view, as well as its inherent ambiguities.

3.7.1 Symbolism

To the Copenhagen group, Everett's formulation of quantum mechanics appeared as a "symbolic limbo" having no thread of connectivity with concrete experimental practice.[166] Everett's interpretation of the wave function seemed to them quite confusing and unjustified, since it endowed the *predictive* symbols of the conventional theory with a *descriptive* connotation which they were not meant to have.[167] In his letter of 1956, Stern wrote:

> Then there is the concept of state in quantum theory. An elementary system does not come with a "ready-made" state. It does not possess a state in the sense of classical physics.[168]

A similar remark was made by Petersen in his discussion with Wheeler:

> Ψ does not pertain to a phys[ical] system in the same way as a dynamical variable. [. . .] Ψ fu[nction] for elec[tron] doesn't have sense until we get something like a prob[ability] dist [ribution] of spots.
>
> So, Q.M. formalism no well defined appli[cation] without exp[erimental] arrangement.[169]

[166] Stern to Wheeler, 1956, *op. cit.* Stern is referring here to "Heisenberg's recent attempt at a theory of elementary particles", which he compares to Everett's proposal.

[167] "[. . .] The entire formalism is to be considered as a tool for deriving predictions, of definite or statistical character, as regards information obtainable under experimental conditions described in classical terms and specified by means of parameters entering into the algebraic or differential equations of which the matrices or the wave-functions, respectively, are solutions. These symbols themselves, as is indicated already by the use of imaginary numbers, are not susceptible to pictorial interpretation." (Bohr 1948, p. 314). Everett outlines Bohr's instrumentalist conception of formalism in the long thesis (1973, p. 110). See Stapp (1972) for a discussion.

[168] Stern to Wheeler, 1956, *op. cit.*

[169] Wheeler, *Notes*, 1956, *op. cit.* When he read this sentence, Everett scrawled in the margin: "Nonsense!".

Indeed, as Groenewold pointed out in 1957, one could figure out an accurate theory of atomic phenomena involving no wave functions at all:

> All physical observable quantities may ultimately be expressed in statistical relations between results of various measurements. These relations may be expressed [. . .] without wave functions (or more general statistical operators).[170]

In his reply to Stern, Wheeler addressed such objections:

> Why in the world talk of a wave function under such conditions for it in no way measures up to the role of the wave function in the customary formulation, that we accept without question?
>
> (a) Nothing prevents one from *considering* a wave function and its time evolution in abstracto; that is, without ever talking about the equipment which originally prepared the system in that state, or even mentioning the many alternative pieces of apparatus that might be used to study that state. (b) A state function as used in this sense has absolutely nothing to do with the state function as used in the customary discussion of the measurement problem, for now *no means of external observation are admitted to the discussion.*

This was a "new physical theory", stemming from "Everett's free volition." Again and again Wheeler stresses the same point:

> The greatest possible confusion will result if the mathematical quantities in Everett's theory, such as the wave function, are thought of as having the purpose that the wave function fulfills in the customary formulation.

And referring to the link between Everett's model and the phenomena:

> The very meaning of the word "consequences" has to be defined within the framework of the theory itself, not by applying to Everett's concept of wave function epistemological considerations that refer to 'wave function' in the completely different of the usual formalism.[171]

Of course, the idea that the state vectors provide a "complete *model* for our world", rather than "expressing the probabilities for the occurrence of individual events observable under well-defined experimental conditions",[172] could hardly appear attractive to Bohr, rooted as it was in a conception of theories that he regarded as a vestige of the classical way of thought. In Bohr's eyes, Everett's attempt to avoid any reflection about the use of concepts in physics, by taking the wave function "as the basic physical entity *without a priori interpretation*", could not produce "a further clarification of the foundations of quantum mechanics."[173] Scientific knowledge, for him, was no less concerned with words than it was with the mathematical symbolism (see Sect. 3.7.4). This point was stressed by Bohr in his discussions with Wheeler, who, as we have seen in Sect. 3.6, after his journey to Copenhagen wrote to Everett that the words that went with the formalism had to be

---

[170] Groenewold to Everett and Wheeler, 1957, *op. cit.*

[171] Wheeler to Stern, 1956, *op. cit.*

[172] The quotations are from Wheeler to Stern, 1956, *op. cit.*, and (Bohr 1948, p. 314) respectively.

[173] Everett (1957b, p. 142), and Aage Petersen to Hugh Everett, 24 Apr 1957, *WP* (Series I—Box Di—Fermi Award #1—Folder Everett).

drastically revised in order to avoid "complete misinterpretation of what physics is about."[174] Even though Wheeler's phrasing seems to call more for the improvement of Everett's prose than for a reflection on the use of concepts, there is little doubt that what Bohr actually wanted to emphasise was the general fact that "one can no more exclude meaning and understanding from physics than one can substitute servomechanisms for physicists."[175]

3.7.2 Relativity

Both Everett and Bohr considered it an important lesson to be learnt from quantum mechanics that physical systems could not be endowed with properties "in the absolute". Yet Everett thought that his relative state formulation was the only way to take the fundamental relativity of properties into account without introducing either subjective or dualistic features in physics. As we have seen, this solution did not put into question the assumption that there must be a correspondence between the state vector of a system and its "objective properties". Bohr's complementarity instead dispensed with the idea that measurements reveal (and state vectors describe) properties which are defined independently of the experimental context. In quantum mechanics, the attribution of properties to a system is consistent with the empirical data only in so far as the observations are confined to a given set of "compatible" observables, i.e. to *certain* experimental contexts. Therefore, from a Bohrian point of view, the fact that state vectors do work as a *meta-contextual* predictive tool prevents us from interpreting them as descriptions of the putative properties of a system. Accordingly, the state vector attributed to a system acquires a physical meaning only when it has been related to the eigenvalues of some observable and to the operations through which the observable is measured.[176] As Petersen put it:

> Only a coord[inate] sys[tem] can give a vector a meaning. Have to know Ψ plus *exp [erimental]* apparatus to make predictions.[177]

---

[174] Wheeler to Everett, 1956 [I], *op. cit.* In his notes (*op. cit.*), Wheeler reports that Petersen, recalling that Everett blamed Bohr for his "conservative" attitude, retorted: "Bohr would say Everett much too class[ical], not in math but in recognize new features. Just as in past formalisms, the whole problem the tough one was to find the right words to express the content of the formalism in acceptable form."

[175] Stern to Wheeler, 1956, *op. cit.* See Sect. 3.7.4 for further discussion.

[176] For Bohr, what is relative (to a given experimental context) is not the property itself, but rather the very possibility of attributing a given property to a system. For a discussion see (Murdoch 1987, Chap. 7). It is telling that, in his epistemological writings, Bohr preferred the term "behaviour" to that of "property" (*Ibid*, p. 135). The meta-contextual connotation that the notion of "property" has in ordinary language must have appeared confusing to Bohr when applied to atomic systems.

[177] Wheeler, *Notes*, 1956, *op. cit.*

Bohr himself repeatedly made this point in his lectures and correspondence, emphasising the analogy with the situation encountered in special relativity.[178] As pointed out by DeWitt, however, also Everett's theory could be put into correspondence with Einstein's approach, although of course for different reasons:

> The conventional interpretation of the formalism of quantum mechanics in terms of an "external" observer seems to me similar to Lorentz's original version (and interpretation) of relativity theory, in which the Lorentz-Fitzgerald contraction was introduced ad hoc. Everett's removal of the "external" observer may be viewed as analogous to Einstein's denial of the existence of any privileged inertial frame.[179]

Everett's theory can be regarded as an attempt to "objectify" the relational aspects of Bohr's approach. The "relativity to the context" implied by Bohr's pragmatic view of formalism is replaced by the "relativity of states", which results in correlations that can be entirely represented *within* the symbolic model of the universe. As we know, the main motive for this move was to neutralise the alleged subjectivist implications of the projection postulate. In the relative state formulation, after a measurement, there is no outcome that is more "actual" than the other a priori possible outcomes: all outcomes are "actual" relative to some state of the universe, and this is supposed to eliminate the need to resort to a "magic process" that projects the state vector onto the subspace corresponding to *the* specific property allegedly revealed by the measurement. However, no "magic process" is required in Bohr's approach either. For Bohr, state vectors are merely predictive symbols that serve to anticipate the results obtained in a well-defined context: if the context undergoes an "objective" change, as it does *after* a result has been recorded, so does the state vector to be used for predicting the results of further observations. This point was emphasised by Groenewold:

> Now one can introduce the statistical operator, which just represents in a very efficient way all the information which already has been obtained and which may be used to calculate the conditional probability (with respect to this information) of other information which still may be obtained or used. Thus also the statistical operator is conditional and depends on the standpoint from which the system is described. It is relative like the coordinate frame in relativity theory. It seems to me that this conditional character has been overlooked in your papers (as well as in many others).[180]

---

[178] See e.g. Murdoch (1987, pp. 145–146).

[179] DeWitt to Everett and Wheeler, 1957, *op. cit.*

[180] Hip Groenewold to Hugh Everett and John A. Wheeler, 11 Apr 1957, *WP* (Series I—Box Di—Fermi Award #1—Folder Everett). For a discussion see (Teller 1981). In the light of these considerations, and in spite of the differences emphasised by both parties in the debate, one could be tempted to point out some connections between Bohr's and Everett's approaches. On the one hand, by taking into account Everett's emphasis on correlations, one might argue that Bohr's interpretation of the state vector requires no projection postulate at all. On the other hand, Bohr's notion of complementarity might be helpful in interpreting Everett's principle of the relativity of states. According to such a principle, the properties possessed by a system at a given instant depend critically on the basis chosen to expand the universal wave function; see Barrett (1999). One may assume the existence of some "internal" mechanism which selects a preferred basis. But as long as this is not done, the arbitrary choice of the preferred basis that determines which sort of

3.7.3 Irreversibility

For the Copenhagen group, the main shortcoming of Everett's theory was that it failed to recognize the fundamental role of irreversibility in physics. Stern wrote to Wheeler:

> Everett does not seem to appreciate the FUNDAMENTALLY irreversible character and the FINALITY of a macroscopic measurement. One cannot follow through, nor can one trace to the interaction between the apparatus and the atomic system under observation. It is not an "uncontrollable interaction", a phrase often used in the literature. Rather, it is an INDEFINABLE interaction. Such a connotation would be more in accord with the fact of the irreversibility, the wholeness of the quantum phenomenon as embodied in the experimental arrangement.[181]

Likewise, in his letter of 1957, after pointing out the necessity of cutting off the "measuring chain", Groenewold remarked:

> But it is extremely fundamental that this cutoff is made after the measuring result has been recorded in a permanent way, so that it no longer can be essentially changed if it is observed on its turn (i.e. if the chain is set forth). This recording has to be more or less irreversible and can only take place in a macrophysical (recording) system. This macrophysical character of the later part of the measuring chain is decisive for the measuring process. I do not think that it can be left out of consideration in its description. It does not seem to act an essential part in your considerations.[182]

From Everett's standpoint, such objections were completely misguided.[183]

> [. . .] one of the fundamental motivations of the paper is the question of how can it be that mac[roscopic] measurements are "irreversible", the answer to which is contained in my theory (see remarks chap. V), but is a serious lacuna in the other theory.

Indeed, as we have seen in Sect. 3.4.2, the way in which the Copenhagen group accounted for the irreversibility of the measurement process was for Everett highly unsatisfactory and mysterious. In Bohr's writings, the fundamental role of irreversibility in physics was often stressed. But, according to Everett, little was said about the origin of this "magic irreversibility".

The arguments put forward by the Copenhagen group about this and other aspects of measurement involved (and sometimes mixed up) two different levels of reflection. The first and more fundamental level implied a *pragmatic-transcendental* argument to the effect that irreversibility is a constitutive feature of measurement, and that it cannot be ensured unless the description of the results is framed within the representation of ordinary "objective" experience. The second

---

(relative) properties are attributed to a system (for instance, a definite value for position, but not for momentum) looks very much like the Bohrian choice between "complementary" contexts. For a discussion see (Bitbol 1998, pp. 286–293).

[181] Stern to Wheeler, 1956, *op. cit.*

[182] Groenewold to Everett and Wheeler, 1957, *op. cit.*

[183] They indicated "rather clearly" that his critics had "had insufficient time to read" his work. This and the following quotations are taken from Everett's notes on Stern's letter, 1956, *ME*.

level implied a *physical* explanation of irreversibility connecting irreversibility with the "reduction" of the state vector, and the reduction of the state vector with the macroscopic nature of the measuring apparatus. Assuming that quantum mechanics should also apply to the macroscopic domain, the former (pragmatic) argument raised a problem of consistency, which the latter (physical) argument was designed to settle.[184]

A detailed analysis of the issue of irreversibility in connection with Everett's work can be found in the correspondence that Rosenfeld had with Belinfante in 1972. It is worth quoting some passages from these letters, with the caveat that they were written many years after Everett's dissertation. The context was then heavily influenced by the controversies of the 1960s, in which Bohr, who died in 1962, took no part. Rosenfeld, who, as he says himself, was doing his best to pull Belinfante out of the pitfall in which he had been precipitated by the reading of Everett,[185] wrote to him:

> . . . I do not think you are right to go on and say that one could do without reducing the state vector, which means physically without carrying the measurement to its completion by recording a permanent mark of its result. You should leave such a heresy to Everett.[186]

In his letters, Rosenfeld explained that the reason why there is "no choice whatsoever about the necessity of applying the [state] reduction" is that "the reduction rule is nothing else than a formal way of expressing the idealized result of the registration": without it "the phenomenon is not well defined."[187] He also stressed that the "reduction rule" did not require an *ad hoc* postulate: it could be deduced (in principle) from thermodynamic considerations that applied to macroscopic systems. Since the registration is necessary, and since it requires state reduction, which can only be established for *macroscopic* systems, Rosenfeld concluded that nobody "can avoid committing himself to accepting the necessity of macroscopic measuring instruments."[188] Indeed, as we have seen (Sect. 2.3), in the early 1960s Rosenfeld supported, against Wigner, the theory of measurement proposed by Daneri, Loinger and Prosperi (1962). In his opinion, such a theory provided a rigorous framework for Bohr's ideas.[189] In the 1950s, however, the Copenhagen group did not oppose Everett's objections with anything like a *theory* of measurement, but merely with a collection of generic statements.

---

[184] See Murdoch (1987, pp. 112–118). See also Sect. 3.7.5.

[185] Léon Rosenfeld to Frederik J. Belinfante, 24 Aug 1972, *RP*.

[186] Rosenfeld to Belinfante, 22 Jun 1972, *op. cit.*

[187] Léon Rosenfeld to Frederik J. Belinfante, 24 Jul 1972. *RP*.

[188] *Ibid.*

[189] "Now, the crux of the problem which worries Wigner so much is that the reduction rule appears to be in contradistinction with the time evolution described by Schrödinger's equation. The answer, which was of course well known to Bohr, but has been made formally clear by the Italians [Daneri, Loinger and Prosperi], is that the reduction rule is not an independent axiom, but essentially a thermodynamic effect, and accordingly, only valid to the thermodynamic approximation." Rosenfeld to Belinfante, 24 July 1972, *op. cit.*

3.7.4 Words

In Bohr's view, the mathematical symbols employed in physics have a meaning only inasmuch as they refer to well-defined measurements. Therefore, the meaningful use of a theory presupposes that one can define unambiguously the experimental setup, in which the measurements are performed, as well as their possible outcomes.[190] This point was made by Petersen during the discussions of 1956:

> [. . .] Math can never be used in phys[ics] until have words. [. . .] What mean by physics is what can both be expressed unambig[uously] in ordinary language. Spots on plate have meaning but not in Everett—he talks of correlations but can never build that up by Ψ fun [ctions].[191]

Stern stressed the same idea in his letter:

> Our formalism must be in terms of possible or idealized experiments whose interpretations thereby involves [sic] the use of concepts intimately connected with our own sphere of experience which we choose to call reality. The epistemological nature of our experiments and the objective nature of the abstract mathematical formalism TOGETHER form the body and spirit of science.

He also illustrated this point by means of an example taken from biology:

> To trace the schizophrenic phenomenon from the basic molecular level to the observational level of its psychological symptomatic manifestations is an aspect of the observation problem. It cannot be traced in the detail of a space-time description.[192]

This example is meant to show that physical theories establish correlations between facts of our experience, the "definition" of which does *not* involve the mathematical constructs of those very theories. Such a remark generalized a typical Copenhagen assertion, which Groenewold summarised as follows:

> Because all observable quantities may ultimately be expressed in statistical relations between measuring results and the latter are represented by essentially macrophysical recordings, the former ones may ultimately be expressed in macrophysical language. That does of course not mean that the formalism, which serves as a tool for calculating these statistical relations could also be expressed in macrophysical language. On the contrary in this field the macrophysical language is liable to loose its original more or less unambiguous meaning.[193]

Besides highlighting the importance of "classical" concepts (i.e. concepts used in ordinary language and classical physics) for describing the experimental context in which atomic phenomena are observed, Bohr also insisted on the need to use such concepts for providing a pictorial description of the phenomena themselves. In both cases Bohr assumed that an account based on classical concepts automatically fulfilled the conditions for an *objective* description. In the former case, as we have

---

[190] See e.g. Stapp (1972).

[191] Wheeler, Notes, 1956, *op. cit.*

[192] Stern to Wheeler, 1956, *op. cit.*

[193] Groenewold to Wheeler and Everett, 1957, *op. cit.*

seen, such conditions were related to the requirements of communicability and repeatability which are constitutive of experimental practice. In the latter case, they were related to the *ob*jectification of phenomena allegedly required by the very concept of *ob*servation.[194] This twofold argument is summarised by Petersen in his letter of 1957:

> There can on [Bohr's] view be no special observational problem in quantum mechanics in accordance with the fact that the very idea of observation belongs to the frame of classical concepts. The aim of [Bohr's] analysis is only to make explicit what the formalism implies about the application of the elementary physical concepts. The requirement that these concepts are indispensable for an unambiguous account of the observations is met without further assumptions [. . .].[195]

As we have pointed out in the discussion about irreversibility, the Copenhagen scientists did not always clearly distinguish the various levels involved in Bohr's argument—the level of language, that of the conditions for the possibility of physics, and that of the content of physical knowledge. This is even more true for Bohr's critics. Everett's reading of Bohr's argument, for example, was that

> [in the Copenhagen interpretation] the deduction of classical phenomena from quantum theory is impossible simply because no meaningful statements can be made without preexisting classical apparatus to serve as a reference frame.[196]

Here Bohr's *transcendental* reasoning, according to which the formulation of a physical problem *presupposes* the specification of the corresponding experimental conditions ("apparatus"), and hence requires a suitable conceptual framework, is presented as a *physical* assumption about the *existence* of a macroscopic world ("phenomena") governed by classical mechanics. That Everett understood Bohr's argument as a postulate implying "that macrosystems are relatively immune to quantum effects" is confirmed by the main criticism that he addressed to the Copenhagen interpretation, namely that it "[adhered] to a 'reality' concept [. . .] on the classical level but [renounced] the same in the quantum domain."[197] Unsurprisingly, Everett regarded such a "postulate" with no sympathy at all ("epistemologically garbage", he annotated on Groenewold's letter). For him, Bohr's conception of formalism, as well as his insistence on the primitive role of classical concepts, imposed arbitrary limits upon the scope of quantum mechanics. Everett contrasted this dogmatic position with the pragmatic view that he advocated with regard to "the constructs of classical physics" (see Sect. 3.5.1), and he claimed that, by showing that classical *physics* can be derived from quantum theory, one could in fact replace "classical" *concepts* by "quantum" ones. In his reply to Petersen, after pointing out that he did not think that his viewpoint could be dismissed "as simply a misunderstanding of Bohr's position", Everett formulated it as follows:

---

[194] See Bitbol (1996b, pp. 256–269) for a critical analysis.

[195] Petersen to Everett, 1957, *op. cit.*

[196] Everett (1973, p. 111). Everett regarded this position as "conservative".

[197] Everett to Petersen, 1957, *op. cit.* See Sect. 3.4.2.

> The basing of quantum mechanics upon classical physics was a necessary provisional step, but now [. . .] the time has come to proceed to something more fundamental. There is a good analogy in mathematics. The complex numbers were first introduced only in terms of the real numbers. However, with sufficient experience and familiarity with their properties, it became possible and indeed more natural to define them first *in their own right* without reference to the reals. I would suggest that the time has come to do the same for quantum theory—to treat it in its own right as a fundamental theory without any dependence on classical physics, and to derive classical physics from it. While it is true that initially the classical concepts were required for its formulation, we now have sufficient familiarity to formulate it without classical physics, as in the case of the complex numbers.

Everett concluded this passage by observing: "I'm sure that you will recognize this as Bohr's own example turned against him".[198] Indeed, from Wheeler's notes, we know that, during their discussions, Petersen had made the following example:

> Bohr (ac[cording] to A[age] P[etersen]) need non rel[ativistic] way to live self into rel [ativistic] world—have to sep[arate] between space [and] time—consider watch; *entrance into* Complex n[umbers] only via real n[numbers]; hence entrance into rel via non rel.[199]

Of course, from a Bohrian standpoint, Everett's hope to *derive* from the theory the conceptual framework *presupposed* by physics was an illusion, since one could not even make sense of the theory without relying on a well-defined experimental practice. As Rosenfeld put it in 1959:

> Everett's work [. . .] suffers from the fundamental misunderstanding which affects all the attempts at 'axiomatizing' any part of physics. The 'axiomatizers' do not realize that every physical theory must necessarily make use of concepts which *cannot*, in principle, be further analysed, since they describe the relationship between the physical system which is the object of study and the means of observation by which we study it: these concepts are those by which we give information about the experimental arrangement, enabling anyone (in principle) to repeat the experiment. It is clear that *in the last resort* we must here appeal to *common experience* as a basis for common understanding. To try (as Everett does) to include the experimental arrangement into theoretical formalism is perfectly hopeless, since this can only shift, but never remove, this essential use of unanalysed concepts which alone makes the theory intelligible and communicable.[200]

With similar arguments in mind, in 1957 Petersen wrote to Everett:

> Of course, I am aware that from the point of view of your model-philosophy most of these remarks are besides the point. However, to my mind this philosophy is not suited for approaching the measuring problem. I would not like to make it a universal principle that ordinary language is indispensable for definition or communication of physical experience,

---

[198] *Ibid.* See also Wheeler (1957, p. 151).

[199] Wheeler, *Notes*, 1956, *op. cit.* Bohr often remarked that the use of imaginary numbers in quantum theory prevents one from interpreting the quantum formalism "as an extension of our power of visualization" (Bohr 1998, p. 86). Also, he liked to mention the discovery of irrational numbers as an example of how concrete problems (e.g. measuring the diagonal of the square) may lead us to extending the use of ordinary concepts (in the example: rational numbers) (Petersen 1985, pp. 301–302).

[200] Léon Rosenfeld to Saul Bergmann, 21 Dec 1959, RP. The letter answered the request for "an opinion about Everett's point of view on the presentation of the principles of quantum mechanics" formulated by Saul M. Bergmann of the Boston Laboratory for Electronics.

> but for the elucidation of the measuring problem [. . .] the correspondence approach has been quite successful.[201]

During the discussions in Copenhagen, Wheeler came to realise that, if Everett's "model philosophy" intended to do away with Bohr's prescriptions about the use of classical concepts, it had to show (without relying on Bohr's pragmatic-transcendental argument) that the general conditions which make experimental activity possible are indeed fulfilled in the world described by the theory.[202] In the words of one of Everett's epigones, the theory was demanded to explain "why the sentient beings we know [. . .] have the particular concepts they do for describing their world" (Vaidman 2002). According to Wheeler, one could thus show that Everett's theory "does not require for its formulation any reference to classical concepts" and is "conceptually self-contained" (Wheeler 1957, pp. 151–152). Along these lines, in his discussions with Petersen, Wheeler had sketched an argument according to which, since human practices (including communication and experimentation) are an outgrowth of (the complex physical processes underlying) biological selection, they could be expected to be described by some process occurring within Everett's "model universe": "Thinking, experimentation and communication—or psychophysical duplicates thereof—are all taken by Everett as going on *within* the model universe.".[203] He wrote to Everett:

> Aage Petersen [. . .] had a tendency to insist that small interaction, small $e^2/\hbar c$, was essential for a world in which one could use normal words. On the contrary, I argued that the world came first—it could have small or large $e^2/\hbar c$, but grant only complex systems, and evolution, and you have systems that *must* find a way to communicate with each other to give mutual assistance in the struggle for existence; in the struggle for survival words would necessarily be invented to deal with a large $e^2/\hbar c$. You don't first give a list of words and then ask what systems are compatible with them; instead, the system comes first, and the words second.[204]

Wheeler's argument was developed in his letter to Stern, in which he concluded:

> The kind of physics that occurs does not adjust itself to the available words; the words evolve in accordance with the kind of physics that goes on.[205]

In the assessment of 1957, we find almost the same sentence. Yet, there is an interesting semantic shift, due to the fact that the term "words", which in the letter stands essentially for "concepts", is replaced by "terminology", and the verb "evolve", which in the letter is clearly related to the evolutionary argument that immediately precedes it, becomes "adjust". Formulated that way, the statement no longer alludes so strongly to a physical explanation of the fact that physicists use

[201] Petersen to Everett, 1957, *op. cit*.

[202] In his paper of 1957 (pp. 151–152), Wheeler says: "The results of the measurements can be spelled out in classical language. Is not such 'language' a prerequisite for comparing the measurements made by different observing systems?".

[203] Wheeler to Stern, 1956, *op. cit*.

[204] Wheeler, Notes, 1956, *op. cit*.

[205] Wheeler to Stern, 1956, *op. cit*.

certain concepts. We can only conjecture that the objections of the Copenhagen group played some role in this reformulation. However, there is no doubt that the idea of providing a naturalized account of the conditions that make physics possible was in contrast to Bohr's doctrine. This is testified by a lapidary remark in Wheeler's notes: "Language second. Very contrary to Bohr, say A[age] P [etersen]."[206]

3.7.5 *Observers*

In the 1950s, the Copenhagen group seems to have regarded the idea of developing a "quantum theory of measurement" (which would apply to measuring devices) as a possible source of confusion. For example, in the above mentioned report of 1957 (see Sect. 3.2.2), Rosenfeld argued:

> Bohr's considerations were never intended to give a 'theory of measurement in quantum theory', and to describe them in this way is misleading, since a proper theory of measurement would be the same in classical and quantal physics, the peculiar features of measurements on quantal systems arising not from the measuring process as such, but from the limitations imposed upon the use of classical concepts in quantum theory. By wrongly shifting the emphasis on the measuring process, one obscures the true significance of the argument and runs into difficulties, which have their source not in the actual situation, but merely in the inadequacy of the point of view from which one attempts to describe it. This error of method has its origin in v. Neumann's book 'Foundations of Quantum Mechanics' [. . .].

In the report, Rosenfeld made some sarcastic remarks on the efforts made by a group of physicists "to develop their own 'theory of measurement' in opposition to what they believed to be the 'orthodox' theory of measurement, as presented by v. Neumann." According to Rosenfeld, these "reformers [. . .] involved themselves in a double misunderstanding, criticizing a distorted and largely irrelevant rendering of Bohr's argument by v. Neumann, and trying to replace it by a 'theory' of their own, based on quite untenable assumptions."[207]

---

[206] Wheeler, Notes, 1956, *op. cit.* See Petersen (1985). For a thorough analysis of the philosophical background of Bohr's doctrine of concepts, see Chevalley (1994). See also Faye (1991), Murdoch (1987).

[207] Rosenfeld, 1957, *op. cit.* Rosenfeld is here alluding to David Bohm and other "young physicists, who, misled partly by v. Neumann's ideas, partly by preconceived philosophical opinions, were unable to understand the real problems underlying the formulation of quantum theory, and [. . .] undertook to reform quantum theory according to their own liking, and to develop, as they put it, a 'causal interpretation' of this theory." However, since the report was written in 1957, it is likely that Everett's work had some role in exacerbating Rosenfeld's irritation.

In the notes he took in Copenhagen, Wheeler reports these words of Petersen:

> Von N[eumann] + Wig[ner] all nonsense; their stuff beside the point; [. . .]Von N[eumann] + Wig[ner]—mess up by including [the] meas[uring] tool in [the observed] system. [. . .] Silly to say apparatus has Ψ-function.[208]

Also, Petersen insisted that, when considering the "paradox outlined by Everett", one must keep in mind the "distinction between Bohr way & the two postulate way to do q[uantum] mech[anics]". It should be stressed, however, that "Bohr way" did not rule out the possibility of treating observers quantum-mechanically.[209] Nothing prevents one from providing a model of the physical process which is supposed to correspond to a measurement. Yet the symbols appearing in such a model acquire a meaning only when one states the set of measurements that can be performed upon the compound system S + O (where S and O are the *physical systems* which represent the "object-system" and the "apparatus", respectively). In other words, *any* formal model presupposes an observer who can perform the experimental operations and interpret the possible outcomes in accordance with a given conceptual and pragmatic framework. As Groenewold put it:

> . . . the observer [. . .] not only "observes" the object system, but also describes it with some theory and "interprets" if you like. . . . I do not see how your automatical observer included in the described combined system also could be used for describing the activities of reading the recorded measuring result and of assigning statistical operators to the object system on the ground of the obtained information.[210]

The "transcendental" role that the observer (or the apparatus) plays within the instrumentalist view of formalism is taken into account by Bohr's *functional* distinction between the apparatus *qua* physical system and the same qua measuring instrument.[211] As Petersen pointed out in his discussion with Wheeler, "QM description of measuring tool prevents its use as a meas[uring] tool."[212] In a letter to Everett, Petersen developed this point:

> I do not understand what you mean by quantized observers. Obviously, one can treat any interaction quantum-mechanically, including the interaction between an electron and a photographic plate, but when utilized as an "observer" the definition of the "state" (position) of the plate excludes considerations of quantum effects. It seems to me that as far as

208 Wheeler, Notes, 1956, *op. cit.*

209 See Bohr (1939). In that paper, Bohr asserted: "In the system to which the quantum mechanical formalism is applied, it is of course possible to include any intermediate auxiliary agency employed in the measuring process." (Bohr 1998, p. 104). In one of the above mentioned letters, referring to Wigner's allusions to a special role played by consciousness in the measuring process, Rosenfeld asserted that the opinion according to which the "recording process is not entirely describable by quantum mechanics" was "simply wrong". (Rosenfeld to Belinfante, 24 Jul 1972, *op. cit.*).

210 Groenewold to Everett and Wheeler, 1957, *op. cit*. The term "super-observer", which Wheeler uses in his paper of 1957 (p. 152), is possibly reminiscent of some analogous remark made during the discussions in Copenhagen.

211 See Murdoch (1987, Chap. 5).

212 Wheeler, *Notes*, 1956, *op. cit.*

> your treatment of many-body systems is consistent with the proper use of the formalism it has nothing to do with the measuring problem.[213]

Nonetheless, the existence of two "complementary" ways of conceiving the apparatus raised an issue of consistency:

> On one hand the combined object and measuring systems are considered from the microphysical quantum mechanical point of view. So far one could not even speak of measurement. On the other hand the later part of the measuring chain and in particular the recording system is regarded from the macrophysical classical point of view. A satisfactory theory of measurement has to relate these two aspects to each other.[214]

A solution to this consistency problem is sketched by Rosenfeld in his letter of 1959:

> The fact, emphasized by Everett, that it is actually possible to set up a wave-function for the experimental apparatus and Hamiltonian for the interaction between system and apparatus is perfectly trivial, but also terribly treacherous; in fact, it did mislead Everett to the conception that it might be possible to describe apparatus + atomic object as a closed system. This, however, is an illusion: the formalism used to achieve this must of necessity contain parameters such as external fields, masses, etc. which are precisely the representatives of the uneliminable residues of unanalysed concepts.[215]

A similar remark had been made by Petersen in 1957:

> There is no arbitrary distinction between the use of classical concepts and the formalism since the large mass of the apparatus compared with that of the individual atomic object permits that neglect of quantum effects which is demanded for the account of the experimental arrangement.[216]

With some reason, Everett found this and similar physical explanations loosely formulated and unconvincing. And since he thought that the conclusions reached by Bohr on the basis of his reflection on the preconditions of physics must ultimately be justified by some physical arguments, this led him to conclude that Bohr's doctrine rested in fact on a "flatly asserted dogma" (see Sect. 3.4.2). Indeed, as we have seen, rendering Bohr's analysis superfluous by exhibiting a self-consistent physical model of the world (including observers) was one of the main goals of the final version of Everett's dissertation. This reflected a concern that Wheeler had already expressed in 1956, when he wrote to Bohr:

> But I am more concerned with your reaction to the more fundamental question, whether there is any escape from a formalism like Everett's when one wants to deal with a situation where several observers are at work, and wants to include the observers themselves in the system that is to receive mathematical analysis.[217]

---

[213] Petersen to Everett, 1957, *op. cit.*

[214] Groenewold to Everett and Wheeler, 1957, *op. cit.*

[215] Rosenfeld to Bergmann, 1959, *op. cit.*

[216] Petersen to Everett, 1957, *op. cit.*

[217] John A. Wheeler to Niels Bohr, 24 Apr 1956, *BSC* (reel 34).

From Stern's letter we know that the idea of providing a naturalized account of the "emergence" of the pragmatic framework presupposed by the instrumentalist interpretation of formalism had been cautiously put forward by Wheeler in a letter of the same year:

> In your letter you ask, "Do we need mathematical models, like those of game theory, that will include the observers, in order to put across to the mathematically minded what is meant by these ideas?" (I take it you mean complementarity and other ideas of quantum theory "as distinct from the mere formalism.")[218]

In the 1957 paper, this proposal was contrasted with the external observation formulation. In such a formulation, the idea that the very possibility of linking the symbolic structure to experience presupposes a pragmatic framework is replaced by a postulate implying that "the 'measuring chain' has to be cut off" and that some physical system has to be left out of the mathematical description whenever an observation takes place.[219] The foregoing analysis should have made clear that Bohr's hostility towards Wheeler's programme was not due to his commitment to such a postulate. Indeed, Petersen wrote to Everett: "I don't think that you can find anything in Bohr's papers which conforms with what you call the external observation interpretation."[220] What made little sense for Bohr was the attempt to restore what Pauli called the "ideal of the detached observer",[221] by postulating an "independent reality" and assuming that physics must describe it. To him, taking this approach was overlooking the analysis of the very conditions which make it possible for an observer engaged in the investigation of experience to describe atomic phenomena objectively.[222]

---

[218] Stern to Wheeler, 1956, *op. cit.* The letter quoted by Stern is now lost.

[219] The quotation is from the letter of Groenewold to Wheeler and Everett (*op. cit.*). The "external observation" reading of Bohr's approach was arguably based on his frequent remarks emphasising "the necessity of describing entirely on classical lines all ultimate measuring instruments which define the external conditions of the phenomenon, and therefore of keeping them outside the system for the treatment of which the quantum of action is to be taken essentially into account." (Bohr 1998, p. 107).

[220] Petersen to Everett, 1957, *op. cit.*

[221] Wolfgang Pauli to Niels Bohr, 15 Feb 1955 (Pauli et al. 1994, p. 43). Pauli uses this expression to denote Einstein's view. Hooker (1991, p. 507) has described such a view as one in which the objectivity of the physical description depends on its ability "to put [us] into the models as *objects* in such a way as to take [us] out of the picture as *subjects*."

[222] Accordingly, Bohr's idea of completeness, like that of objectivity, had little to do with the possibility of providing an all-encompassing model of the universe, including observers. What counted, instead, was the ability to answer all the possible questions that can be concretely framed in an experimental context. As Hooker (1991, p. 507) puts it: "To be Bohr-objective is to achieve simultaneously both an empirically adequate, exhaustive and symbolically unified description of the phenomena we can produce and an accurate portrayal of the conditions under which such phenomena are accessible to us." Hence "*Bohr-objectivity* cannot consist in removing the knowing subject from the representation of reality—precisely to the contrary." From Bohr's point of view, the "restrictions" that the instrumentalist interpretation of formalism allegedly imposed upon the scope of quantum theory did not deprived us of any portion of physical knowledge. On the contrary, they were (in a Kantian sense) *constitutive* of knowledge. For an analysis of the Kantian

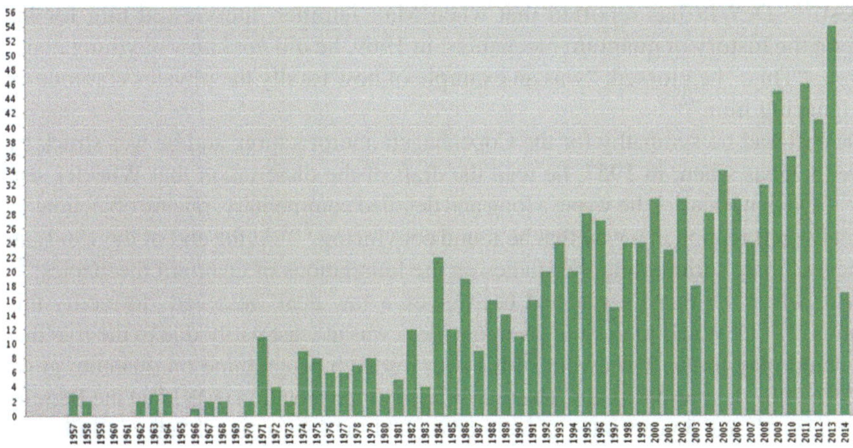

Picture 3.2 Citations of Everett's 1957 paper, from 1957 to 2014—Source of the data: Web of Science

3.8 Epilogue

Contrary to Wheeler's hopes, after obtaining his PhD, Everett continued to collaborate with the Pentagon and did not return to academic research.[223] In 1962 he was invited to present the relative-state formulation at a conference on the foundations of quantum mechanics held at the Xavier University of Cincinnati, before an audience including Furry, Wigner, Dirac, Aharanov, Rosen, and Podolsky (a short account of the conference appeared in *Physics Today*[224]). But except for this and other sporadic signs of interest, the impact of Everett's work was

aspects of Bohr's philosophy see for example (Honner 1987; Murdoch 1987; Faye 1991; Kaiser 1992; Chevalley 1994).

[223] See Wheeler to Everett, 1956 [I], *op. cit.*; (Byrne 2010).

[224] Werner (1964). At the conference, Everett was invited to outline his approach, which he did, insisting particularly on the "deduction" of the standard probabilistic interpretation. In reply to questions about the status of branches, Everett examined the case in which an observer performs a sequence of measurements on an ensemble of identical systems. In this case, he argued, each "element" of the resulting superposition of states "contains the observer as having recorded a particular definite sequence of results of observation". He concluded that any such element can be identified as "what we think of as an experience", and that "it is tenable to assert that all the elements simultaneously coexist." To the remark of Podolsky: "It looks like we would have a non-denumerable infinity of worlds", Everett answered: "Yes." (Proceedings of the Conference on the Foundations of Quantum Mechanics, Xavier University, Cincinnati, 1962; deposited at the American Institute of Physics.)

modest.[225] DeWitt has reported that when Max Jammer interviewed him for his book on the history of quantum mechanics, in 1969, he did not know anything about Everett. "This," he glossed, "was an example of how totally the physics community was ignoring him."[226]

DeWitt had no sympathy for the Copenhagen interpretation, and he was struck by Everett's ideas when, in 1957, he read the draft of the dissertation that Wheeler sent him.[227] On that occasion he wrote a long and detailed commentary, raising objections to which Everett replied in a way that he found convincing.[228] At the end of the 1960s, in the new climate surrounding the studies on the foundations of quantum mechanics,[229] DeWitt, who "felt that Everett had been given a raw deal" resolved "to rectify this situation".[230] DeWitt's interest in Everett's ideas was at least partly due to the role that they could play in the framework of his own research programme on quantum gravity.[231] In 1967, he presented the "Everett–Wheeler interpretation (EWI)" at the Battelle Rencontres,[232] and 3 years later he lectured on it at the International School of Physics "Enrico Fermi", in the framework of a course on the foundations of quantum physics organised by Bernard d'Espagnat. In 1970 Physics Today published a paper in which DeWitt contrasted his many-worlds version of the EWI with both the Copenhagen interpretation and the mentalistic approach advocated by Wigner. The paper gave rise to a lively debate, which marked the beginning of the "rediscovery" of Everett's work.

Everett took no part in that debate. In 1971, he consented to the publication of the long version of the thesis in a small book edited by DeWitt and his student Neill Graham "with the proviso that [he] would not have to devote any effort to editing, proof reading, etc."[233] In 1977, Wheeler, who was then at the University of Texas in Austin, invited Everett for a conference. There Everett met DeWitt for the first and

---

[225] Shikhovtsev (2003) mentions in particular an invitation by Wheeler to give a seminar at Princeton in 1959. Everett's paper was cited in the philosophical works of Margenau (1963), Shimony (1963), and Petersen (1968). It was not cited in the famous papers on the measurement problem that Wigner wrote in that period (Wigner 1961, 1963). In 1963, referring to Everett in a letter, Wigner observed: "The state vector, as he imagines it, does not convey any information to anyone, and I don't see what its role is in the framework of science as we understand it." (Eugene Wigner to Abner Shimony, 24 May 1963, *WigP* (Box 94, Folder 1). The limited impact of Everett's work is discussed by Freire Jr. (2004) based on the statistics of the citations that it received in the decade that followed the publication. See also Picture 3.2.

[226] DeWitt interview, *op. cit*, p. 7.

[227] "I read it and I was stunned, I was shocked." (DeWitt interview, *op. cit.*, p. 7). However, for a more detailed analysis of DeWitt's ideas, see Hartz (2013).

[228] Everett to DeWitt, 1957; *op. cit.*; DeWitt interview, *op. cit.*, p. 7.

[229] See Freire Jr. (2004).

[230] DeWitt to Shikhovtsev, [w/d], *op. cit.*; Bryce S. DeWitt to Olival Freire, pers. comm., 29 Jun 2002.

[231] The paper in which DeWitt presented the famous Wheeler–DeWitt equation relies on Everett's approach in order to provide an interpretive framework for "the state functional of the actual universe" (DeWitt 1967).

[232] DeWitt (1968).

[233] Hugh Everett to Bill Harvey, 20 Jun 1977, *EP* (Series I–8). The book was published in 1973.

last time.[234] Everett's ideas sparked the interest of some of Wheeler's students who attended the conference. David Deutsch, who was among them, has reported that Everett appeared quite sympathetic to the many-worlds interpretation.[235] However, answering a letter of that year in which he was explicitly asked if he advocated such an interpretation, Everett said laconically: "I certainly approve of the way Bryce DeWitt presented my theory, since without his efforts it would never have been presented at all."[236] And referring in another letter to the title of DeWitt's and Graham's book, *The Many-Worlds Interpretation of Quantum Mechanics*, he said: "This of course was not my title as I was pleased to have the paper published in any form anyone chose to do it in!" And he added: "I, in effect, had washed my hands of the whole affair in 1956."[237] Indeed, Everett made little effort to promote and develop his ideas, and showed himself reluctant to go beyond generic comments in private correspondence either.[238]

There are some hints that Wheeler's attitude after the publication of Everett's dissertation was not very supportive.[239] As we have seen, Wheeler's admiration for Bohr did not prevent him from attaching great importance to Everett's unorthodox ideas, and from believing that it was indeed possible to get "his great master" and his young student to agree.[240] Consequently, the reception of Everett's work in Copenhagen must have left him rather disappointed. In his interview, DeWitt recalled that when the EWI was brought to the knowledge of the wider public by his own paper in *Physics Today*, Wheeler "promptly disowned Everett." DeWitt added that he asked Wheeler why he did not "accept Everett more", but never got a satisfactory answer from him.[241] The circumstance pointed out by DeWitt is confirmed by the *incipit* of a letter which Everett received in 1977 from Jean-Marc Lévy-Leblond:

---

[234] DeWitt interview, *op. cit.*, p. 15.

[235] Shikhovtsev (2003).

[236] Everett to Harvey, 1977, *op. cit.*

[237] Everett to Lévy-Leblond, 1977, *op. cit.*

[238] DeWitt asserted many years later: "Everett always took the attitude—and I got this from Charlie Misner as well—that he was not really strongly committed to this." (DeWitt interview, *op. cit.*, p. 15.) DeWitt confirmed this opinion in a recent letter, arguing that Everett "was lackadaisical and couldn't care less if other physicists would accept his views." (DeWitt to Shikhovtsev, [w/d], *op. cit.*) It is likely that the reception of his ideas in Copenhagen diminished Everett's original enthusiasm. In any case, even in his last years, Everett maintained that the relative state formulation was the "simplest" and the "only completely coherent approach" "to come to grips with the paradoxes of the measurement process", and that the alternative proposals were "highly tortured and unnatural" and "by far more artificial and unsatisfactory." Everett to Jammer, 1973, *op. cit.*; Everett to Raub, 1980, *op. cit.*

[239] For instance, in a paper about cosmology of 1962, in which he mentioned the "so-called 'universal wave function'", Wheeler (1962) cited his own assessment, but not Everett's paper.

[240] Everett interview, *op. cit.*, p. 8.

[241] DeWitt interview, *op. cit.*, p. 7; DeWitt to Shikovtsev, [w/d], *op. cit.*

Dear Dr. Everett,

I obtained your address through the kindness of Prof. Wheeler,who suggested that I directly ask your opinion on what I believe to be a crucial question concerning the 'Everett & no-longer-Wheeler' (if I understood correctly!) interpretation of Qu. Mech.[242]

Everett himself alludes to Wheeler's ambiguity in a letter of 1980:

Dr. Wheeler's position on these matters has never been completely clear to me (perhaps not to John either). He is, of course, heavily influenced by Bohr's position (he was a student of Bohr) which essentially regards the entire formalism as merely a calculating device, and does not worry any further about "reality". It is equally clear that, at least sometimes, he wonders very much about that mysterious process, "the collapse of the wave function". The last time we discussed such subjects at a meeting in Austin several years ago he was even wondering if somehow human consciousness was a distinguished process and played some sort of critical role in the laws of physics.[243]

As is apparent from this passage, Wheeler's attitude towards Everett's work was not as clear-cut as described by DeWitt. Everett reported an anecdote according to which, during the meeting in Austin, Wheeler told him that he mostly believed his interpretation, but reserved Tuesdays once a month to disbelieve it.[244] In 1977, being requested to give an opinion on a paper dealing with the EWI, Wheeler answered that he "still [felt] it[was] one of the most important contributions made to quantum mechanics in recent decades". He added nonetheless that he had "difficulty subscribing to it today." As he had done with Lévy-Leblond, he asked the author to "change the reference from Everett–Wheeler to Everett interpretation".[245] (A copy of the letter was forwarded to Everett, who scrawled on the term "difficulty": "Only on Tuesday"!) To be sure, Wheeler continued to pay attention to Everett's ideas, and never gave up the hope to work with him again.[246] The papers he published in the 1970s and 1980s reflect his effort to reach a satisfactory understanding and an appropriate generalization of the Copenhagen view. From

[242] Jean-Marc Lévy-Leblond to Hugh Everett, 17 Aug 197[7], *EP*. In a lecture reported in the proceedings of the School "Enrico Fermi" of 1977, Wheeler says: "Imaginative Everett's thesis is, and instructive, we agree. We once subscribed to it. In retrospect, however, it looks like the wrong track" (Wheeler 1979a, p. 396).

[243] Everett to Raub, 1980, *op. cit.* Wheeler's temporary interest for Wigner-like approaches coincided with his efforts to clarify the question as to whether Bohr's views did involve any reference to consciousness [see Wheeler's letters to Aage Bohr in Freire Jr. (2007) and Chapter 4, this book. See also Wheeler and Zurek (1983, p. 207) and Wheeler (1981)].

[244] Everett interview, *op. cit.*, p. 8.

[245] John A. Wheeler to Paul Benioff, 7 Jul 1977; and 7 Sep 1977, *EP*.

[246] According to DeWitt, "one of the very first things he did when he arrived [at the University of Texas] was actually to invite and pay for Everett to come." (DeWitt interview, *op. cit.*, p. 15.) Furthermore, according to Shikhovtsev (2003), Wheeler planned to bring Everett back to theoretical physics in the framework of a project which aimed to create a working group devoted to the quantum theory of measurement at the Institute for Theoretical Physics in Santa Barbara, but the whole project was eventually abandoned.

such papers, it is apparent that the time elapsed since the discussions of 1956 had not erased his doubts, and that Everett's work had not completely lost its appeal for him.[247]

Concluding Remarks

The epilogue of the Everett affair seems to support the idea that as late as in the 1950s the Copenhagen school still exerted a decisive influence, which could go as far as undermining the career of a brilliant physicist in the US. The interpretive model of the "dictatorial imposition" (Jammer 1974, p. 250) is nonetheless too crude to account for all the aspects of the Everett episode. Indeed, our analysis suggests that the mechanisms which ensured the supremacy of the Copenhagen view (and led to its decline a few years after Bohr's death, in the new climate of which Everett was a forerunner) were actually subtler than they are habitually depicted to be (Howard 2004).

Urged by Wheeler (who was a dedicated Bohrian, but did not belong to the inner circle of Bohr's collaborators), the Copenhagen scientists did not refuse to debate the non-conventional proposal of Wheeler's pupil. Admittedly, the objections raised in Copenhagen were very general, and they resulted only partly from a rigorous appraisal of the merits and shortcomings of Everett's work. But this reflected the fact that what bothered Bohr was not so much the technical aspects of Everett's project as the very concept of physical knowledge which underlay it. The existence of such a chasm in the very premises of Everett's and Bohr's interpretations of the quantum formalism was manifestly not apparent to Wheeler. He was one of the very few "missionaries of the Copenhagen Spirit" (Heilbron 2001) in America, but his understanding of some aspects of the Bohrian gospel was neither firm nor unequivocal. This explains at once his doubts on the Copenhagen approach to measurement, and his belief that these doubts could be solved without abandoning the framework of Bohr's view. The discussions that Wheeler had with the Copenhagen group were pretty frank, and, notwithstanding his caution, he did not hesitate to put forward arguments which could sound heretical. When it became clear that they were given no importance whatsoever in Copenhagen, he curbed his

(continued)

[247] Wheeler's idea of a "participatory universe" (Wheeler and Zurek 1983, pp. 182–183) can be said to have inspired a number of attempts to "go beyond" Bohr's view of measurement along the lines of the relative state formulation [see e.g. Omnès (1992), Rovelli (1996), Zurek (1998)]. In some of these approaches, the explicit inclusion of the observer in the quantum description of the universe is supposed to enable one to dismiss the postulate of projection; see Barrett (1999). Furthermore, in order to demonstrate the "emergence of a classical world from a quantum universe" (a definitely Everettian idea), the advocates of such approaches have sometimes put forward evolutionary arguments reminiscent of those sketched by Wheeler in the discussion with the Copenhagen group; see Vaidman (2002) for a list of references, and Bitbol (1996b, pp. 414–418) for a discussion.

enthusiasm for Everett's ideas. But his veneration for Bohr could not remove the tension between his firm belief that Bohr's approach provided indeed a deep insight into quantum physics and the feeling that it missed something crucial, and had to be amended. That this situation was a source of inner trouble for him is suggested by his wavering attitude in the 1970s, as well by his reluctance to mention the events of 1956 in later recollections.

We can contrast this attitude with that of Everett, who never bothered too much about the relationship between his ideas and the Copenhagen view. Everett was an exponent of the new American generation growing up in an intellectual and scientific context which had little to do with that of the German-speaking Europe between the two wars: his attitude prefigures that of many physicists and philosophers of the 1960s, for whom Bohr came to represent a positivism out of date. Everett pointed out what he considered to be the limitations of Bohr's approach and straightforwardly ascribed them to Bohr's dogmatic and conservative stance. There was no effort on his part to reach a deeper understanding of the philosophical background of complementarity, and no hesitation to seek a formulation of quantum mechanics in which Bohr's reflections on the nature of scientific knowledge could be simply bypassed.

References

Assmus, A.: The Americanization of molecular physics. Hist. Stud. Phys. Biol. Sci. **23**, 1–34 (1992)

Bacciagaluppi, G., Valentini, A.: Quantum Theory at the Crossroads: Reconsidering the 1927 Solvay Conference. Cambridge University Press, Cambridge (2009)

Barrett, J.A.: The Quantum Mechanics of Minds and Worlds. Oxford University Press, Oxford (1999)

Becker, L.: That von Neumann did not believe in a physical collapse. Br. J. Philos. Sci. **55**(1), 121–135 (2004)

Bell, J.S.: Speakable and Unspeakable in Quantum Mechanics: Collected Papers on Quantum Philosophy. With an Introduction by Alain Aspect. Cambridge University Press, Cambridge (2004)

Beller, M.: Quantum Dialogue—The Making of a Revolution. The University of Chicago Press, Chicago (1999a)

Beller, M.: Jocular Commemorations the Copenhagen Spirit. Osiris **14**, 252–273 (1999b)

Ben-Dov, Y.: Everetts Theory and the many-worlds interpretation. Am. J. Phys. **58**(9), 829–832 (1990)

Bitbol, M.: Schrödinger's Philosophy of Quantum Mechanics. Kluwer Academic Publishers, Dordrecht (1996a)

Bitbol, M.: Mécanique Quantique - Une Introduction Philosophique. Flammarion, Paris (1996b)

Bitbol, M.: L' Aveuglante Proximité Du Réel—Anti-Réalisme Et Quasi-Réalisme En Physique. Flammarion, Paris (1998)

Bitbol, M.: Physique et Philosophie de l'Esprit. Flammarion, Paris (2000)

Bohm, D.: Quantum Theory. Prentice-Hall, New York (1951)

Bohm, D.: A suggested interpretation of the quantum theory in terms of hidden variables—I & II. Phys. Rev. **85**(2), 166–179 (1952). 180–193

Bohr, N.: The causality problem in atomic physics. The New Theories of Physics, pp. 11–45. International Institute of Intellectual Cooperation, Paris (1939) (Reprintedin: Bohr, N. (ed.) Causality and complementarity: supplementary papers. In: Faye, J., Folse, H. (eds.) The Philosophical Writings of Niels Bohr, vol. 14, pp. 94–121. Ox Bow Press, Wooldbridge, CT (1998). Page numbers refer to the reprint.)

Bohr, N.: On the notions of causality and complementarity. Dialectica **2**, 312–319 (1948)

Bohr, N.: Discussion with Einstein on epistemological problems in atomic physics. In: Schilpp, P.A. (ed.) Albert Einstein—Philosopher–Scientist, pp. 199–242. The Library of the Living Philosophers, Evanston (1949)

Bohr, N.: Essays, 1958-1962, on Atomic Physics and Human Knowledge. Interscience Publishers, New York (1963)

Bohr, N.: Causality and Complementarity: Supplementary Papers. In: Bohr, N., Faye, J., Folse, H.J. (eds.) The Philosophical Writings of Niels Bohr. Ox Bow Press, Woodbridge, CT (1998)

Bohr, N., French, A.P., Kennedy, P.J.: Niels Bohr: A Centenary Volume. Harvard University Press, Cambridge, MA (1985)

Born, M.: The interpretation of quantum mechanics. Br. J. Philos. Sci. **IV**(14), 95–106 (1953)

Brush, S.G.: The chimerical cat: philosophy of quantum mechanics in historical perspective. Soc. Stud. Sci. **10**(4), 393–447 (1980)

Butterfield, J.N.: Some worlds of quantum theory. In: Russell, R., Polkinghorne, J. (eds.) Quantum Mechanics. Scientific Perspectives on Divine Action, vol. 5, pp. 111–140. Vatican Observatory Publications, Vatican City (2002)

Byrne, P.: The many worlds of Hugh Everett. Sci. Am. **297**(6), 98–105 (2007)

Byrne, P.: The Many Worlds of Hugh Everett III: Multiple Universes, Mutual Assured Destruction, and the Meltdown of a Nuclear Family. Oxford University Press, New York (2010)

Camilleri, K.: Constructing the myth of the Copenhagen interpretation. Perspect. Sci. **17**(1), 26–57 (2009)

Cartwright, N.: Philosophical problems of quantum theory: the response of American physicists. In: Krüger, L., Gigerenzer, G., Morgan, M.S. (eds.) The Probabilistic Revolution, vol. 2, pp. 407–435. MIT Press, Cambridge, MA (1987)

Cassinello, A.: La interpretación de los muchos universos de la mecánica cuántica. Apuntes históricos. Arbor - Ciencia Pensamiento y Cultura **148**(584), 47–68 (1994)

Chevalley, C.: Niels Bohr and contemporary philosophy. In: Faye, J., Folse, H.J. (eds.) Boston Studies in the Philosophy of Science, pp. 33–55. Kluwer Academic, Dordrecht (1994)

Chevalley, C.: Mythe et philosophie: la construction de "Niels Bohr" dans la doxographie. Physis, Rivista Internazionale di Storia della Scienza **34**(3), 569–603 (1997)

Chevalley, C.: Why do we find Bohr obscure? In: Greenberger, D., Reiter, W.L., Zeilinger, A. (eds.) Epistemological and Experimental Perspectives on Quantum Mechanics, pp. 59–73. Springer, Dordrecht (1999)

Daneri, A., Loinger, A., Prosperi, G.M: Quantum theory of measurement and ergodicity conditions. Nucl. Phys. **33**, 297–319 (1962) (Reprinted in: Wheeler, J.A., Zurek, W.H. (eds.) Quantum Theory and Measurement, pp. 657–679. Princeton University Press, Princeton (1983)

DeWitt, B.S.: Quantum theory of gravity. I. Canonical theory. Phys. Rev. **160**(5), 1113–1148 (1967)

DeWitt, B.: The Everett–Wheeler interpretation of quantum mechanics. In: DeWitt, C.M., Wheeler, J.A. (eds.) Battelle Rencontres 1967. Lectures in Mathematics and Physics, pp. 318–332. W.A. Benjamin, New York (1968)

DeWitt, B.S.: Quantum mechanics and reality. Phys. Today **23**(9), 155–165 (1970)

DeWitt, B.S., Everett, H., Graham, N.: The Many-Worlds Interpretation of Quantum Mechanics: A Fundamental Exposition. Princeton University Press, Princeton, NJ (1973)

DeWitt-Morette, C., Rickles, D. (eds.): The Role of Gravitation in Physics: Report from the 1957 Chapel Hill Conference. Edition Open Access, Berlin (2011)

Dicke, R.H., Wittke, J.P.: Introduction to Quantum Mechanics. Addison-Wesley, Reading, MA (1960)

Dirac, P.A.M.: The Principles of Quantum Mechanics. Clarendon Press, Oxford (1958)

Everett, H.: On the foundations of quantum mechanics. PhD dissertation, Princeton University (1957a)

Everett, H.: Relative state formulation of quantum mechanics. Rev. Mod. Phys. **29**(3), 454–462 (1957b) [Reprinted in: Wheeler, J.A., Zurek, W.H. (eds.), Quantum Theory and Measurement (pp. 315–323). Princeton University Press, Princeton (1983). Page numbers refer to the reprint]

Everett, H.: The theory of the universal wave function. In: DeWitt, B.S., Graham, N. (eds.) The Many-Worlds Interpretation of Quantum Mechanics, pp. 3–140. Princeton University Press, Princeton, NJ (1973)

Everett, H.: Objective vs subjective probability. In: Everett, H., Barrett, J.A., Byrne, P. (eds.) The Everett Interpretation of Quantum Mechanics: Collected Works 1955-1980 with Commentary, pp. 57–60. Princeton University Press, Princeton (2012[1955a])

Everett, H.: Quantitative measure of correlation. In: Everett, H., Barrett, J.A., Byrne, P. (eds.) The Everett Interpretation of Quantum Mechanics: Collected Works 1955-1980 with Commentary, pp. 61–63. Princeton University Press, Princeton (2012[1955b])

Everett, H. (2012 [1955c]). Probability in Wave Mechanics. In: Everett, H., Barrett, J.A., Byrne, P. (eds.) The Everett Interpretation of Quantum Mechanics: Collected Works 1955-1980 with Commentary, pp. 64–70. Princeton University Press, Princeton

Everett, H., Barrett, J.A., Byrne, P.: The Everett Interpretation of Quantum Mechanics: Collected Works 1955-1980 with Commentary. Princeton University Press, Princeton (2012)

Faye, J.: Niels Bohr: His Heritage and Legacy: An Anti-Realist View of Quantum Mechanics. Kluwer Academic, Dordrecht (1991)

Faye, J.: Copenhagen interpretation of quantum mechanics. In: Zalta, E.N. (ed.) The Stanford Encyclopedia of Philosophy (Summer 2002 edition). Stanford (2002), Available online: http://plato.stanford.edu/archives/sum2002/entries/qm-copenhagen/

Feyerabend, P.K.: Problems of microphysics. In: Colodny, R.G. (ed.) Frontiers of Science and Philosophy, pp. 189–283. George Allen, London (1964)

Forman, P.: "Weimar Culture, Causality, and Quantum Theory, 1918-1927: Adaptation by German Physicists and Mathematicians to a Hostile Intellectual Environment (1971) [Reprinted in Forman, P. et al. (eds.) Weimar Culture and Quantum Mechanics: Selected Papers by Paul Forman and Contemporary Perspectives on the Forman Thesis, 2011]." Hist. Stud. Phys. Sci. **3**, 1–115

Forstner, C.: The early history of David Bohm's quantum mechanics through the perspective of Ludwik Fleck's thought-collectives. Minerva **46**(2), 215–229 (2008)

Freire Jr., O.: The historical roots of "foundations of quantum mechanics" as a field of research (1950–1970). Found. Phys. **34**(11), 1741–1760 (2004)

Freire Jr., O.: Science and exile: David Bohm, the cold war, and a new interpretation of quantum mechanics. Hist. Stud. Phys. Biol. Sci. **36**(1), 1–34 (2005)

Freire Jr., O.: Orthodoxy and heterodoxy in the research on the foundations of quantum physics: E.P. Wigner's case. In: Santos, B.d.S. (ed.) Cognitive Justice in a Global World: Prudent Knowledges for a Decent Life, pp. 203–224. Lexington Books, Lanham, MD (2007)

Freire Jr., O.: "On the connections between the dialectical materialism and the controversy on the quanta." Jahrbuch Für Europäihce Wissenschaftskultur—Yearbook of the European Culture of Science **6**, 195–210 (2011)

Freire Jr, O.: "Book review: Hugh Everett III. The Everett interpretation of quantum mechanics—collected works 1955–1980 with commentary, Barrett, J.A., Byrne, P. (eds.)" Stud. Hist. Philos. Mod. Phys. **46**, 263–264 (2014)

Freitas, F.: Os estados relativos de Hugh Everett III: Uma analise historica e conceitual. Master dissertation, Universidade Federal da Bahia (2007)

George, A.: Louis de Broglie, physicien et penseur. Albin Michel, Paris (1953)

Graham, L.R.: The Soviet reaction to Bohr's quantum mechanics. In: Feshbach, H., Matsui, T., Oleson, A. (eds.) Niels Bohr and the World: Proceedings of the Niels Bohr Centennial Symposium, pp. 305–317. Harwood Academic, New York (1988)

Hartz, T.: As heterodoxias quânticas e o olhar do historiador: uma história dos usos dos argumentos de Niels Bohr acerca da medição de campos quânticos (1930–1970). PhD dissertation, Universidade Federal da Bahia and Universidade Estadual de Feira de Santana (2013)

Heilbron, J.: The earliest missionaries of the Copenhagen spirit. In: Galison, P., Gordin, M., Kaiser, D. (eds.) Science and Society—The History of Modern Physical Science in the Twentieth Century, vol. 4, pp. 295–330. Routledge, New York (2001)

Heisenberg, W.: The development of the interpretation of the quantum theory. In: Pauli, W. (ed.) Niels Bohr and the Development of Physics; Essays Dedicated to Niels Bohr on the Occasion of his Seventieth Birthday, pp. 12–29. McGraw-Hill, New York (1955)

Home, D., Whitaker, M.A.B.: Ensemble interpretations of quantum-mechanics—a modern perspective. Phys. Rep. Rev. Sect. Phys. Lett. **210**(4), 223–317 (1992)

Honner, J.: The Description of Nature: Niels Bohr and the Philosophy of Quantum Physics. Oxford University Press, Oxford (1987)

Hooker, C.A.: Projection, physical intelligibility, objectivity and completeness—the divergent ideals of Bohr and Einstein. Br. J. Philos. Sci. **42**(4), 491–511 (1991)

Howard, D.: Einstein on locality and separability. Stud. Hist. Phil. Sci. **16**(3), 171–201 (1985)

Howard, D.: Who invented the "Copenhagen interpretation"? A study in mythology. Philos. Sci. **71**, 669–682 (2004)

Institut International de Physique Solvay: Electrons et photons—Rapports et discussions du Cinquième Conseil de Physique tenu à Bruxelles du 24 au 29 octobre 1927. Gauthier-Villars, Paris (1928) [English translation in Bacciagaluppi and Valentini 2009]

Jacobsen, A.: Léon Rosenfeld's Marxist defense of complementarity. Hist. Stud. Phys. Biol. Sci. **37**(Supplement), 3–34 (2007)

Jammer, M.: The Conceptual Development of Quantum Mechanics. McGraw-Hill, New York (1966)

Jammer, M.: The Philosophy of Quantum Mechanics—The Interpretations of Quantum Mechanics in Historical Perspective. Wiley, New York (1974)

Kaiser, D.: More roots of complementarity—kantian aspects and influences. Stud. Hist. Phil. Sci. **23**(2), 213–239 (1992)

Kaiser, D.: Cold War requisitions, scientific manpower, and the production of American physicists after World War II. Hist. Stud. Phys. Biol. Sci. **33**, 131–159 (2002)

Kent, A.: Against many-world interpretation. Int. J. Mod. Phys. **A5**, 1745–1762 (1990). An updated version is available online: http://arxiv.org/abs/gr-qc/9703089S

Körner, S.: Observation and Interpretation—A Symposium of Philosophers and Physicists. Butterworths, London (1957)

Kragh, H.: Quantum Generations: A History of Physics in the Twentieth Century. Princeton University Press, Princeton, NJ (1999)

Kuhn, T.S.: The Structure of Scientific Revolutions. The University of Chicago Press, Chicago (1970)

Lakatos, I., Feyerabend, P., Motterlini, M.: For and Against Method: Including Lakatos's Lectures on Scientific Method and the Lakatos-Feyerabend Correspondence. University of Chicago Press, Chicago (1999)

Lehner, C. (1997). Quantum Mechanics and Reality: An Interpretation of Everett's Theory. PhD dissertation, Stanford University

London, F., Bauer, E.: La théorie de l'observation en mécanique quantique, pp. 217–259. Hermann, Paris (1939) [English translation in Wheeler & Zurek 1983]

Margenau, H.: Advantages and disadvantages of various interpretations of the quantum theory. Phys. Today **7**(10), 6–13 (1954)

Margenau, H.: Philosophical problems concerning the meaning of measurement in physics. Philos. Sci. **25**(1), 23–33 (1958)

Margenau, H.: Measurement and quantum states—I & II. Philos. Sci. **30**, 1–16 (1963). 138–157
Mehra, J., Rechenberg, H.: The Historical Development of Quantum Theory, vol. 6(Part 2). Springer, New York (2001)
Murdoch, D.: Niels Bohr's Philosophy of Physics. Cambridge Cambridgeshire; Cambridge University Press, New York (1987)
Olwell, R.: Physical isolation and marginalization in physics—David Bohm's cold war exile. ISIS **90**, 738–756 (1999)
Omnès, R.: Consistent interpretations of quantum-mechanics. Rev. Mod. Phys. **64**(2), 339–382 (1992)
Osnaghi, S., Freitas, F., Freire Jr., O.: The origin of the Everettian heresy. Stud. Hist. Philos. Mod. Phys. **40**(2), 97–123 (2009)
Paty, M.: The nature of Einsteins objections to the copenhagen interpretation of quantum-mechanics. Found. Phys. **25**(1), 183–204 (1995)
Pauli, W.: Niels Bohr and the Development of Physics; Essays Dedicated to Niels Bohr on the Occasion of His Seventieth Birthday. McGraw-Hill, New York (1955)
Pauli, W., Enz, C.P., Meyenn, K.V.: Writings on Physics and Philosophy. Springer, Berlin (1994)
Petersen, A.: Quantum Physics and the Philosophical Tradition. Belfer Graduate School of Science, Yeshiva University, New York (1968)
Petersen, A.: The philosophy of Niels Bohr. In: French, A.P., Kennedy, P.J. (eds.) Niels Bohr, A Centenary Volume, pp. 299–310. Harvard University Press, Cambridge, MA (1985)
Rédei, M., Stöltzner, M.: John von Neumann and the Foundations of Quantum Physics. Kluwer Academic, Dordrecht (2001)
Rosenfeld, L.: Measuring process in quantum mechanics. Suppl. Prog. Theor. Phys. 222–231 (1965)
Rovelli, C.: Relational quantum mechanics. Int. J. Theor. Phys. **35**(8), 1637–1678 (1996)
Rozental, S.: Niels Bohr; His Life and Work as Seen by His Friends and Colleagues. Wiley, Amsterdam (1967)
Scheibe, E.: The Logical Analysis of Quantum Mechanics. Pergamon, Oxford (1973)
Schilpp, P.A., Einstein, A.: Albert Einstein, Philosopher-Scientist. Library of Living Philosophers, Evanston, IL (1949)
Schrödinger, E.: Are there quantum jumps? I and II. Br. J. Philos. Sci. **3**, 109–123 (1952). 233–242
Schrödinger, E.: The meaning of wave mechanics. In: George, A. (ed.) Louis de Broglie—Physicien et Penseur, pp. 16–32. Albin Michel, Paris (1953)
Schrödinger, E.: Might perhaps energy be a merely statistical concept. Nuovo Cimento **9**(1), 162–170 (1958)
Schweber, S.S.: The empiricist temper regnant—theoretical physics in the United-States 1920-1950. Hist. Stud. Phys. Biol. Sci. **17**, 55–98 (1986)
Shimony, A.: Role of observer in quantum theory. Am. J. Phys. **31**(10), 755–773 (1963)
Sopka, K.R.: Quantum Physics in America, 1920-1935. Arno Press, New York (1980)
Stapp, H.P.: Copenhagen interpretation. Am. J. Phys. **40**(8), 1098–1116 (1972)
Stapp, H.P.: Quantum theory and the place of mind in nature. In: Faye, J., Folse, H. (eds.) Niels Bohr and contemporary philosophy. Boston studies in the philosophy of science, pp. 245–252. Kluwer, Dordrecht (1994)
Stöltzner, M.: What John von Neumann thought of the Bohm interpretation. In: Greenberger, D., Reiter, W.L., Zeilinger, A. (eds.) Epistemological and Experimental Perspectives on Quantum Mechanics, pp. 257–262. Springer, Dordrecht (1999)
Teller, P.: The projection postulate and Bohr's interpretation of quantum mechanics. In: Asquith, P., Giere, R. (eds.) PSA 1980: Proceedings o fthe 1980 Biennial Meeting of the Philosophy of Science Association, vol. 2, pp. 201–223. Michigan State University, East Lansing (1981)
Vaidman, L.: The many-worlds interpretation of quantum mechanics. In: Zalta, E.N. (ed.) The Stanford Encyclopedia of Philosophy. Stanford (2002). Available online: http://plato.stanford.edu/archives/sum2002/entries/qm-manyworlds

Von Neumann, J.: Mathematical Foundations of Quantum Mechanics. Princeton University Press, Princeton, NJ (1955)

Werner, F.G.: The foundations of quantum mechanics. Phys. Today **17**(1), 53 (1964)

Wheeler, J.A.: A septet of sibyls—aids in the search for truth. Am. Sci. **44**(4), 360–377 (1956)

Wheeler, J.A.: Assessment of Everett's relative state formulation of quantum theory. Rev. Mod. Phys. **29**(3), 463–465 (1957). Reprinted in: Wheeler, J.A., Zurek, W.H. (eds.) Quantum theory and measurement, pp. 324–325 Princeton University Press, Princeton (1983). Page numbers refer to the reprint

Wheeler, J.A.: The Universe in the light of general relativity. Monist **47**, 40–76 (1962)

Wheeler, J.A.: Frontiers of time. In: Toraldo di Francia, G. (ed.) Problems in the Foundations of Physics. Proceedings of the International School of Physics E. Fermi, pp. 395–497. North Holland, Dordrecht (1979a). Partially reprinted in: Wheeler, J.A., Zurek, W.H. (eds.) Quantum theory and measurement, pp. 182–213. Princeton University Press, Princeton (1983)

Wheeler, J.A.: Mercer street and other memories. In: Tauber, G.E. (ed.) Albert Einstein's Theory Of General Relativity, pp. 182–184. Crown, New York (1979b)

Wheeler, J.A.: Not consciousness but the distinction between the probe and the probed as central to the elemental quantum act of observation. In: Jahn, R.G. (ed.) The Role of Consciousness in the Physical World, pp. 87–111. Westview, Boulder (1981)

Wheeler, J.A.: Physics in Copenhagen in 1934 and 1935. In: French, A.P., Kennedy, P.J. (eds.) Niels Bohr: A Centenary Volume, pp. 221–226. Harvard University Press, Cambridge, MA (1985)

Wheeler, J.A., Ford, K.W.: Geons, Black Holes, and Quantum Foam: A Life in Physics. Norton, New York (1998)

Wheeler, J.A., Zurek, W.H.: Quantum Theory and Measurement. Princeton University Press, Princeton, NJ (1983)

Wigner, E.: Remarks on the mind-body question. In: Good, I.J. (ed.) The Scientist Speculates: An Anthology of Partly-Baked Ideas, pp. 284–302. Heinemann, London (1961). Reprinted in Wheeler, J.A., Zurek, W.H. (eds.) Quantum Theory and Measurement, pp. 168–181. Princeton University Press, Princeton (1983)

Wigner, E.P.: Problem of measurement. Am. J. Phys. **31**(1), 6–15 (1963)

Zurek, W.H.: Decoherence, einselection and the existential interpretation (the rough guide). Philos. Trans. R. Soc. A Math. Phys. Eng. Sci. **356**(1743), 1793–1821 (1998)

Chapter 4
The Monocracy is Broken: Orthodoxy, Heterodoxy, and Wigner's Case

Abstract From the 1950s awareness of the existence of a problem with measurement in quantum theory grew among physicists. Framed in von Neumann's terms, it concerns the two kinds of evolution of the quantum states. In the early 1960s the debate on measurement was further stirred up by Eugene Wigner and by Léon Rosenfeld. Wigner held that the mind may be responsible for measurement. He also supported a number of younger physicists who began to tackle the measurement problem, such as Abner Shimony and Michael Yanase. Rosenfeld presented the results from the Italian physicists Adriana Daneri, Angelo Loinger, and Giovanni Prosperi as the crowning of Bohr's complementarity. The Italians had suggested that measurements should be understood as thermodynamic amplifications in the measurement device after it interacts with the quantum system, in line with a hint from Bohr that measurement implies irreversibility. Rosenfeld and Wigner embraced the conflict with a number of papers crossing the Atlantic criticizing each other. In addition to the quantum controversy their background fuelled the controversy, with Wigner supporting the US in the atomic race and Rosenfeld a Marxist. As a result of the battle, the Copenhagen monocracy was broken. Physicists began to speak of the Copenhagen school and the Princeton school as two variants of orthodoxy in quantum mechanics.

4.1 Introduction[1]

Dealing with Eugene Wigner's ideas on the measurement procedure in quantum physics and unearthing the controversy that pitted him against supporters of the interpretation of complementarity, I will show how Wigner and his followers contributed to the defeat of a seemingly unshakeable consensus. Indeed, as a result of the quarrel between Wigner and Léon Rosenfeld, with a number of papers

[1] This Chapter is an enlarged version of my "Orthodoxy and Heterodoxy in the Research on the Foundations of Quantum Physics: E.P. Wigner's Case", published in Boaventura de Sousa Santos (Org.). *Cognitive Justice in a Global World: Prudent Knowledges for a Decent Life*. Lanham, MD, USA: Lexington Books, p. 203–224, 2007.

O. Freire Junior, *The Quantum Dissidents*, DOI 10.1007/978-3-662-44662-1_4

crossing the Atlantic criticizing each other, the Copenhagen monocracy was broken. Physicists began to speak of the Copenhagen school and the Princeton school as two variants of orthodoxy in quantum mechanics. In addition to the quantum controversy their ideological background fuelled the controversy, with Wigner supporting the US in the atomic race and Rosenfeld a Marxist. Ironically, although Wigner intended to defend what seemed to him to be orthodoxy, he himself ultimately became heterodox.

Wigner's conjectures on the role of consciousness in physical phenomena were not fruitful and were discarded, and today they form part of the *history* of physics rather than physics proper. However, his ideas and actions left an indelible mark on the physics of the second half of the twentieth century. The current wide use of the term quantum measurement problem, which indicates the existence of such a problem, is largely due to Wigner, who was one of the first to use it. This is our main interest insofar as it is related to the vicissitudes suffered by the foundations of quantum physics until it became a regular field of research in physics. Wigner formulated his ideas in opposition to the "Copenhagen monocracy,"—paraphrasing the historian Max Jammer—which still had a stronghold on the interpretation of quantum physics in the 1960s. He stressed the unsolved status of the measurement problem, effectively defended his ideas, and supported those who were willing to investigate the foundations of quantum physics.[2] He thus contributed to the creation of a new field of research in physics, that of the foundations of quantum physics, which attributed a higher scientific status to the old controversy on the interpretations and foundations of this theory. This new field has had to deal with important theoretical, experimental, and philosophical issues with significant repercussions in recent decades.

In addition to being of interest to the history of the quantum controversy, Wigner's case is of general interest for the history of physics. The way in which he dealt with controversies in science and with young scientists involved in such controversies has a reach which goes far beyond the case of the quanta. Furthermore, the current varying opinions about his contribution to the measurement problem in quantum physics may lead us to reflect about anachronism in history of science and the obstacles it puts in the way of a better public image of science.

4.2 Measurement Problem Before Wigner

What would be later called the "quantum measurement problem" was present, albeit implicit, from the inception of quantum theory in around 1927. The problem may be summarized as follows: The evolution of the state describing a quantum

[2] Jammer's (1974, p. 250) words were: "In the early 1950s the almost unchallenged monocracy of the Copenhagen school in the philosophy of quantum mechanics began to be disputed in the West."

system, let us say, the electron's spin projection in a given direction, is ruled by Schrödinger's equation. This means that the two possible states, spin "up" and spin "down," are expressed in the quantum states and are preserved while time evolves. If one measures this electron spin, one will discover it to be either up or down. How did it happen that a state which contains a superposition of two possibilities became just one? Physicists first christened this evolution the "reduction of the wave-packet," an expression which is reminiscent of the wave formulation of quantum theory. A more sophisticated analysis of this process was suggested by Niels Bohr, assuming that measurements require macroscopic devices and appealing to the complementarity view suggested by himself. Bohr suggested that such devices had to be treated within the framework established by classical physics, not because one could not treat them from a quantum point of view, but because they had to be treated classically so that measurement results could be compared to those of other researchers. As communication is a requirement to attain objectivity, and communication requires ordinary language refined by concepts from classical physics (e.g., concepts indicated by words such as "work," "force," etc.), the classical treatment of measurement devices is a condition for preserving objectivity in scientific research.[3]

The other major solution to the measurement problem was suggested by John von Neumann as part of his work to lay rigorous mathematical foundations for the mathematical formalism of quantum physics. He began to work on this subject just after the elaboration of the mathematical formalism of the quantum theory, around 1925–1927, which was the time of completion of quantum physics. In 1927 he wrote a trilogy of papers which would be the basis for his book *Mathematische Grundlagen der Quantenmechanik*, published in 1932.[4] Considering that the "transformation theory," formulated by Dirac and independent contributions by Pascual Jordan and Fritz London, were the "definitive form" of quantum mechanics, von Neumann departed from it due to its lack of mathematical rigor. According to him, in the opening of his 1932 book, "it should be emphasized that the correct structure need not consist in a mathematical refinement and explanation of the Dirac method, but rather that it requires a procedure differing from the very beginning, namely, the reliance on the Hilbert theory of operators" (Von Neumann 1955, p. ix). Von Neumann not only based his presentation on the mathematical structure of Hilbert vector spaces and Hermitian operators, but also extended it beyond its "classical limits." From then on, matrix mechanics, wave mechanics, and transformation theory should be considered as manifestations of Hilbert space vectors. Von Neumann's and Jordan's work has been dissected by historians Anthony Duncan and Michael Janssen, who argued, "So, rather than following the Jordan-Dirac

[3] For the purposes of this text, I would like to emphasize that Bohr's complementarity treats measurement devices according to classical physics, not according to quantum physics. For a standard, comprehensive description of complementarity, see Bohr's (1949) report of his discussions with Einstein.

[4] For von Neumann's biographies, see Macrae (1992) and Heims (1980).

approach and looking for ways to mend its mathematical shortcomings, von Neumann, as indicated in the passage from his 1932 book quoted above, adopted an entirely new approach. He generalized Hilbert's spectral theory of operators to provide a formalism for quantum mechanics that is very different from the one proposed by Jordan and Dirac" (Duncan and Janssen 2013, p. 194).[5] In the search for the consistency of his mathematical scheme von Neumann used it to deal with measurement in quantum physics and he diverged from Bohr's solution. The milestone in von Neumann's treatment of measurement was the introduction of a distinction between two kinds of time evolution of quantum states. The first one, "discontinuous, non-causal and instantaneously acting experiments or measurements," occurs during the measurement processes. The second one, "continuous and causal," is governed by the Schrödinger equation. In addition, von Neumann treated measuring devices quantum mechanically, instead of treating them classically as suggested by Bohr. This choice leads to the transfer of the singular superposition of quantum states from the system under scrutiny to the combination of system and measuring apparatus. In mathematical terms, this transfer is represented by the inner product between the two Hilbert vectors, one related to the system and the other related to the measurement device. As no such measurement device described by such a bizarre superposition has been seen, it raises the questions: how, where, and when does this superposition become a vector with just one component, which is an eigenstate of the physical property of interest? After all, what we obtain after measurements is related to vectors and probabilities rather than to superposition of vectors. Von Neumann solved the problem appealing to the distinction between the two kinds of evolution of quantum states and the role of the cognizant subject, that is, the individual observer. He recalled the general epistemological view that in any measurement there ultimately is a moment in which "we must say: and this is perceived by the observer," and framed his answer in the requirement of the psycho-physical parallelism, which means "it must be possible so to describe the extra-physical process of the subjective perception as if it were in reality in the physical world," i.e., "to assign to its parts equivalent physical processes in the objective environment, in ordinary space" (Von Neumann 1955, p. 419). It should be noted that at that moment von Neumann remarked that Bohr had been the first to link this dual description (two kinds of evolution) to the psycho-physical parallelism (Von Neumann 1955, p. 420 footnote 207), thus diluting the implicit different in their approaches to the measurement issues in quantum physics.[6]

[5] On Hilbert's and von Neumann's early axiomatic activity in the field of quantum mechanics, see Lacki (2000) and references therein.

[6] That von Neumann appealed to the psycho-physical parallelism and refrained from attributing a physical role for the mind in the quantum measurement processes has not been acknowledged by some commentators. For an example of this misreading, see Jammer (1974, pp. 480–482). The psycho-physical parallelism was articulated by Gustav Theodor Fechner as part of the debates on the mind-body issue in the second half of the nineteenth century. According to Heidelberger, it was seen "as compatible with science and science's materialistic inclination, without necessitating

In the 1930s, therefore, there was no awareness of the differences between Bohr and von Neumann as regards measurements in quantum mechanics. As late as 1955, in the preface to the English translation he himself had revised, von Neumann dedicated a paragraph to setting forth his criticisms towards attempts to complete quantum physics with "hidden variables," which was an indirect reference to Bohm's 1952 work. However, he did not waste time contrasting his and Bohr's approach to measurement in quantum theory (Von Neumann 1955, p. x). Von Neumann's work imposed itself as the rigorous mathematical presentation of quantum physics but contemporary criticism with the publication of his work did not exploit the distinctions between his and Bohr's approach. This had been the case with the two now well-known *Gedankenexperiments* created by Einstein and Schrödinger in 1935, the EPR experiment and Schrödinger's cat, respectively.[7] We will see the EPR experiment in more detail in Chap. 6. As for Schrödinger's cat experiment, we deal with it here as it is more closely related to the measurement problem and to the paradox of Wigner's friend. It appeared in a paper in which the Austrian physicist, motivated by the publication of the EPR paper, raised the stakes against what he called the "reigning doctrine" of blurred reality. As he suggested, imagine a device, a steel chamber in which there is an atomic sample with a 50 % probability of decaying after a certain time. If the atoms decay they will trigger a Geiger counter and this will trigger a hammer that will hit and break a bottle storing a lethal gas. In addition there is a cat inside the chamber. Schrödinger argued that quantum physics will describe the whole setting as a superposition of dead and live cats, that is, quantum theory does not say that the cat is either alive or dead. However, as the cat is in fact not in a suspended state, Schrödinger concludes, quantum theory is incomplete as it is unable to say if the cat is dead or alive (Schrödinger 1983).[8]

In the twilight of the European interwar period two events highlighted the measurement issues and Bohr's and von Neumann's approaches, but it was too late for immediate consequences of these events. In 1939, Fritz London and Edmond Bauer wrote a concise essay attempting to explain von Neumann's theory of measurement for not highly mathematically-skilled readers. Fritz London, one of the creators of quantum chemistry, was then fully dedicated to the study of superconductivity and superfluidity, which were seen by him as the exemplary quantum macroscopic phenomena. He had arrived in Paris in 1936 escaping Nazi Germany three years earlier. London and Bauer were more emphatic than von

recourse to crude materialism." On its origins, its German-speaking cultural background, including its influence on physicists such as Einstein, Bohr, and von Neumann, see Heidelberger (2003).

[7] Einstein et al. (1935) and Schrödinger (1983).

[8] According to Schrödinger's (1983, p. 157) own conclusions, "It is typical of these cases that an indeterminacy originally restricted to the atomic domain becomes transformed into macroscopic indeterminacy, which can then be resolved by direct observation. That prevents us from so naively accepting as valid a 'blurred model' for representing reality. In itself it would not embody anything unclear or contradictory. There is a difference between a shaky or out-of-focus photograph and a snapshot of clouds and fog banks."

Neumann about the role of consciousness in quantum measurement. As commented by London's biographer, "von Neumann did not include the consciousness of the observer in the measuring chain. The novelty of the London-Bauer treatment was the explicit claim that the reduction of the wave function was the result of the conscious activity of the human mind" (Gavroglu 1995, p. 171). Through quantum formalism, after coupling an apparatus and an object, they noticed that "a coupling, even with a measuring device, is not yet a measurement," and went on to draw the bold conclusion: "a measurement is achieved only when the position of the pointer has been *observed*. [. . .] We note the essential role played by the consciousness of the observer in this transition from a mixture to a pure case" (London and Bauer 1939, p. 41, their emphasis). To make it more explicit they coupled three systems: object x, apparatus y, and observer z, and noted that the quantum description (superposition of pure states) would remain unchanged, except for the description from the point of view of the observer. According to them:

> The observer has a completely different impression. For him it is only the object x and the apparatus y that belong to the external world, to what he calls 'objectivity.' By contrast he has *with himself* relations of a very special character. He possesses a characteristic and quite familiar faculty which we can call the 'faculty of introspection.' He can keep track from moment to moment of his own state. By virtue of this 'immanent knowledge' he attributes to himself the right to create his own objectivity—that is, to cut the chain of statistical correlations [. . .] by declaring 'I am in [this] state." (London and Bauer 1939, p. 42, their emphasis)[9]

London and Bauer were fully aware of the implications of this step and dedicated one section of their work to the "scientific community and objectivity," arguing that such an appeal to the consciousness would not lead to solipsism. They maintained that the quantum case was, in these circumstances, related to the problem of the "determination of the necessary and sufficient conditions for an object of thought to possess objectivity and to be an object of science," which was an important philosophical problem discussed by philosophers such as Malebranche, Leibniz, Bolzano, and, more recently, Husserl and Cassirer (London and Bauer 1939, pp. 50–51). At the time, London and Bauer's approach to the measurement problem was not considered to be subjectivist and the book had a laudatory preface by the physicist Paul Langevin, who was a paragon of French rationalism with a realistic view and materialistic inclinations (Bensaude-Vincent 1987; Freire Jr. 1993).[10] However, as we will see, it was only in the 1960s, in Wigner's hands that the boldness of their statements would be revived.

In 1938, the Institut International de Coopération Intellectuelle (IICI 1939) held the conference *Les nouvelles théories de la physique* in Warsaw which Bohr and

[9] For this English translation of London and Bauer's original paper, which was in French, see Wheeler and Zurek (1983).

[10] To trace London and Bauer's philosophical concerns with quantum physics is not easy, according to Gavroglu (1995, p. 175), as they never wrote anything else, before or after this book, on the philosophical aspects of quantum physics. On London's influence from both philosophy and psychology; see Gavroglu (1995, p. 179).

von Neumann attended. Bohr presented a report on the quantum theory, which was followed by a short presentation by von Neumann (Institut international de coopération intellectuelle 1939, pp. 11–48). The proceedings of the conference were published with the transcription of the debates following the presentations. Some of the differences which would flourish later, mainly by commentators of Bohr and von Neumann, were recorded in a very subtle manner. Von Neumann presented his proof against the existence of additional variables in quantum mechanics and his works on the new kind of logic he thought quantum mechanics would require. Bohr praised the mathematical skills presented by von Neumann but remarked that the same aspects had been covered by him in a simpler way. On the need for a new logic for quantum physics, Bohr stated that he had preferred to stick to the logical forms of everyday life. Von Neumann emphasized the arbitrary distinction between the observer and the system implying that the former could be treated through quantum mechanics while Bohr noted that the distinction between phenomenon and observer is naturally set when the everyday language necessary to describe experiments is adopted. Finally, H. Kramers (Institut international de coopération intellectuelle 1939, p. 102) suggested a difference between Bohr's more physical approach, and von Neumann's rather mathematical one, which was criticized by von Neumann and accepted by Bohr.[11]

In the 1950s there was a growing interest in the measurement process in quantum physics, mainly in the German-speaking world of physics. Slowly, Bohr's view on the non-eliminable role of the classic concepts began to be articulated as a physical insight, not only a philosophical approach. It took the form of the assumption of the need for macroscopic devices for the measurement processes and the role of irreversible thermodynamic amplification in such devices. The first hint came from Pascual Jordan (1949), one of the creators of matrix mechanics, with Heisenberg and Max Born, who suggested that macroscopicity and irreversibility should be taken as essential features of measurements in quantum theory. The idea was further developed by Gunther Ludwig (1953), who maintained that the transitions from the quantum description of microscopic bodies to the description of macroscopic measurement devices could have their physics explained. As we know the former are described by a superposition of eigenstates while the latter are not. He suggested taking into account that after the interaction between the microscopic body and the measurement device a thermodynamic amplification of the signal in the latter would happen, from where the irreversibility of quantum measurements comes. Amplification here means, according to Jordan (1949, p. 271), "an *avalanche process* set off by the microphysical object of investigation." For Ludwig measurement was over as a thermodynamically irreversible process in the

[11] For a survey of other contributions on the quantum measurement process between the 1920s and the early 1960s, see Jammer (1974, pp. 470–521).

macroscopic device. Later on, the Italian physicists (Daneri et al. 1962) would be more precise as for them the measurement device was considered a macroscopic body in a thermodynamic metastable state. It was triggered by small perturbations coming from the interaction with the microsystem being measured. Then the measurement device would evolve towards a thermodynamic stable state. Ludwig's and Jordan's approaches implied that collapses of wave packets, the first kind of evolution of the quantum states, were a thermodynamic effect and that quantum mechanics was not applicable to macroscopic bodies. Thermodynamic amplification as a programmatic idea was supported by the philosopher Paul Feyerabend and the physicist H.J. Groenewold and criticized by the physicist G. Süssmann at the Colston Symposium in Bristol 1957, the first meeting after World War II to resume the debate on the foundations of quantum physics (Körner 1957, pp. 121–147). Süssmann argued along the lines of von Neumann's measurement treatment, which implied considering the discontinuous and non-causal evolution of the quantum state—the quantum jump—an independent assumption in the mathematical formalism of quantum theory. As we will see, the work by the Italians Daneri, Loinger, and Prosperi is a mathematical development of Jordan's and Ludwig's programmatic ideas. However, while scholars may find cues in Bohr's writings on the role of irreversibility and macroscopicity in quantum measurements there was no clear-cut endorsement of this by the Danish physicist. Therefore the physicists who got involved in dealing with the quantum measurement processes in the 1950s were aware of the different approaches to the problem but there was no heated dispute between Bohr's and von Neumann's partisans on this issue. Rather, there was a fair and friendly debate among physicists and philosophers featuring Ludwig, Feyerabend, and Süssmann, among others. Moreover, even referring to quantum measurement as a problem was not common in the 1950s. Feyerabend (1957), for instance, while aligned with the conceptual ideas of the role played by irreversibility and macroscopic bodies in quantum measurements, was at variance with what he realized as Bohr's instrumentalistic philosophical ideas and pleaded for a "realistic interpretation of the formalism of quantum mechanics" (Körner 1957, p. 129). A brief inspection of the proceedings of the Colston Symposium (Körner 1957) would reveal the contrast between the calm debate on the measurement problem and the heated one on the causal interpretation. In addition, the very term "quantum measurement problem" was not yet used. Feyerabend (1957) spoke "on the quantum-theory of measurement" and Süssmann (1957) presented "an analysis of measurement."

Von Neumann died in 1957 and Bohr in 1962. In the 1960s, it would be up to other physicists to perpetuate the cleavage in the quantum measurement treatment.

Picture 4.1 Casual portrait of Eugene Wigner (1902–1995) taken at a meeting in Lindau. AIP Emilio Segre Visual Archives, Segre Collection

4.3 Enter Wigner

Complementarity had faced great challenges coming from outside the circle of the founding fathers of quantum mechanics, as we saw in Chaps. 2 and 3 while analyzing David Bohm's and Hugh Everett's stories. Now, another major challenge came from within. Eugene P. Wigner was born in 1902 in Budapest, where he graduated in chemical engineering. Early on, at his Lutheran high school, he met John von Neumann and became his friend and an admirer forever.[12] After a stay in

[12] Wigner considered von Neumann's mathematical work on the foundations of quantum mechanics "more important than any of these inventions [computing machine and implosion bomb]." See

Berlin, Wigner together with von Neumann emigrated to the United States in order to jointly develop a mathematical physics program at Princeton in the 1930s. In the early 1960s, Wigner's prestige was approaching its zenith. He was recognized early on for his use of the theory of groups in quantum mechanics and recognition increased with his contributions to nuclear physics, including his participation in the Manhattan Project (Mehra 1993).[13]

From the late thirties, Wigner began to play a role beyond physics proper, motivated by the military implications of recent discoveries in nuclear physics. He and his Hungarian colleague Leo Szilard suggested that Albert Einstein write the famous letter to President Roosevelt calling for the development of the U.S.'s nuclear program (Doncel et al. 1984). For the many roles he played in the Manhattan Project, he was consequently awarded the title of "the founder of nuclear engineering" (Weinberg 2002). After the war, Wigner's involvement with defense matters did not wane. In the late fifties, he was one of the "Princeton three," along with John Archibald Wheeler and Oskar Morgenstern, who urged the American government to build an enormous national laboratory dedicated to defense research, an initiative that failed but eventually led to the setting up of JASON, a group of academic physicists who advised the U.S. Department of Defense on defense matters.[14] Wigner assumed responsibility for promoting the role of civil defense in the Cold War context and even built a nuclear fallout shelter in his own home.[15]

In the early 1960s Wigner decided to intensify his public involvement beyond physics, publishing papers on the philosophy of science and dealing with the measurement problem of quantum mechanics. This central issue in the foundations of quantum physics would be of interest not only to physicists but to other audiences as well, particularly philosophers. As we shall see, Wigner believed that the measurement problem was part of the philosophy of physics, which in turn he saw as an integral part of physics itself, a view that many of his colleagues did not share. It should be noted that from the 1930s, working with von Neumann, Wigner was interested in and contributed to measurement issues in quantum mechanics (Shimony 1997). In the early 1950s he resumed the subject showing

---

E.P. Wigner, interviewed by W. Aspray, 04 Dec 1984, American Institute of Physics, College Park, MD.

[13] There is no professional biography on Wigner. His recollections are in Wigner and Szanton (1992). Hargittai (2006) drew from the same source to describe a biographical picture of the Hungarian-born physicists Theodore von Kármán, Leo Szilard, Eugene P. Wigner, John von Neumann, and Edward Teller. A concise biographical note is Westfall (2008). In another biographical note it was said that "Wigner's deep interest in the foundations of quantum mechanics, especially the quantum theory of measurement, persisted longer than any of his other interests" (Seitz et al. 1998). The relationship between Wigner and Michael Polanyi is exploited in Nye (2011); on their discussions on epistemology, see Jha (2011).

[14] On the "Princeton three", see Aaserud (1995). On Jason, see Finkbeiner (2006) and Moore (2008).

[15] *Trenton Evening Times*, 6 November 1961: "Princeton Scientist Who Did Work On Atom Bomb Has Own Shelter". See Eugene Wigner Papers [hereafter WigP], Box 97, Folder 1, Manuscripts Division, Department of Rare Books and Special Collections, Princeton University Library.

how quantum formalism exhibits limitations of measurability (Wigner 1952). In the same year, with G. Wick and A. Wightman, they further extended such limitations introducing the concept of a superselection rule. None of these papers, however, were as influential as his 1960s works.[16]

Between 1961 and 1963, Wigner published the two papers that would become the centerpieces of his views on the foundations of quantum mechanics. He revisited the distinction first emphasized by von Neumann between two kinds of evolution of quantum states. Additionally, but still following von Neumann, he treated measuring devices quantum mechanically, instead of treating them classically as suggested by Bohr. The latter choice leads, as we have seen, to the transfer of the singular superposition of quantum states from the system under scrutiny to the combination: system plus the measuring apparatuses. After all, what we get after measurements is related to vectors and probabilities rather than to superposition of vectors. Wigner emphasized this point and arrived at the same conclusion as London and Bauer's: in order to eliminate this superposition one needs to admit that the analysis of quantum measurement leads eventually to the role of the observer's introspection, i.e., when the information enters the mind of the observer.

Conjecturing that the mind plays an essential role in the description of quantum measurements was one of Wigner's distinctive features when approaching the measurement problem. According to him, "when the province of physical theory was extended to encompass microscopic phenomena, through the creation of quantum mechanics, the concept of consciousness came to the fore again: it was not possible to formulate the laws of quantum mechanics in a fully consistent way without reference to the consciousness."[17] He presented his arguments in two steps. The first, and less incisive, was that the quantum state changes every time the observer obtains new information from observations. While in classical mechanics you also have to observe to obtain the initial conditions and establish the classic state, when you get them and solve the equations of motion, the new information is no longer relevant to change the state. In the second step, he strengthened his case, presenting an idealized experiment in order to demonstrate the difference between quantum descriptions of measurements with and without human observers. Nowadays the argument related to Wigner's idealized experiment is known as "Wigner's friend" (Wigner 1961).[18] Wigner suggests you observe an object quantum described by a linear combination of two states, helped by a friend. Your friend observes the object, hence to him/her it is in one of the two states and no longer in a

---

[16] For a technical presentation of Wigner's papers on quantum measurements, see Shimony (1997) and G. Emch, "Annotation," in (Wigner 1995, pp. 1–28).

[17] Wigner (1961), cited from (Wigner 1995, p. 248).

[18] Indeed Hugh Everett was the first to write the argument we now call "Wigner's friend," in the long version of his doctoral thesis. To get his PhD degree in 1957, he submitted an abridged version of the dissertation, without this argument. The full dissertation only was published in 1973. However, it is uncertain who was the first to conceive this argument as Everett interacted with Wigner at Princeton in the mid-1950s. See Chap. 3; Osnaghi et al. (2009, pp. 104–105) and Everett et al. (2012, pp. 14, 29–32).

linear combination of the two. Before s/he tells you the result of the observation, there will be a conflict between your description of the object (linear combination of the two states) and that given by your friend (one of the two states). Accepting your quantum description as the right one, you must admit that your "friend was in a state of suspended animation before he answered" your question. This is a paradoxical conclusion. So for Wigner quantum theory is unable to embrace measurements or it does not produce consistent results if human observers are included in the quantum description. Thus, if quantum theory is to encompass not only inanimate bodies, but also life and mind, it needs to be modified, and Wigner suggested looking explicitly for a non-linear equation of motion.[19] This was indeed the originality of Wigner's approach in these papers as, in fact, he was suggesting a true research program: to acknowledge the existence of a measurement problem and to solve it changing the standard quantum physics mathematical formalism. For Wigner this was necessary in order to include life and mind in the scope of physical theories.[20] As for lexicon, Wigner was, as far as I know, to title a paper considering the quantum measurement issue as a "problem" (Wigner 1963).

Furthermore, Wigner's arguments entailed a more sociological and historical issue: to define the orthodox interpretation of quantum mechanics, and to identify its protagonists. Introducing himself as a supporter of the orthodox, standard view of quantum mechanics, he wrote: "The standard view is an outgrowth of Heisenberg's paper in which the uncertainty relation was first formulated. The far-reaching implications of the consequences of Heisenberg's ideas were first fully appreciated, I believe, by von Neumann, but many others arrived independently at conclusions similar to this. There is a very nice little book, by London and Bauer, which summarizes quite completely what I shall call the orthodox view" (Wigner 1963). Bohr's paper on complementarity is only referred to in a footnote. In Wigner's account, therefore, Bohr and complementarity occupy a behind the scenes role in the quantum story, and Heisenberg and von Neumann become its chief protagonists. Historians of science know the role played by the creation of disciplinary histories. These are "attempts to create discrete and unified histories of scientific disciplines, complete with founding fathers, fundamental innovations and

[19] In his 1961 paper, he wrote a section under the heading "Non-linearity of Equations as Indicators of Life." Later, Wigner (1995[1973]) kept the same stance: "it seems unlikely [...] that the superposition principle applies in full force to beings with consciousness. If it does not, or if the linearity of the equations of motion should be invalid for systems in which life plays a significant role, the determinants of such systems may play the role which proponents of the hidden variable theories attribute to such variables. All proofs of the unreasonable nature of hidden variables are based on the linearity of the equations."

[20] Wigner's conjecture about the role of mind in quantum physics was strongly intertwined with his metaphysical and epistemological beliefs. He kept a dualistic view about mind and matter and maintained the former was primary. He criticized mechanistic approaches to the question of life because, for him, the phenomenon of consciousness entreats us to admit the existence of biotonic laws, that is, laws of nature not contained in the laws of physics (Wigner 1995[1972], 1997a, b). I will not, however, extend my analysis of his broader philosophical views here. For a discussion on such issues, see (Esfeld 1999). Thanks to Ron Anderson for bringing this paper to my attention.

so forth" (Christie 1990, p. 11). One may think of Wigner's account as the disciplinary history of the research on the foundations of quantum theory. I think he had broader aims and interpret this excerpt as a dispute over the intellectual heritage of the founding fathers of quantum mechanics. Wigner wrote this text after von Neumann's and Bohr's deaths, and while scientists and historians in the U.S. were involved in one of the largest projects ever to collect and store records which were significant in the creation and evolution of a scientific theory and which would come to be known as the *Archives for the History of Quantum Physics* (Kuhn 1967).

Wigner's papers drew both support and opposition. Abner Shimony, who had a PhD from Yale in Philosophy and was doing his second PhD, this time in Physics at Princeton under the supervision of Wigner, was very impressed by it: "I found your paper on the mind-body problem extremely stimulating. It is one of the few treatments of the problem which considers the mind-body relationship to be a legitimate subject for scientific investigation, without achieving this scientific status for the problem by reducing it to behavioristic or materialistic considerations."[21] M. Satosi Watanabe (1910–1993), a Japanese physicist, who had studied in Europe with de Broglie and Heisenberg and was interested in foundational issues and information, also reacted very favorably to Wigner's suggestion about the role of consciousness in physical processes. We find in their correspondence hints regarding the subsequent opposition to Wigner's ideas from Rosenfeld. Rosenfeld's Marxist motivation can also be traced here. Apparently, Wigner had underestimated the ideological backdrop of the quantum controversy. He wrote to Watanabe, "Do you know of any political background that has come into the open in these discussions? I am under the happy impression that we can keep the discussion on these subjects free from politics and am not aware of anyone having brought in any doctrine into the argument." Watanabe's reply was premonitory of the Wigner-Rosenfeld dispute: "... I have indeed had quite a few experiences myself of being exposed to shameless attacks by Marxists in Japan for what they call my bourgeois idealism. In spite of the fact that Marxism is not a mechanical materialism, they are dead against giving any kind of independent reality to consciousness. There are Marxists who are quite broad-minded (like Prof. Rosenfeld) in many respects, but they usually become quite emotional when the topic touches upon their basic dogmas. (I was rather disappointed by the partisan emotion which tainted Prof. Rosenfeld's paper on Statistical Mechanics which was published in Poland. Even a broadminded Marxist like Prof. Rosenfeld acts like this.")[22]

---

[21] Letter from Abner Shimony to Wigner, May 1, 1961. WigP, Box 94, folder 1. Shimony would always defend that the mind, or cognitive faculties, should be investigated in relation with the quantum measurement problem. "A possibility that seems to me largely to have been neglected in the literature on the measurement problem [...] is that the locus of reduction is the macromolecules of the sensory and cognitive faculties." He conducted experiments on the subject (Hall et al. 1977) and maintains that quantum mechanics may bridge the gap between psychology and natural sciences (Shimony 1993, pp. 74 and 319).

[22] Wigner to Watanabe, 30 Aug 1961; Watanabe to Wigner, 15 Dec 1961; *WigP*, Box 63, folder 12, and box 71, folder 1, respectively.

Picture 4.2 Léon Rosenfeld (1904–1974). AIP Emilio Segre Visual Archives

It was up to Rosenfeld to oppose Wigner in defense of complementarity. Rosenfeld had been Bohr's assistant since the 1930s, and a physicist who was very sensitive to epistemological matters.[23] Rosenfeld and Wigner had, however, quite different stands on a number of issues. Politically, Wigner was very conservative—he was a follower of the Republican Party and was supportive of U.S. foreign policy to the point of receiving a telegram from President Richard Nixon thanking him for his support in the Vietnam War effort.[24] In contrast, Rosenfeld had been engaged in Marxist philosophy since the thirties. Rosenfeld's Marxism was closer to Western Marxism than it was to Soviet Marxism, to use terms introduced by Perry Anderson (1976) in order to make sense of Marxist trends in the twentieth century.[25] In the late 1940s and 1950s Rosenfeld participated actively in organizations and movements, such as the "World Federation of Scientific Workers," "Science for Peace," and the "Manchester University Socialist Society." His political record led him to doubt whether he would be granted a visa to visit the U.S., in Cold War times.[26] To preserve what seemed to him to be a dialectical feature of complementarity, Rosenfeld criticized both the Soviet and

[23] On Rosenfeld, see his comprehensive biography by Anja Jacobsen (2012) and his collected papers on epistemology (Rosenfeld et al. 1979).

[24] R. Nixon to Wigner, 22 Jun 22, 1970: "Encouragement is always gratifying, but I particularly appreciated your very thoughtful letter and I want you to know how pleased I was to hear from you. Your support for our policies toward Southeast Asia means a great deal to America's fighting men, and needless to say, it means a great deal to me." *WigP*, Box 97, folder 3.

[25] Anderson's distinction is driven to label those Marxist intellectuals, such as Lukacs, Korsch, Bloch, and Adorno, who kept their distance from the Soviet Marxism and the Western Communist parties related to it. Anderson's categories are not trivial since Antonio Gramsci, the leader of Italian Communism, is considered for his works part of Western Marxism. Roughly used, however, they are useful for understanding Rosenfeld's Marxism.

[26] See letters to L. Rosenfeld, from J.A. Wheeler [March 27, 1952]; R.E. Marshak [September 24, 1954], and A. Roberts [December 22, 1955]. Léon Rosenfeld Papers, Niels Bohr Archive, Copenhagen [Hereafter *RP*].

Marxist physicists, like D. Blokhintsev and D. Bohm, who were themselves critics of complementarity, and physicists like Heisenberg, who leaned towards idealism, as we have discussed in Chap. 2. So, for a number of reasons—political, ideological, and philosophical—Rosenfeld could not accept a view like Wigner's which assigned a central role to the mind in physical phenomena.

Wigner and Rosenfeld also displayed significant differences in their approach to the measurement problem, which could also be referred to as different scientific styles. For Wigner, following von Neumann, dissecting the mathematical formalism of quantum physics in order to exhibit its axiomatic structure was a necessary step in grasping the theory's full implications. That is not to say that for Wigner axiomatic theories were necessary for all research in physics, because in other fields, nuclear physics for instance, his approach was phenomenological.[27] But Rosenfeld, possibly following Bohr, always emphasized his distrust of the reach of any axiomatic treatment of physical theories, or, at least, his distrust in von Neumann's axiomatic approach to quantum mechanics. Even before his dispute with Wigner, Rosenfeld had written: "the 'axiomatizers' do not realize that every physical theory must necessarily make use of concepts which *cannot*, in principle, be further analyzed, since they describe the relationship between the physical systems which is the object of study and the means of observation by which we study it: these concepts are those by which we give information about the experimental arrangement, enabling anyone (in principle) to repeat the experiment. It is clear that *in the last resort* we must here appeal to *common experience* as a basis for common understanding."[28] A little earlier, in a report on quantum theory textbooks requested by a publisher, he had been more explicit in his distance from von Neumann's approach: "v. Neumann's book 'Foundations of Quantum Mechanics' [...], though excellent in other respects, ha[d] contributed by its unhappy presentation of the question of measurement in quantum theory to create unnecessary confusion and raise spurious problems."[29] Last but not least, Rosenfeld maintained that complementarity was the great epistemological lesson of quantum theory, and for this reason, he could not accept Wigner's position according to which Bohr's complementarity played no role in the orthodox interpretation of quantum theory.

---

[27] I am thankful to Sam Schweber for his discussion on this issue. Commenting on the founding fathers of quantum mechanics, Schweber (1996) wrote: "Wigner stands out by being, on the one hand, the theorist who had perhaps the greatest affinity to pure mathematics and, on the other, probably the most phenomenologically inclined among them." For Wigner's insertion in the mathematical physics tradition, see Schweber (2014).

[28] L. Rosenfeld to Saul M. Bergmann, December 21st, 1959. *RP*. The subject of the letter concerns Everett's approach to quantum physics. Emphases are in the original.

[29] L. Rosenfeld. "Report on: Louis de Broglie, La théorie de la mesure en mécanique ondulatoire (Paris: Gauthier-Villars)," 1957, *RP*.

4.4 The Heated Dispute: Wigner Versus Rosenfeld and the Italians

Rosenfeld's strategy for criticizing Wigner's view was to give strong praise to certain work, by writing, "these misunderstandings [i.e. that the translation of Bohr's argument into the formal language of the theory should present unrecognized difficulties], which go back to the deficiencies in von Neumann's axiomatic treatment, have only recently been completely removed by the very thorough and elegant discussion of the measuring process in quantum mechanics carried out by Daneri, Loinger and Prosperi" (Rosenfeld 1965). This paper had been published in *Nuclear Forces*, a journal edited by Léon Rosenfeld. These Italian physicists had used the ergodic theorem, which states that under certain conditions the time average of a function equals its space average, to explain quantum measurements as a thermodynamic amplification of a signal triggered by the interaction between quantum systems and measurement devices (Daneri et al. 1962). Indeed, the Italian physicists had quantum mechanically treated both the system and the interaction between the system and the measurement device, but, after the interaction ended, they considered the measurement device as evolving according to classical statistical physics, which was compatible with Bohr's requirement that the measurement devices should be considered classical bodies.[30]

The Italian paper came from a research tradition devoted to applying the ergodic theorem to problems in statistical mechanics and quantum theory and was inspired by the Italian theoretical physicist Piero Caldirola.[31] What had been a rather technical issue in the quantum measurement problem became a controversial issue due to its conjunction with Rosenfeld's praise, which raised the stakes of the paper. If Rosenfeld's point of view about the reach of the Italian work were accepted, Wigner's claims would be considered ungrounded. The dispute lasted throughout the second half of the 1960s and it was marked by bitter arguments, even though it dealt with rather technical content, i.e., to determine whether the Italian work was a rigorous solution or just an approximation.

Wigner was particularly upset by the Italian physicists' subsequent paper. In this paper they (Daneri et al. 1966) cited Wigner, Shimony, Moldauer, Yanase, and Jauch's analyses of the measurement problem, stating that "none of them gives new substantial contributions to the subject; therefore we shall not discuss them in detail, but we shall limit ourselves to the sketchy comments reported in footnotes" (Daneri et al. 1966, pp. 120–121). In fact, this was a paper that in some places dedicated more room to footnotes (roughly 80 %) than to the main text itself. In addition, Daneri, Loinger, and Prosperi drew the lines of the battle as they criticized Wigner, Shimony, Moldauer, Yanase, and Jauch's analyses for no "new substantial contributions" and aligned their paper with Bohr, Jordan, and Ludwig's approaches

[30] For a more detailed discussion of Daneri, Loinger, and Prosperi's paper, see Chap. 5.

[31] See Chap. 5 and Pessoa Jr. et al. (2008).

to the quantum measurement problem, as well as counting on Rosenfeld as their main cheerleader (Daneri et al. 1966, p. 120).

Wigner's reaction was to write to Josef M. Jauch suggesting a common response, together with Yanase, Wigner's former student. Jauch was at that time leading what would be later known as the Geneva school dedicated to the axiomatic foundations of quantum physics.[32] Wigner acknowledged that he was particularly irritated not by the attack on him but by its significance for young researchers like Abner Shimony and Michael Yanase, his former doctoral students. Thus he wrote,

> I just finished reading the article of Daneri Loinger and Prosperi in the July issue of Nuovo Cimento and am really a bit irritated by it. First of all, it is not good taste to say about a set of articles that they do not make substantial contributions to a subject. Needless to say, I am less concerned about myself than about other people who are much younger than I am and whose future careers such statements may hurt.[33]

Wigner also told Jauch of his disagreement with Rosenfeld's support of the paper by the Italian physicists. While referring to macroscopic systems with states not described by classical mechanics, he was probably thinking of phenomena such as superconductivity and superfluidity, which had been studied just through the use of quantum mechanics. Wigner wrote to Jauch,

> I am also saddened by Rosenfeld's endorsement of the article which, after all, considers it axiomatic that macroscopic systems have only states which can be described by classical mechanics. This is, of course, in conflict with quantum mechanics, but this is never mentioned in the article except by the explicit agreement with the work of Ludwig, who is entirely explicit on this subject.[34]

The letter to Jauch was a typical maneuver in search of allies, as Wigner was not in complete agreement with Jauch, notwithstanding the fact that the latter was trying to refine von Neumann's mathematical treatment. Jauch did not agree with Wigner's conjecture on the role of mind in the measurement process and believed that the changes he himself had introduced in von Neumann's treatment had transformed the difference between the two kinds of evolution of the state vector into a pseudo-problem. According to him,

> It is shown that the two ways of the change of state vectors can be understood without introducing von Neumann's 'ultimate observer' and without abandoning the linear law of the time evolution of states. Consciousness or even the macroscopic nature of the measuring device is not an essential requirement for a measurement. (Jauch 1964, p. 293)

The maneuver was acknowledged by Wigner in a letter to Shimony, "enclosed is a preliminary manuscript of an article which attempts a reconciliation of the views of Jauch and ourselves. It is a response to what I consider to be a rather intemperate criticism by Daneri, Loinger and Prosperi."[35] Looking for allies, Wigner went so far

[32] For references to the Geneva school, see Arthur (1981).

[33] Wigner to Jauch, 06 September 1966. *WigP*, Box 94, folder 7.

[34] Ibid.

[35] Wigner to Shimony, 16 Dec 1966. *WigP*, Box 71, folder 3.

as to accept a suggestion from Shimony and propose that Jauch include in their joint paper a favorable citation of a paper by David Bohm and Jeffrey Bub. However, Jauch could not accept this because in his own work with Constantin Piron, whose doctoral thesis had been co-supervised by Ernst Stueckelberg and himself, he was trying to reinforce von Neumann's proof against hidden variables (Jauch and Piron 1963), while Bohm and Bub's work was an open criticism to this approach. [36]

The joint paper by Jauch, Wigner, and Yanase (1967) was a piece exhibiting both conceptual precision and diplomatic skills. After an explanation of the measurement problem, following von Neumann's standard presentation of measurement in quantum physics, they went on to address the criticisms voiced by Daneri, Loinger, Prosperi, and Rosenfeld. Instead of merely rebutting them, Jauch, Wigner, and Yanase acknowledged the paper by the Italian physicists as "a useful contribution to the theory of measurement" insofar as they showed that certain macroscopic bodies used as measurement devices can evade the quantum superposition of states. In addition, the Italian physicists had acknowledged that microscopic systems play only the role of triggering the thermodynamic amplification in the measurement processes while the amplification happens exclusively in the macroscopic device after the interaction with the microscopic systems. However, Wigner, Jauch, and Yanase pointed out that not all measurements follow the scheme of a microscopic system triggering a thermodynamic amplification in the measurement system. They used as counter-example the case of "negative-result measurements" which had already been suggested by Mauritius Renninger.[37] Finally, they looked for a common ground with the work of the Italian physicists noting that both Jauch's macrostates and the Italian classical states of the macroscopic apparatus do not obey the Schrödinger equation in their evolution, thus both of them may be considered "steps in the direction of a generalization of the quantum-mechanical description of physical systems" (Jauch et al. 1967 on 151).

Afterwards, Rosenfeld (1968) and Loinger (1968) replied to Jauch, Wigner, and Yanase's paper, however, the next round did not take place in papers published in journals. Instead, Wigner waited for a special gathering, the 1970 Varenna summer school dedicated to the foundations of quantum mechanics, to have a live debate, as we shall see. It can be said that the very existence of this school reflected the changing mood among physicists concerning the status of research on foundations

[36] The suggestion is in Shimony to Wigner, 1 Jan 1967. *WigP*, Box 83, folder 7. Shimony's suggestion was to cite (Bohm and Bub 1966a). Jauch's reaction was: "the second major point which I should modify refers to your remarks on Bohm and Bub. This concerns me perhaps more directly because their second paper (Bohm and Bub 1966b) is entitled as a 'refutation' of the paper by Piron and myself [...] but what concerns me more in connection with the problem on measurement is that the model which they propose in the first paper has absolutely no predictive value." Wigner then accepted Jauch's restrictions. Jauch to Wigner, 13 Oct 1966, Wigner to Jauch, 25 Oct 1966, *WigP*, Box 71, folder 3. On Bub's work as a doctoral student of Bohm, see Chap. 2.

[37] For a review on negative-result measurement, both theory and experiments, since Epstein and Renninger, see Whitaker (2000). For the relationships between Renninger's negative-result measurements, the paper by Jauch, Wigner, and Yanase, and the work of the physicist Klaus Tausk, see Chap. 5.

of quantum theory (Freire Jr. 2003a). In addition to these shifting intellectual trends, we must remember that Rosenfeld was not in an easy position during this dispute. In 1952, as mentioned in Chap. 2, he had derided David Bohm, the leader of the first round of dissent, labeling him a "tourist" or a "dilettante" in the field of the foundations of quantum mechanics,[38] but he could not deal with the 1963 Nobel Prize winner, Eugene Wigner, in the same way.

The Varenna courses, organized by the Italian Society of Physics, had been held regularly since 1953 in the summer in Varenna on Lake Como. The 1970 course was dedicated to the theme "Foundations of Quantum Mechanics" following a suggestion from Franco Selleri supported by Toraldo di Francia, then the president of the society. The motivation and background for holding such a course will be discussed in Chap. 6. For the moment however, we need a few pieces of information from it. The course was held under the direction of Bernard d'Espagnat. There were 84 participants, and its proceedings (d'Espagnat 1971) reveal a diversified spectrum of subjects, such as measurement, hidden variables and non-locality, and interpretations. People from different perspectives about the quantum issues, such as Wigner, Jauch, Shimony, d'Espagnat, Bell, de Broglie, Prosperi, Selleri, and Bohm were invited. d'Espagnat suggested some diplomatic rules to be followed in the invitation letter so as to guarantee a peaceful and creative atmosphere in which to discuss scientific controversies.[39] As remarked by the historian Anja Jacobsen, Rosenfeld's biographer, Rosenfeld was invited to speak about the measurability of quantum fields and accepted. However, after reading the invitation letter and the list of speakers, he withdrew and sent Jørgen Kalckar, a younger physicist from the Niels Bohr Institute, in his place. Indeed, through this maneuver Rosenfeld avoided endorsing an event he was against the very existence of. In addition, Rosenfeld's retraction reflected the ongoing battle between Wigner, Bohm, and Jauch, on the one hand, and himself, on the other hand. This was expressed in his refusal letter with the following words: "Wigner's talk [in a recent event in Trieste] convinced me that no dialogue is possible and, furthermore, that he is not looking for it. As for Bohm, I just say that I am tired with his last words; and you know what I think of Jauch's axiomatic prestidigitations."[40] According to Jacobsen, "d'Espagnat regretted Rosenfeld's decision" but denied "that the direction the meeting had taken was pointing at Rosenfeld any way" (Jacobsen 2012, pp. 307–308). As a matter of fact, the intellectual climate concerning the foundations of quantum mechanics was one of far more openness than in the 1950s and early

[38] Rosenfeld (1953, p. 56) and letter from L. Rosenfeld to N. Bohr, 14 Jan 1957. Archives for the History of Quantum Physics, American Philosophical Society, Philadelphia, PA (*AHQP*, hereafter), Bohr Scientific Correspondence, reel 31. For the context, see Chap. 2.

[39] These diplomatic rules will be presented in Chap. 6. They were previously analyzed in (Freire Jr. 2003a, 2004).

[40] "L'exposé de Wigner m'a convaincu qu'aucun dialogue n'était possible, et que d'ailleurs il ne le souhaitait nullement; quant à Bohm, je dirais seulement que ses prêches des derniers jours me fatiguent; et vous savez ce que je pense des prestidigitations axiomatiques de Jauch." Rosenfeld to d'Espagnat, 23 December 1969, RP.

1960s and Rosenfeld's refusal to attend also reflected the decline of his influence among the physics milieu interested in the foundations of quantum physics.

The dispute involving Wigner, Rosenfeld, Jauch, Yanase, and the Italian physicists Daneri, Loinger, and Prosperi eventually ended at the Varenna summer school, which Wigner succeeded in transforming into an agreement on the need for a research program on quantum measurement processes. Wigner gave the keynote talk at Varenna (d'Espagnat 1971, pp. 1–19), and Prosperi spoke about "macroscopic physics and the problem of measurement in quantum mechanics," in a section dedicated to "Measurement and basic concepts" (d'Espagnat 1971, pp. 97–122). An informal discussion between Wigner and Prosperi ensued from these lectures. Assisted by Shimony and d'Espagnat,[41] Wigner reconstructed his arguments and had them published (d'Espagnat 1971, pp. 122–126). It is worth taking a detailed look at Wigner's conclusions because we can discern in them two distinctive features of Wigner's approach to the foundations of quantum physics: his diplomatic and open-minded attitudes, and his consideration of the philosophy of physics as part of physics. He divided his conclusions into two, the first related to "the philosophical problem," and the second about "questions of physical theory." He explained that the main issue at the center of both Prosperi's and of his own concerns was related to the knowledge of the "reason for the statistical, that is probabilistic, nature of the laws of quantum-mechanical theory." In other words, how can one understand that quantum predictions are not "uniquely given by the inputs" even though equations of quantum and classical physics are deterministic? He suggested one might answer this question in different ways, and cleverly framed Prosperi's and his own responses on the same side. This type of answer implies that "the possible reason for the probabilistic nature of quantum theory's conclusions concerning the outcomes of measurements is that the theory cannot completely describe the process of measurement, that some part of the process is not subject to the equations of quantum mechanics." The difference between Wigner's and Prosperi's views resided in "the *area* to which quantum mechanics is inapplicable." For Prosperi, probability is necessary for the translation of the quantum-mechanical description into the classic description because this translation is not unique. Wigner says that for him and von Neumann quantum mechanics does not "apply to the functioning of the mind" as "the conscious content of the mind is not uniquely given by its state vector." Finally, arguing on more scientific grounds, Wigner remarked that Prosperi and collaborators were using phrases such as "macroscopic variables" and "macroscopic objects" without giving a precise definition of these terms. He remembered examples of phenomena with macroscopic bodies but which

[41] "I received a letter from d'Espagnat telling me about your suggestion that the discussion between Prosperi and myself be included in the Proceedings of our conference in Varenna." Wigner to Shimony, 09 Oct 1970; Wigner to Shimony, 21 Jan 1971, Box 1, Folder 07-B, Shimony to Wigner, 02 Feb 1971, Box 1, Folder 7, Abner Shimony Papers (*AS* hereafter), Archives of Scientific Philosophy, Special Collections Department, University of Pittsburgh. "It would indeed be a service to people interested in foundations of quantum mechanics for you to reconstruct your discussion with Prosperi." Shimony to Wigner, [w/d 1971], *WigP*, Box 72, folder 2.

exhibit quantum features, such as permanent currents in superconductors—superconductivity—and spontaneous magnetization in different directions, besides the observable difference between dextro and levorotatory sugar, which is based on a quantum relation of microscopic phases. In his conclusion, Wigner once more looked for areas of agreement between the two physicists and presented a proposal for a genuine research program. His point of departure was that Prosperi's premises could not be rigorously formulated (at least not at that time) and their formulation would entail a significant modification of the theory current at that time. Then, Wigner argued, the convergence resided in the conclusion, common to Prosperi and Wigner's views, namely "the inapplicability of quantum mechanics to some part of the measurement process has to be postulated or admitted." Surely, Wigner had been postulating this for some time, and he was asking Prosperi to admit it. If Rosenfeld had been present at the Varenna school, one might suppose that he would have not accepted Wigner's suggestion because for him, according to Bohr's views, concepts such as "macroscopic variables" or "macroscopic bodies" should be admitted without previous definitions since one gains nothing when trying to axiomatize, or to define all theoretical terms. As close as his views were to Bohr's, Prosperi thought differently and did not insist on the thesis maintained by Daneri, Loinger, and himself. This thesis had implied that their work should be seen as an accomplishment of quantum theory. He did not publish any additional reports on his own arguments and returned from Varenna to Milan convinced that the measurement problem was still unsolved.[42]

4.5 The Orthodoxy Splits

It is important to consider how Wigner's contemporaries interpreted his dispute with Rosenfeld and the Italian physicists. The dispute was seen as evidence of the existence of a controversy on the foundations of quantum physics. Thus, Otto R. Frisch, in a colloquium held in 1968, said: "I understand that at present there exists a controversy, roughly speaking between a group of people which includes Wigner as the best known person and another group centered on Milan in Italy, and that these two have different views on how this reduction happens" (Frisch 1971, p. 14). For the first time in the literature, the name "Princeton school" was used to differentiate Wigner's views from the Copenhagen school. According to Ballentine (1970, p. 360), there were "several versions of the Copenhagen interpretation" and, "although both claim orthodoxy, there now seems to be a difference of upholds between what may be called the Copenhagen school represented by Rosenfeld, and the Princeton school represented by Wigner."[43] Since then, the labeling

---

[42] G. Prosperi, July 3, 2003, Milan, interviewed by the author.

[43] Neither Wigner nor Rosenfeld, however, held Ballentine's views in high esteem. "Ballentine, whom I had the honour to meet at your old place, Vancouver, last April, looked to me as a

Copenhagen and Princeton schools has become widely used in the literature (Home and Whitaker 1992; Leggett 1987). The monocracy of the Copenhagen school, a term used by Max Jammer, was thus broken, from the inside. I do not want to say that Wigner was the sole driving force in breaking that monocracy. Other factors also contributed to changing physicists' attitudes to research in foundations of quantum mechanics, but I will not discuss them here as they will be dealt with in the following chapters. What I am saying is that Wigner made a major contribution in this direction, which is not always recognized today.

4.6 Wigner's Style of Intellectual Leadership

The portrait of Wigner as simply a controversial actor in the creation of the field of foundations of quantum mechanics is not completely fair. He engaged in a variety of activities and had a kind of non-dogmatic but highly influential leadership style. He put together a group of students to work on the subject, including Abner Shimony, who had a PhD in Physics with a dissertation on the foundations of statistical mechanics, and Michael Yanase, a Jesuit priest whose dissertation dealt with the measurement problem. We have already seen how he mobilized Yanase to join the debate with the Italians. Shimony (1993, p. xii) provides a very impressive testimony of the role Wigner played in his career: "I am most deeply grateful to Eugene Wigner, [who] encouraged my later work on foundations of quantum mechanics. The preponderance of the physics community at that time accepted some variant of the Copenhagen interpretation of quantum mechanics and believed that satisfactory solutions had already been given to the measurement problem, the problem of Einstein-Podolsky-Rosen, and other conceptual difficulties. My decision to devote much research effort to these problems would have been emotionally more difficult without Wigner's authority as one of the great pioneers and masters of quantum mechanics." The bulk of the correspondence exchanged between Shimony and Wigner on philosophical matters suggests that Wigner also benefited from this intellectual relationship because in fact Shimony acted informally as Wigner's assistant on philosophical matters.[44]

Wigner was also supportive of senior physicists working on topics of foundations of quantum physics, such as Bernard d'Espagnat, Henry Margenau, and John

rejuvenation of Everett himself, just as bumptious and probably no less stupid. I was giving a general lecture [...] and at the end Ballentine came to me and said, 'I am very embarrassed because I expected that I would strongly disagree with you and I find what you said is in agreement with my views'", Rosenfeld to F. J. Belinfante, 22 June 1972, *RP*. "Did you see Ballentine's article in the Rev. Mod. Phys.? It does show how difficult the communication is between physicists and philosophers, and how much more the latter believe in the meaningful nature of words which we consider ill defined. We do need people like yourself to establish a modicum of mutual understanding." Wigner to Shimony, 21 January 1970, *AS*, Box 1, Folder 07-B.

44 Most of this correspondence, of meaningful philosophical value, is deposited at the Eugene W. Papers (*WigP*) and Abner Shimony Papers (*AS*).

Archibald Wheeler. Bernard d'Espagnat was already a senior-level high-energy physicist when in the 1960s he decided to resume a project begun in his younger days, to philosophize on the problems of contemporary physics.[45] D'Espagnat found in Wigner a dialoguer, even if he did not completely share his views, and in Rosenfeld an ironic and bitter critic, albeit friendly, when he accepted some of Wigner's positions. As early as 1964, d'Espagnat criticized Jauch for his idea that mixture and pure state are in the same "equivalence class," and supported Wigner: "This is a matter into which I always took a great interest and I found your article in AJP very illuminating." Later, Rosenfeld praised d'Espagnat's book (1965), but not the paper in which d'Espagnat (1966) suggested a generalization of Wigner's point of view, according to which "the framework of the orthodox theory of (ideal) measurements" means that these cannot as a rule be described by means of linear quantum mechanical laws. In February 1966, d'Espagnat wrote to Rosenfeld, "I am thankful [. . .] because you had the kindness of approving my book." Four months later, Rosenfeld wrote (8 July 1966), "your last work 'Two Remarks on the Theory of Measurement' seems to indicate you need reinvigorate yourself in the pure air of Copenhagen [d'Espagnat had been there in 1954, and had received an invitation to return from Rosenfeld, in January 1966]. There is nothing as such to heal you from this Wignerite crisis you seem to be suffering from, which I hope is a light one."[46]

Unlike d'Espagnat, Margenau was a seasoned veteran in the field of foundations of physics as he had been criticizing the complementarity view since the 1930s. In the early 1960s, he welcomed Wigner's analysis of quantum measurement, which motivated him to resume his own ideas and to present them in a clearer and more concise way:[47]

> I have read your illuminating paper in the American Journal of Physics. This, together with thoughts about other materials from your pen and further recent publications, has prompted me to put together what I consider a simple and consistent theory of measurement [. . .]. I believe I have not made sufficiently clear in the past what I regard as important, for I really think that my basic concepts do not differ from your version.

They engaged in a published debate with Hillary Putnam over the conceptual structure of quantum mechanics and planned to write a book together, but this project was never seriously initiated.[48] In the late 1960s, Wigner accepted

[45] Bernard d'Espagnat, interviewed by the author, Paris, 26 Oct 2011, Center for History of Physics, American Institute of Physics, College Park, MD.

[46] "Je tiens à vous remercier pour [. . .] l'approbation que vous avez la gentilesse d'y exprimer à l'égard de mon livre," d'Espagnat to Wigner, 18 Feb 1964, *WigP*, Box 94, Folder 1. D'Espagnat to Rosenfeld, 26 Feb 1966. "Votre dernier travail 'Two Remarks on the Theory of Measurement' semble indiquer que vous avez besoin de vous retremper dans l'air pur de Copenhague, [. . .] Il n'y a rien de tel comme cure de cette wignérite dont vous paraissez subir une atteinte, que j'espère légère," Rosenfeld to d'Espagnat, 8 July 1966, *RP*.

[47] Margenau to Wigner, 21 Jan 21, 1963. Henry Margenau Papers [*MP* hereafter], Manuscripts and Archives, Yale University Library, box 1, folder 12.

[48] See Margenau and Wigner (1962). On the debate with Putnam, see Santos and Pessoa Jr. (2011). For the book they planned, as suggested by Wigner, see the letter from Margenau to Wigner, 4 Oct 4, 1974 (*WigP*, box 56, folder 13); idem, 26 Dec 1974 (*WigP*, box 72, folder 3); and the letter from

Margenau's invitation to be a member of the editorial board of a new journal, *Foundations of Physics*, designed to foster research of "disciplined speculations suggestive of new basic approaches in physics,"[49] including those concerning the foundations of quantum mechanics. Wigner not only accepted this invitation but assumed the editorial responsibilities of the journal, suggesting papers and influencing the choice of editor who would replace Margenau upon retirement.[50]

Picture 4.3 (L-R): John Wheeler, Eugene Wigner. AIP Emilio Segre Visual Archives, Wheeler Collection

Wigner to Margenau, 28 Dec 28, 1974 (*MP*, box 1, folder 12). Later, however, Wigner apparently did not follow Margenau's admission of extrasensory perception and remained skeptical about Margenau's essays on blending science and religion. See letter from Margenau to Wigner, 27 May 1988; and Wigner's to Margenau, 30 June 1988 (*WigP*, box 56, folder 13). Documents from Margenau's views on extrasensory perception are in *WigP* (box 56, folder 13), and *MP* (box 1, folder 6).

[49] See *Foundations of Physics*, 1970, 1, "editorial preface."

[50] Letter from Wigner to Robert Ubell, 24 Sep 1974, *WigP*, box 72, folder 3.

A final example, and perhaps the most significant example of Wigner's influence on certain contemporaries, is John Archibald Wheeler, a physicist who was well known for his insights, both sound and speculative, in fields as diverse as cosmology and quantum physics. The two men were very close not only in science, but also in political matters and in defense-related research. As we have seen, Wigner and Wheeler were two of the "Princeton three" who were involved at the outset in the JASON project. In foundations of quantum mechanics, however, Wheeler's views were very close to Bohr's, but Wigner was so influential that by the mid-1970s he had began to doubt what Bohr's real opinion was on the role of consciousness in the quantum measurement process. Haunted by this doubt, he wrote the following to Niels Bohr's son, Aage Bohr:

> I have the impression, perhaps mistaken, that your father at one time thought that for the making of an observation it only took in the end an irreversible account of amplification; but that later on he changed his position to something closer to the idea that no observation is an observation unless and until it enters the consciousness. However, I am not able to find anything to document this supposed change of view and my understanding of the history may be quite wrong.[51]

On receiving no response from the younger Bohr, he asked a friend who was in Copenhagen, John Hopfield, to answer a list of some questions after consulting Aage Bohr. This included the following item: "Niels Bohr did change position from (a) 'Measurement requires irreversible act of amplification' to (b) something closer to Wigner's 'a measurement is not a measurement until the result has entered the consciousness' YES ___; NO ___; QUESTION ILL DEFINED ___." This time, however, it did not take long for Aage Bohr to send a reply corroborating the accuracy of Rosenfeld's interpretation of Niels Bohr's views. Aage Bohr wrote:

> [...] our reactions can be deduced from the answers to the questionnaire which you have formulated so cleverly that no evasion is possible. Let me just add that it is quite true that my father strongly emphasized that for an unambiguous description it is essential to include the detection device in the definition of a quantum phenomenon and even advocated that one reserved the word 'phenomenon' for processes that are 'closed' in this sense. However, I do not think he meant this to imply that the act of observation need have any effect on the processes which generated the phenomenon in question.[52]

It is worth concluding these comments about Wigner's style of intellectual influence with a remark about a characteristic that the reader has surely noted, namely, the non-dogmatic manner in which he dealt with subjects related to the foundations of quantum mechanics. Shimony was well situated to demonstrate this because his point of view on the role of the mind in quantum mechanics, different as it was from Wigner's, did not impinge on their close collaboration.[53] According to

---

[51] Wheeler to Aage Bohr, 25 Feb 1977; John Wheeler Papers, Series V, Notebook October 1976–April 1977, American Philosophical Society, Philadelphia, PA.

[52] Wheeler to John Hopfield, 2 May 1977; Aage Bohr to Wheeler, 16 May 1977. Ibid.

[53] In his first published paper on the foundations of quantum mechanics, Shimony (1963) analyzed these two proposals for interpreting the quantum state evolution during measurements: von Neumann's and Bohr's approaches. His point of view on the former was that "although this

Shimony (2004, p. 60), one of the salient features of Wigner's contribution to the measurement problems in quantum mechanics was "freedom from dogmatism, open-mindedness towards new ideas, [...] and in general an exploratory attitude regarding the frontiers of physics, other sciences, and of philosophy." Still, according to Shimony, "consequently, it is a historical error and a misunderstanding of his work, to speak of 'The Wigner solution to the measurement problem' without attention to his exploratory attitude."

One last example is related to Wigner's reaction to the approach suggested by H. Dieter Zeh (1970). This approach was critical both of Wigner's and of the Italian physicists' approach because both admitted the validity of Schrödinger's equation to describe the measurement devices, and according to Zeh, measurement devices are not closed systems to which such an equation could be applied. While Zeh had difficulty publishing his paper elsewhere, Wigner, upon receiving a preprint version of Zeh's paper, supported its publication in the first volume of *Foundations of Physics* and opened his Varenna keynote talk with six possible solutions to the measurement problem, Zeh's solution being the last. According to Zeh, a preliminary version (in German) of this paper had been rejected by several journals in 1967, "the usual answer being that 'quantum theory does not apply to macroscopic objects,'" a kind of answer based on Bohr's and Rosenfeld's point of view.[54]

Comparing Wigner's and Rosenfeld's styles as far as foundational issues are concerned, it may be said that Wigner intentionally supported a wide debate on the foundations of quantum mechanics, while Rosenfeld only unintentionally contributed to the opening of such a debate as he found there were no problems to be solved in these foundations.

Epilogue and Conclusion: Orthodoxy Becomes Heterodoxy

After the clash with Wigner, Rosenfeld continued working on the measurement processes in quantum theory. Privately, he acknowledged that the work of the Italian physicists had shortcomings concerning its expressions, as he wrote to F.J. Belinfante in the early 1970s.[55]

> I agree with your mild criticism of the Italian physicists whose method is, as you say, not entirely rigorous and also rather complicated. In fact this was the motivation for me to give the simplified and I think also more strictly correct exposition of their argument, which you quote.

(continued)

interpretation appears to be free from inconsistencies, it is not supported by psychological evidence and it is difficult to reconcile with the inter-subjective agreement of several independent observers."

[54] Wigner to Margenau [editor of *Foundations of Physics*], 31 March 1970, "I am really very glad that Zeh's paper was accepted." *MP*, box 1, folder 12. Zeh (1970) thanks Wigner for his support of his paper. For more information about the refusal of this paper, see Freire Jr. (2009, p. 281) and Chap. 8.

[55] L. Rosenfeld to F.J. Belinfante, 22 June 1972, *RP*.

Still in the early 1970s he began a collaboration with the Belgian physical chemist Ilya Prigogine and his colleague Claude George to better formulate the macroscopic character of the measurement devices from the physical point of view (George et al. 1972). They did not use the ergodic argument used by the Italian physicists. Instead they used the kinetic approach of statistical mechanics to deal with the time evolution of large atomic systems. According to Rosenfeld's biographer, Anja Jacobsen (2012, p. 312), in doing so "Rosenfeld's interpretation of the measurement process therefore appears to have deviated from Bohr's since Bohr had maintained that macroscopic things such as measuring devices should be described classically." This approach did not get a wider audience. However, macroscopicity would later become relevant in the studies on the processes called decoherence. Indeed coupling between quantum systems and their environments would be the key concept behind decoherence and one may roughly associate such environments to the macroscopic features of measurement devices. While decoherence did not solve the measurement problem (Pessoa Jr. 1998), it constituted a step towards its understanding thus vindicating, to a certain extent, some aspects of Rosenfeld's approach to quantum measurement processes.

Coming back to Wigner, let us conclude with three remarks: on Wigner's self-awareness of the role he played in the foundations of quantum mechanics; on the success of his ideas and action; and on a very different question, anachronism in the history of science. Shimony had the insight to record Wigner's feelings about the attitudinal changes he underwent. These changes may also help us understand changes in the *Zeitgeist* of physicists in the 1960s and early 1970s with respect to the foundations of quantum mechanics. By attempting to defend what he considered to be the "quantum orthodoxy," he in fact helped to legitimize heterodoxy on this subject, and he himself became a dissident. In Shimony's (1997, p. 412) words: "Wigner recognized with some relish a similarity between the 'heterodox' view that quantum mechanics is only approximate in the physical world and the 'orthodox' view that a reduction of the wave packet occurs only when there is a registration upon the consciousness of an observer." Shimony concluded, citing Wigner: "Both points of view come to the conclusion that the validity of quantum mechanics' linear laws is limited."

During the 1970s, the community working on the foundations of quantum mechanics was mainly occupied with another subject, Bell's inequalities and their experimental tests. Wigner was not as interested in this subject as in measurement problems, but he continued to play an active role until his intellectual vigor began to fade. However, physicists continued to work on the measurement problem research program and in the 1980s and 1990s it matured into the decoherence approach with its first experimental results in

(continued)

the middle of the 1990s. Where Wigner saw a role for the mind in quantum measurements, the current trend is to look for an exchange of information between the experimental devices and the environment (Zurek 1991; Haroche 1998). Wigner followed such developments and recanted both from his stance on the role of mind in quantum measurements and from the defense of a non-linear change in Schrödinger's equation. He stated that, as suggested by D. Zeh, macroscopic systems cannot be isolated from outside effects; according to his words "this shows that the probabilistic phenomenon enters not only when a living being observes, as I believed some time ago, but already if any macroscopic system plays a role" (Wigner 1995[1983], p. 136).[56] Wigner had already become more skeptical of the non-linear proposal. As he wrote to Shimony, in 1977, "there is only one point of 'Abner's views' with which I do not agree. It is the implication that all is needed is to make the equations of motion nonlinear. I believe that much more fundamental changes will be necessary—as they were when a description of electromagnetism was introduced or when microscopic physics, that is quantum mechanics, was created."[57] Today, Wigner's conjecture about the role of the mind in the quantum measurement process is no longer part of physics, but rather part of the history of physics. Nevertheless, the question persists and from time to time physicists devote some time to building technical arguments against it (Brandt 2002). In contrast, Wigner's research program—to understand from a physical point of view what quantum measurement is—has flourished and is part of physics; indeed it remains an open question in the foundations of quantum mechanics. Furthermore, in order to create this subfield of physics, foundations of quantum physics, it was necessary to break what Jammer called the "Copenhagen monocracy." As stated by the French physicist Alain Aspect (2004), a leader in the field of foundations of quantum physics and not at all a critic of the complementarity view, "questioning the 'orthodox' views, including the famous Copenhagen interpretation, might lead to an improved understanding of the quantum mechanics formalism, even though that formalism remained impeccably accurate." Wigner made major contributions to achieving this goal. As a token of recognition for his contributions, the "first comprehensive meeting on the [foundations of quantum theory] to be held in the United States" was dedicated to Eugene P. Wigner (Greenberger 1986, p. xiii).[58]

(continued)

[56] I am thankful to Frederik Santos for discussions about Wigner's withdrawal concerning the role of mind and non-linearity in quantum physics; see his dissertation (Santos 2010, pp. 53–57). Wigner's recantation is also noted by Michael Nauenberg (2007, p. 1614).

[57] Eugene Wigner to Abner Shimony, 12 Oct 1977, *WigP*, Box 83, folder 7.

[58] Incidentally, the homage speech to Wigner, given by Arthur Wightman (1986), is, as far as I know, the sole account of Schrödinger's cat experiment from the perspective of cats.

Current accounts of Wigner's contributions to the understanding of the foundations of quantum physics involve distinct perspectives. We have just seen statements from people who valued his contributions highly. Let us now examine opposing perspectives. Indeed, some contemporary physicists fail to appreciate Wigner's insight into the role of the mind in the measurement problem of quantum theory. They read Wigner's work neglecting to take into account the role he played in the context of the 1960s. The first version of this paper (Freire Jr. 2003b) appeared in a volume organized by the sociologist Boaventura de Sousa Santos as part of a late battle of the science wars which plagued culture and science in the 1990s. The paper was motivated by a question formulated by the Portuguese physicist António M. Baptista as part of his criticisms towards Santos. Baptista (2002, pp. 63–74) asked where a serious physicist had said consciousness plays a role in measurements. He had also written that Wigner speculated outside the boundaries of the natural sciences. Baptista was not alone in this. Earlier, the Physics Nobel Prize winner Murray Gell-Mann (1994, p. 155) had written: "... many sensible, even brilliant commentators have written about the alleged importance of human consciousness in the measurement process. Is it really so important?"

As we can conclude from this chapter, deprecating Wigner's contribution to the foundations of quantum mechanics constitutes an anachronistic reading of events. It is a reading based on the current state of art of the problems and it is a misreading of what actually happened at that time, the 1960s. A reference to consciousness was a legitimate issue in the emergence of the quantum measurement problem as a problem in physics. Furthermore, through his ideas and action Wigner was instrumental in enhancing physicists' awareness of the foundational issues in quantum physics. Judging any historically significant work by contemporary standards constitutes an anachronism and anachronism does not facilitate our understanding of how science really works as it gives rise to a distortion in the practice of science. It produces distorted images of an idealized science.[59]

One of the aims of the study of the history of science is to rectify anachronistic perceptions of science because, according to Lucien Febvre (1982), historians should prevent the sin of all sins—the unforgivable sin, anachronism. Historians know, however, that there is a tension implied. According to Marc Bloch and Lucien Febvre, the creators of new

(continued)

[59] At the time Santos invited me to participate in the book he was organizing, I thought the intrinsic historic worth of Wigner's case would be enough to justify its inclusion in a book organized to criticize the hubris of contemporary scientism and to suggest, instead, the role of prudent knowledge for a decent life, which was the theme of Santos' book (2003). Parts of the book in Portuguese were later translated into English; see Santos (2007) and Freire Jr. (2007). For Baptista's replica, see Baptista (2004, pp. 88–95).

perspectives for the historical disciplines, historians should ask questions about the past, and these questions may be provoked by contemporary questions. Thus, still according to Febvre, the metaphor of sin for anachronism should be further extended to its ultimate limits; as the original sin, anachronism may be a source of knowledge (Dumoulin 1986).[60]

References

Aaserud, F.: Sputnik and the Princeton 3—the national-security laboratory that was not to be. Hist. Stud. Phys. Biol. Sci. **25**, 185–239 (1995)

Anderson, P.: Considerations on Western Marxism. NLB, London (1976)

Arthur, R.T.W.: Quantum-mechanics, a half century later—Lopes, JL, Paty, M. Philos. Sci. **48**(1), 156–161 (1981)

Aspect, A.: Introduction. In: Bell, J.S. (ed.) Speakable and Unspeakable in Quantum Mechanics. Cambridge University Press, New York (2004)

Ballentine, L.E.: Statistical interpretation of quantum mechanics. Rev. Mod. Phys. **42**(4), 358–381 (1970)

Baptista, A.M.: O discurso pós-moderno contra a ciência—obscurantismo e irresponsabilidade, 2nd edn. Lisboa, Gradiva (2002)

Baptista, A.M.: Crítica da razão ausente. Gradiva, Lisboa (2004)

Bensaude-Vincent, B.: Langevin 1872–1946 science et vigilance. Belin, Paris (1987)

Bloch, M.: The Historian's Craft. Vintage Books, New York (1953)

Bohm, D., Bub, J.: A proposed solution of measurement problem in quantum mechanics by a hidden variable theory. Rev. Mod. Phys. **38**(3), 453–469 (1966a)

Bohm, D., Bub, J.: A refutation of proof by Jauch and Piron that hidden variables can be excluded in quantum mechanics. Rev. Mod. Phys. **38**(3), 470–475 (1966b)

Bohr, N.: Discussion with Einstein on epistemological problems in atomic physics. In: Schilpp, P.A. (ed.) Albert Einstein—Philosopher–Scientist, pp. 199–242. The Library of the Living Philosophers, Evanston (1949)

Brandt, H.E.: Deconstructing Wigner's density matrix concerning the mind-body question. Found. Phys. Lett. **15**(3), 287–292 (2002)

Christie, J.R.R.: The development of the historiography of science. In: Olby, R.C., Cantor, G.N., Christie, J.R.R. (eds.) Companion to the History of Modern Science, pp. 5–22. Routledge, London (1990)

D'Espagnat, B.: Two remarks on the theory of measurement. Suppl. al Nuovo Cimento **1**, 828–838 (1966)

[60] There is further significance in quoting Marc Bloch as there are some parallelisms between Bloch's and Santos' intellectual *démarches*. In "A Discourse on the Sciences," Boaventura de Sousa Santos (2001) took into account what seemed to him to be lessons from the natural sciences in order to reflect on the changing paradigms of social sciences. Marc Bloch (1953), in a beautiful but unfinished essay about the historian's craft, written sometime before being killed by the Nazis on June 16, 1944, on the order of Klaus Barbie, affirmed that our mental environment was not the same anymore. Quantum physics, relativity theory, and the kinetic theory of gases soundly changed the ideas we had formed about science, making it more flexible, he wrote. Bloch added that we were then better prepared to admit that historical knowledge, even without Euclidian proofs or immutable laws of repetition, could nevertheless aim to be named scientific. For an analogous argument, see Gaddis (2002).

d'Espagnat, B.: Foundations of Quantum Mechanics—Proceedings of the International School of Physics "Enrico Fermi". Academic, New York (1971)

Daneri, A., Prosperi, G.M., Loinger, A.: Quantum theory of measurement and ergodicity conditions. Nucl. Phys. **33**(2), 297–319 (1962) [reprinted in Wheeler, J.A., Zurek, W.H. (eds.) Quantum Theory and Measurement, pp. 1657–1679. Princeton University Press (1983)]

Daneri, A., Loinger, A., Prosperi, G.M.: Further remarks on relations between statistical mechanics and quantum theory of measurement. Nuovo Cimento B **44**(1), 119–128 (1966)

Doncel, M.G., Michel, L., Six, J.: Interview de Eugene P. Wigner sur sa vie scientifique. Arch Int Hist Sci **34**(112), 177–217 (1984)

Dumoulin, O.: Anachronisme. Dictionnaire des sciences historiques, pp. 34–35. A. Burguière (ed). Presses Universitaires de France, Paris (1986)

Duncan, A., Janssen, M.: (Never) Mind your p's and q's: Von Neumann versus Jordan on the foundations of quantum theory. Eur Phys J H **38**(2), 175–259 (2013)

Einstein, A., Podolsky, B., Rosen, N.: Can quantum-mechanical description of physical reality be considered complete? Phys. Rev. **47**, 777–780 (1935)

Esfeld, M.: Wigner's view of physical reality. Stud. Hist. Philos. Mod. Phys. **30**(1), 145–154 (1999)

Everett, H., Barrett, J.A., Byrne, P.: The Everett interpretation of quantum mechanics : collected works 1955-1980 with commentary. Princeton University Press, Princeton (2012)

Febvre, L.P.V.: The problem of unbelief in the sixteenth century, the religion of Rabelais. Harvard University Press, Cambridge, MA (1982)

Feyerabend, P.K.: On the quantum-theory of measurement. In: Körner, S. (ed.) Observation and Interpretation in Physics, pp. 121–130. Butterworths Scientific Publications, Dover, NY (1957)

Finkbeiner, A.: The Jasons: The Secret History of Science's Postwar Elite. Viking, New York (2006)

Freire Jr., O.: L'interprétation de la mécanique quantique selon Paul Langevin. La Pensée **292**, 117–134 (1993)

Freire Jr., O.: A story without an ending: the quantum physics controversy 1950-1970. Sci. Educ. **12**(5–6), 573–586 (2003a)

Freire Jr., O.: O debate sobre a imagem da ciência—a propósito das ideias e da ação de E. P. Wigner. Conhecimento Prudente para uma vida Decente—'Um Discurso sobre as Ciências' revisitado. B. S. Santos. Porto, Edições Afrontamento, pp 481–506 (2003b)

Freire Jr., O.: The historical roots of "foundations of quantum mechanics" as a field of research (1950–1970). Found. Phys. **34**(11), 1741–1760 (2004)

Freire Jr., O.: Orthodoxy and heterodoxy in the research on the foundations of quantum physics: E.P. Wigner's case. In: Santos, B.d.S. (ed) Cognitive Justice in a Global World: Prudent Knowledges for a Decent Life, pp. 203–224. Lexington Books, Lanham, MD (2007)

Freire Jr., O.: Quantum dissidents: research on the foundations of quantum mechanics circa 1970. Stud. Hist. Philos. Mod. Phys. **40**(4), 280–289 (2009)

Frisch, O.R., Frisch, O.R.: The conceptual problem of quantum theory from the experimentalist's point of view. In: Bastin, T. (ed.) Quantum Theory and Beyond—Essays and Discussions Arising From a Colloquium, pp. 13–21. Cambridge University Press, London (1971)

Gaddis, J.L.: The Landscape of History: How Historians Map the Past. Oxford University Press, New York (2002)

Gavroglu, K.: Fritz London a Scientific Biography. Cambridge University Press, Cambridge (1995)

Gell-Mann, M.: The Quark and the Jaguar: Adventures in the Simple and the Complex. W.H. Freeman, New York (1994)

George, C., Rosenfeld, L., Prigogine, I.: Macroscopic level of quantum-mechanics. Nature **240** (5375), 25–27 (1972)

Greenberger, D. (ed.): New techniques and ideas in quantum measurement theory, vol. 480. Annals of the New York Academy of Sciences, New York (1986)

Hall, J., Kim, C., Mcelroy, B., Shimony, A.: Wave-packet reduction as a medium of communication. Found. Phys. **7**(9–10), 759–767 (1977)

Hargittai, I.: The Martians of Science: Five Physicists Who Changed the Twentieth Century. Oxford UniversityPress, Oxford, New York (2006)

Haroche, S.: Entanglement, decoherence and the quantum/classical boundary. Phys. Today **51**(7), 36–42 (1998)

Heidelberger, M.: The Mind-body problem in the origin of logical empiricism: Herbert Feigl and psychophysical parallelism. In: Parrini, P., Salmon, M.H., Salmon, W.C. (eds.) Logical Empiricism: Historical and Contemporary Perspectives, pp. 233–262. University of Pittsburgh Press, Pittsburgh, PA (2003)

Heims, S.J.: John Von Neumann and Norbert Wiener: From Mathematics to the Technologies of Life and Death. MIT Press, Cambridge, MA (1980)

Home, D., Whitaker, M.A.B.: Ensemble interpretations of quantum-mechanics—a modern perspective. Phys Rep Rev Sect Phys Lett **210**(4), 223–317 (1992)

Institut international de coopération intellectuelle: Les Nouvelles théories de la physique. I.I.C.I, Paris (1939). Varsovie, 30 mai-3 juin 1938

Jacobsen, A.: Léon Rosenfeld—Physics, Philosophy, and Politics in the Twentieth Century. World Scientific, Singapore (2012)

Jammer, M.: The Philosophy of Quantum Mechanics—The Interpretations of Quantum Mechanics in Historical Perspective. Wiley, New York (1974)

Jauch, J.M.: Problem of Measurement in Quantum Mechanics. Helv Phys Acta **37**(4–5), 293–316 (1964)

Jauch, J.M., Piron, C.: Can hidden variables be excluded in quantum mechanics. Helv Phys Acta **36**(7), 827–837 (1963)

Jauch, J.M., Wigner, E.P., Yanase, M.M.: Some comments concerning measurements in quantum mechanics. Nuovo Cimento B **48**(1), 144–151 (1967)

Jha, S.: Wigner's "Polanyian" epistemology and the measurement problem: the Wigner-Polanyi dialog on tacit knowledge. Phys. Perspect. **13**(3), 329–358 (2011)

Jordan, P.: On the process of measurement in quantum mechanics. Philos. Sci. **16**, 269–278 (1949)

Körner, S.: Observation and interpretation—a symposium of philosophers and physicists. Butterworths, London (1957)

Kuhn, T.S.: Sources for history of quantum physics; an inventory and report [by] Thomas S. Kuhn [and others]. American Philosophical Society, Philadelphia (1967)

Lacki, J.: The early axiomatizations of quantum mechanics: Jordan, Von Neumann and the continuation of Hilbert's program. Arch. Hist. Exact Sci. **54**(4), 279–318 (2000)

Leggett, A.J.: Reflections on the quantum measurement paradox. In: Hiley, B., Peat, D. (eds.) Quantum implications—essays in honour of David Bohm, pp. 85–104. Routledge, London (1987)

Loinger, A.: Comments on a recent paper concerning quantum theory of measurement. Nucl. Phys. A **A108**(2), 245–249 (1968)

London, F., Bauer, E.: La théorie de l'observation en mécanique quantique. Hermann, Paris (1939). English translation in Wheeler & Zurek 1983, pp. 217–259

Ludwig, G.: Der Messprozess. Zeitschrift Fur Physik **135**(5), 483–511 (1953)

Macrae, N.: John von Neumann. Pantheon Books, New York (1992)

Margenau, H., Wigner, E.P.: Comments on Putnam comments—discussion. Philos. Sci. **29**(3), 292–293 (1962)

Mehra, J., Wigner, E.P.: A biographical sketch. In: Wigner, P. (ed.) The Collected Works of Eugene Paul Wigner, Part A, pp. 3–14. Springer, Berlin (1993)

Moore, K.: Disrupting Science: Social Movements, American Scientists, and the Politics of the Military, 1945-1975. Princeton University Press, Princeton (2008)

Nauenberg, M.: Critique of "quantum enigma: physics encounters consciousness". Found. Phys. **37**, 1612–1627 (2007)

Nye, M.J.: Michael Polanyi and His Generation: Origins of the Social Construction of Science. The University of Chicago Press, Chicago, London (2011)

Osnaghi, S., Freitas, F., Freire Jr., O.: The origin of the Everettian heresy. Stud. Hist. Philos. Mod. Phys. **40**(2), 97–123 (2009)

Pessoa Jr., O.: Can the decoherence approach help to solve the measurement problem? Synthese **113**, 323–346 (1998)

Pessoa Jr., O., Freire Jr., O., De Greiff, A.: The Tausk controversy on the foundations of quantum mechanics: physics, philosophy and politics. Phys. Perspect. **10**(2), 138–162 (2008)

Rosenfeld, L. (1953). L'évidence de la complementarité. In: George, A. (ed.) Louis de Broglie—physicien et penseur, pp. 43–65. Editions Albin Michel, Paris [A slightly modified English version of this paper is Strife about complementarity, Science progress, 163 (1953), 1393–1410, reprinted in Robert Cohen and John Stachel (eds.). Selected papers of Léon Rosenfeld (Dordrecht, D. Reidel, 1979)

Rosenfeld, L.: Measuring process in quantum mechanics. Suppl. Prog. Theor. Phys., 222–231 (1965)

Rosenfeld, L.: Questions of method in consistency problem of quantum mechanics. Nucl. Phys. A **A108**(2), 241–244 (1968)

Rosenfeld, L., Cohen, R.S., Stachel, J.J.: Selected Papers of Léon Rosenfeld. D. Reidel Publishing Company, Dordrecht (1979)

Santos, B.S.: Um discurso sobre as ciências, 12th edn. Afrontamento, Porto (2001)

Santos, B.S. (ed.): Conhecimento Prudente para uma Vida Decente—'Um Discurso sobre as Ciências' revisitado. Edições Afrontamento, Porto (2003)

Santos, B.S.: Cognitive Justice in a Global World: Prudent Knowledges for a Decent Life. Lexington Books, Lanham, MD (2007)

Santos, F.M.: Na fronteira entre a Ciência e a Filosofia: Reflexões Filosóficas de Eugene Wigner. Master Dissertation, Universidade Federal da Bahia (2010)

Santos, F.M., Pessoa Jr., O.: Delineando o Problema da Medição na Mecânica Quântica: o debate de E. P. Wigner e H. Margenau versus H. Putnam. Scientiae Studia **9**, 625–644 (2011)

Schrödinger, E.: The present situation in quantum mechanics. In: Wheeler, J.A., Zurek, W.H. (eds.) Quantum Theory and Measurement, pp. 152–167. Princeton University Press, Princeton, NJ (1983). Original publication in Naturwissenchaften 123, 807–812, 823–828, 844–849, 1935

Schweber, S.S.: Wigner's vast output: four of the eight planned volumes. Physics Today October, pp. 65–66 (1996)

Schweber, S.S.: Writing the biography of Hans Bethe: contextual history and Paul Forman. Phys. Perspect. **6**, 179–217 (2014)

Seitz, F., Vogt, E., Weinberg, A.: Eugene Paul Wigner. Biogr. Mem. Natl Acad. Sci. **74**, 364–388 (1998)

Shimony, A.: Role of observer in quantum theory. Am. J. Phys. **31**(10), 755–773 (1963)

Shimony, A.: Search for a Naturalistic World View. Cambridge University Press, Cambridge (1993). 2 vols

Shimony, A.: Wigner on Foundations of Quantum Mechanics. The Collected Works of Eugene Paul Wigner, Part A—Vol III, Particles and Fields; Foundations of Quantum Mechanics. Springer, Berlin (1997)

Shimony, A.: Wigner's contributions to the quantum theory of measurement. Acta Phys. Hung. B **20**, 59–72 (2004)

Süssmann, G.: An analysis of measurement. In: Körner, S. (ed.) Observation and Interpretation: A Symposium of Philosophers and Physicists, pp. 131–147. Butterworths, London (1957)

Von Neumann, J.: Mathematical Foundations of Quantum Mechanics. Princeton University Press, Princeton, NJ (1955)

Weinberg, A.M.: Eugene Wigner, nuclear engineer. Phys. Today **55**(10), 42–46 (2002)

Westfall, C.: Wigner, Eugene Paul. In: Koertge, N. (ed.) New Dictionary of Scientific Biography. 8, pp. 293–297. Thomson Gale, New York (2008)

Wheeler, J.A., Zurek, W.H.: Quantum Theory and Measurement. Princeton University Press, Princeton, NJ (1983)

Whitaker, M.A.B.: Theory and experiment in the foundations of quantum theory. Prog Quantum Electron **24**, 1–106 (2000)

Wightman, A.: E.P. Wigner: an introduction. In: Greenberger, D. (ed.) New techniques and ideas in quantum theory measurement. The New York Academy of Sciences, New York (1986). xv–xvii

Wigner, E.P.: Die Messung Quantenmechanischer Operatoren. Zeitschrift Fur Physik **133**(1-2), 101–108 (1952). Reprinted in Wigner (1995, pp. 1147–1154)

Wigner, E.P.: Remarks on the mind-body question. In: Good, I.J. (ed.) The Scientist Speculates: An Anthology of Partly-Baked Ideas, pp. 284–302. Heinemann, London (1961). Reprinted in Wheeler, J.A., Zurek, W.H. (eds.) Quantum Theory and Measurement, pp. 168–181. Princeton University Press, Princeton (1983)

Wigner, E.P.: Problem of measurement. Am. J. Phys. **31**(1), 6–15 (1963)

Wigner, E.P.: Philosophical Reflections and Syntheses. Springer, Berlin (1995)

Wigner, E.P.: The Probability of the Existence of a Self-Reproducing Unit. The Collected Works of E. P. Wigner, Part A – Vol III. Springer, Berlin (1997a)

Wigner, E.P.: Are We Machines? The Collected Works of E.P. Wigner, Part A—Vol III. Springer, Berlin (1997b)

Wigner, E.P.: The place of consciousness in modern physics. In: Wigner, E.P. (ed.) Philosophilcal Reflections and Syntheses, pp. 261–267. Springer, Berlin (1995[1972])

Wigner, E.P: Epistemological perspective on quantum theory. In: Wigner, E.P. (ed) Philosophical Reflections and Syntheses, pp. 55–71. Springer, Berlin (1995[1973])

Wigner, E.P.: The limitations of determinism. In: Wigner, E.P. (ed.) Philosophical Reflections and Syntheses, pp. 133–138. Springer, Berlin (1995[1983])

Wigner, E.P., Szanton, A.: The Recollections of Eugene P. Wigner as Told to Andrew Szanton. Plenum, New York (1992)

Zeh, H.D.: On the interpretations of measurement in quantum theory. Found. Phys. **1**, 69–76 (1970). Reprinted in Wheeler and Zurek, Quantum Theory and Measurement, pp. 342–349

Zurek, W.H.: Decoherence and the Transition from Quantum to Classical. Phys. Today **44**(10), 36–44 (1991)

Chapter 5
The Tausk Controversy on the Foundations of Quantum Mechanics: Physics, Philosophy, and Politics

Abstract In 1966 the Brazilian physicist Klaus Tausk (1927-2012) circulated a preprint from the International Centre for Theoretical Physics in Trieste, Italy, criticizing Adriana Daneri, Angelo Loinger,and Giovanni Maria Prosperi's theory of 1962 on the measurement problem in quantum mechanics. A heated controversy ensued between two opposing camps within the orthodox interpretation of quantum theory, represented by Léon Rosenfeld and Eugene P. Wigner. The controversy went well beyond the strictly scientific issues, however, reflecting philosophical and political commitments within the context of the Cold War, the relationship between science in developed and Third World countries, the importance of social skills, and personal idiosyncrasies.

5.1 Introduction

Klaus Stefan Tausk was born in Graz, Austria, on April 11, 1927, and emigrated as a youth with his Jewish parents to São Paulo, Brazil, in 1938. He is virtually unknown among physicists and historians of physics today, although he was one of the protagonists in a controversy that helped to establish the field of the foundations of quantum mechanics: In 1966, while working as a doctoral student at the International Centre for Theoretical Physics (ICTP) in Trieste, Italy, he circulated a preprint (Tausk 1966), based upon some original arguments, in which he criticized a paper that Adriana Daneri, Angelo Loinger, and Giovanni Maria Prosperi had published in 1962 (Daneri et al. 1962). The ensuing heated controversy went well beyond strictly scientific issues, with a number of prominent theoretical physicists, including Léon Rosenfeld, David Bohm, Josef Maria Jauch, Eugene P. Wigner, and John S. Bell taking sides in it. Tausk's work was eventually neglected and ultimately forgotten, even by those who used it to advance their own interpretations of quantum mechanics. Tausk's failure to be recognized for his achievement, in our view, can be attributed to his careless and aggressive

This chapter is based on the paper with the same title, co-authored by Osvaldo Pessoa Jr., Olival Freire Jr., and Alexis De Greiff, *Physics in Perspective* 10 (2008) 138–162. The original text had its format fitted to the editorial guidelines for this book. References were updated. Acknowledgements are in the original paper.

O. Freire Junior, *The Quantum Dissidents*, DOI 10.1007/978-3-662-44662-1_5

style of writing and to his inadequate social skills in communicating his ideas. It also was conditioned by the circumstances surrounding the ongoing debate on the foundations of quantum mechanics, in particular, the lack of respect for this field in the eyes of most physicists at the time, and by the controversial reputation of the ICTP owing to its questionable publication policy. The historian and philosopher of science Ernan McMullin has emphasized that scientific controversies involve much more than logical problems concerning hypotheses and evidence; they are social conflicts involving personality traits and other historical contingencies.[1] The Tausk controversy took place within the context of particular scientific, historical, philosophical, and political circumstances: The foundations of quantum mechanics had become controversial among physicists for a number of reasons, including the issues raised by the Bohr-Einstein debate of 1935, by Soviet criticisms of Bohr's principle of complementarity in the 1950s, and especially by David Bohm's "causal interpretation" of quantum mechanics of 1952 (Bohm 1952), which offered a deterministic picture based upon "hidden variables,"[2] an interpretation that was set within the political context of the Cold War, as he have discussed in Chap. 2. This impinged upon the Tausk controversy, as we shall see, but it also was affected by a different kind of politics, one that reflected attitudes of physicists working in scientific centers in developed countries toward those working in Third World countries, a tension that was mediated by the concrete institutional setting of the ICTP in Trieste, Italy.

5.2 Scientific Background

Tausk's preprint focused on the "measurement problem," one of the central problems in the foundations of quantum mechanics.[3] Setting aside the more heterodox proposals such as Bohm's causal interpretation of 1952 and Hugh Everett's relative-state interpretation of 1957,[4] by the late 1950s there were two orthodox points of view that divided theoretical physicists on the measurement problem.

On one side were physicists such as John von Neumann, Georg Süssmann, Josef Maria Jauch,[5] and Eugene P. Wigner, who described the measurement apparatus

[1] McMullin (1987, pp. 51–54 and 59–61). See also "Controversies," *Science in Context* 11 (2) (1998), 147–325, and Collins and Pinch (1993).

[2] See Cushing (1994, pp. 42–75).

[3] Jammer (1974, pp. 470–521) and d'Espagnat (1989, pp. 159–229).

[4] Bohm (1952) and Everett (1957).

[5] Jauch was born in Lucerne, Switzerland, on September 20, 1914, received his *Diplom* at the Federal Institute of Technology (*Eidgenössische Technische Hochschule*) in Zurich in 1938, and his PhD degree in theoretical physics at the University of Minnesota in 1939. He then returned to Zurich as an Assistant in theoretical physics (1940–1942), but then left again for the United States, where he was an Instructor and Assistant Professor of Physics at Princeton University (1942–1945), a research physicist at Bell Telephone Laboratories (1945–1946), and Associate and Full

used in quantum-mechanical experiments in an exact way (that is, without approximations) as a quantum system. Sometimes called the "Princeton school,"[6] they applied the Schrödinger equation (or another equivalent equation describing a unitary state evolution) to the composite system consisting of apparatus and quantum object, and concluded that such a description is insufficient to account for all aspects of the measurement process—a formal result that was an example of what became known as an "impossibility proof," which von Neumann first derived in 1932,[7] and which served to justify his "projection postulate" describing the discrete change of state as an indeterministic process that accompanies a measurement. The projection postulate thus was regarded as an independent principle, to be added to the five fundamental axioms (or six, if one considers indistinguishable particles) of nonrelativistic quantum mechanics (d'Espagnat 1989, pp. 14–29). The impossibility proof, as reformulated by Wigner (1963), prohibited the reduction of the projection postulate to the other fundamental axioms.

On the other side of the postwar debate were physicists such as Niels Bohr, Pascual Jordan, Günther Ludwig, Paul Feyerabend, H.S. Green, Adriana Daneri, Angelo Loinger, Giovanni Maria Prosperi, and Léon Rosenfeld, who argued that the measurement process can be described adequately by a statistical mechanics of quantum processes, which would amount to a *thermodynamic approach*. These physicists were closely allied to the orthodox Copenhagen interpretation and its central concept of complementarity,[8] but they proposed to modify it by introducing certain approximations in the limit of large numbers. Some, such as Jordan (1949), clearly pointed out the statistical hypothesis that was being used and proposed to simply substitute it for the projection postulate. Others, such as Daneri, Loinger, and Prosperi in 1962, argued that the approximations involved no fundamental physical principle, so the projection postulate could be eliminated and reduced to the other fundamental axioms. This clashed with the impossibility proof and stirred up the debate on the measurement problem, beginning with the papers that Feyerabend and Süssmann presented at the Colson Research Society Symposium at the University of Bristol, England, in 1957.[9]

The postwar thermodynamic approach arose as an "objectivist" alternative to the "idealistic" views that were widespread in the 1930s. Bohr clearly reflected this change when he stressed in 1958 that a measurement could be made in the absence of a conscious observer, "based on registrations obtained by means of suitable amplification devices with irreversible functioning...."[10] The idea was that a

Professor of Physics at the University of Iowa (1946–1959) before returning to his home country permanently in 1960 as Professor of Physics at the University of Geneva.

[6] Ballentine (1970, p. 360). See a more extensive discussion on these two fields about the measurement processes in quantum mechanics in Chap. 4.

[7] Von Neumann (1932, pp. 157–173); English translation in Von Neumann (1955).

[8] Jammer (1974, pp. 86–107 and 197–211).

[9] P.K. Feyerabend, "On the quantum-theory of measurement," and G. Süssmann, "An analysis of measurement," in Körner (1957, pp. 121–130 and 131–136).

[10] Niels Bohr, "Unity of Knowledge," in Bohr (1958, pp. 67–82, quote on p. 73).

measurement is an objective thermodynamic process. The problem that was left open was how to describe mathematically, in the most satisfactory way, the irreversible amplification process that leads from a microscopic event to a macroscopic registration.

Daneri, Loinger, and Prosperi (hereafter DLP), who were working in the Milan section of the Italian *Istituto Nazionale di Fisica Nucleare*, published the most ambitious theory of the thermodynamic-amplification approach in 1962 in the journal *Nuclear Physics* edited by Léon Rosenfeld (Daneri et al. 1962).[11] They divided the measurement process into two stages. First, the microscopic quantum object interacts with the apparatus as prepared in a "metastable" state, which produces a nonequilibrium state. Second, amplification takes place, which involves certain restrictions known as "ergodicity conditions" and which, as defined by Léon van Hove (1959), were weaker than those used earlier. They guaranteed that the system would return to equilibrium, according to the expected behavior of the measurement apparatus, *in the limit of an infinite amount of time*. Rosenfeld approved of DLP's theory, emphasizing the importance of the second, amplification stage.[12]

DLP's theory was the culmination of a series of investigations that Milanese and Pavian theoretical physicists, such as Loinger, Prosperi, Pietro Bocchieri, and Antonio Scotti, had undertaken since the end of the 1950s on the ergodic hypothesis and its applications in statistical mechanics and quantum mechanics. Specifically, their search for a more realist solution to the measurement problem, in opposition to von Neumann's, was inspired by the Italian theoretical physicist Piero Caldirola, who helped to popularize DLP's theory in 1965, which subsequently was widely cited in the literature.[13]

The thermodynamic approach gradually declined in importance, however, for two main reasons. First, Wigner's arguments of 1963, which as noted above were based upon the impossibility proof, undermined it (Wigner 1963). Second, Tausk's argument of 1966, as well as Jauch, Wigner, and Mutsuo M. Yanase's of 1967, which were based upon "negative-result measurements," as had been discussed by physicist Mauritius Renninger at the University of Marburg, Germany, in 1960, also undermined it.[14] Consider the following example. Imagine an experiment in which a quantum-mechanical object (we will call it a "particle" but will not require it to have a well-defined position) strikes with equal probability one of two detectors placed in paths *A* and *B*. Now suppose that the detector in path *A* is removed and the particle is sent to the apparatus. If after a certain time the observer sees no signal at the detector in path *B* (assuming that the detectors are perfectly

---

[11] Reprinted in Wheeler and Zurek (1983).

[12] Rosenfeld (1965, pp. 225, 230); reprinted in Rosenfeld et al. (1979, pp. 536–546, especially pp. 539–540 and 545).

[13] Caldirola (1965) and Garuccio and Leone (2002, pp. 66–68 and 78–90). Until January 2014, DLP's paper was cited 180 times in the literature; see the ISI Web of Science.

[14] Tausk (1966), Jauch et al. (1967, pp. 150–151) and Renninger (1960).

efficient), the observer then would conclude that the particle traveled along path A, which amounts to a state reduction or collapse. No amplification occurred, however, which clearly shows that amplification is not a necessary condition for state reduction or collapse (although in practice it might be a sufficient condition).

Although Tausk's 1966 argument, which was based upon such negative-result measurements, was seen by many as a knock-out argument against DLP's theory following Jauch, Wigner, and Yanase's paper of 1967, Loinger defended it in 1968, showing that it did not require amplification (Loinger 1968). Their formalism required only that a coupling had to exist between quantum object and detector, a situation that Robert H. Dicke clarified much later, in 1981 (Dicke 1981).

5.3 Tausk in Trieste

Tausk studied physics at the University of São Paulo, Brazil, from 1947 to 1951, and later worked there on cosmic-ray experiments with the Czech physicist Kurt Sitte in 1953–1954.[15] He also became acquainted with David Bohm, who worked there from October 1951 to January 1955, although Tausk later claimed that he was not influenced by Bohm's causal interpretation, because he did not have an adequate understanding of quantum mechanics at the time.[16] Tausk then interrupted his studies for a few years, beginning graduate research in 1958, which included a year in Hamburg (1959–1960) to work with Harry Lehman on quantum-field theory. Tausk also met Georg Süssmann in Hamburg, who was on a visit from Frankfurt, and who was doing significant work on measurement theory in quantum mechanics (Süssmann 1958).

Returning to São Paulo in 1962, Tausk read a paper of 1960 by Hitoshi Wakita on the measurement problem in quantum mechanics (Wakita 1960), which stimulated his interest in the subject, and he also came across Renninger's paper of 1960 on negative-result measurements. Renninger had used the thought experiment noted above to criticize the Copenhagen interpretation of quantum mechanics, denying that every measurement produces an uncontrollable disturbance on the observed object (Renninger 1960).[17] Tausk too then began to question the Copenhagen interpretation and to work on the measurement problem.

---

[15] On Tausk and Sitte work on cosmic-rays, see Andrade (2004). Sitte was born in Reichenberg, Bohemia (Liberec, Czechoslovakia) on December 1, 1910, and received his PhD degree in physics at the German University of Prague in 1933. As a non-Jew but outspoken left-wing anti-Nazi, he was arrested in Prague immediately after the German invasion of Czechoslovakia in March 1939, was imprisoned in the Dachau and Buchenwald concentration camps, and was liberated in April 1945. After the war, he had appointments at the Universities of Edinburgh and Manchester (1946–1948) and at Syracuse University (1948–953) before accepting a visiting professorship at the University of São Paulo, Brazil (1953–1954) and subsequently an appointment at the Technion in Haifa, Israel. Later he was imprisoned in Israel convicted of espionage favoring the USSR.

[16] Osvaldo Pessoa Jr., interviews of Klaus S. Tausk, 1991 and 1999. On Bohm, see Chap. 2.

[17] For a discussion of Renninger's work, see Jammer (1974, pp. 495–496).

In 1965 Tausk wrote to Abdus Salam, Director of the International Centre for Theoretical Physics (ICTP), in Trieste, Italy, presenting himself as a doctoral student of the renowned Brazilian theoretical physicist Mario Schönberg, and received a scholarship to work at the ICTP. The ICTP had been created in June 1963 as a division of the International Atomic Energy Agency (IAEA) with the support of UNESCO.[18]

The ICTP was in a delicate situation at the time, because it had been created over the opposition of India, the Soviet Union, the United States, and most of the developed countries. The Swedish physicist Sigvard Eklund, Director of the IAEA, was a friend of Rosenfeld, who during the negotiations to create the ICTP had proposed that it be located in Copenhagen, not Trieste. Rosenfeld and his Danish colleagues felt that the IAEA should support regional institutions such as NORDITA in Copenhagen, because they were skeptical about supporting a center for theoretical physics in Trieste whose goal was to create a scientific elite in Third World countries.[19]

Tausk spent just over a year at the ICTP in Trieste, from the middle of 1965 until the end of September 1966. He had applied to the ICTP to carry out research on quantum-field theory, but he actually continued his studies on the measurement problem. Toward the end of his stay, he finished writing a paper entitled "Relation of Measurement with Ergodicity, Macroscopic Systems, Information and Conservation Laws" (Tausk 1966), in which he criticized the aforementioned paper by Daneri et al. (1962), as well as the orthodox Copenhagen interpretation of quantum mechanics, especially the version of it that Werner Heisenberg had published in 1958 (Heisenberg 1958, pp. 44–58). He also criticized the preprint of a new paper by the Italian trio that circulated in February 1966 (Daneri et al. 1966). Tausk wrote his paper as a thesis to be submitted to the International Advanced School of Physics, a division of the ICTP under the directorship of Luciano Fonda.[20] It began to circulate among physicists in August 1966.

As a scientist working at the ICTP, Tausk had the right to request that his paper be typed and fifty copies printed, without any refereeing, as an internal report of the ICTP. Contrary to the usual procedure, however, Tausk added an official ICTP cover to each copy of the report. He soon apologized to Rosenfeld for this breach in procedure, blaming it on his "ignorance of the regulations, a series of misunderstandings and to the absence of part of the staff from the Centre at the time...."[21]

Tausk distributed his report as a preprint to a number of theoretical physicists, including Süssmann in Frankfurt, Germany; the Argentinian Daniele Amati, who

[18] United Nations Educational, Scientific, and Cultural Organization.

[19] Nordisk Institut for Teoretisk Fysik (NORDITA). On the creation of the ICTP, see Greiff (2002).

[20] On Tausk's research proposal, see Luciano Fonda, "Report [to Salam] on the Fellows of the Centre," February 10, 1967, D.1713, International Advanced School of Physics, ICTP, Trieste.

[21] Klaus Tausk to Rosenfeld, 10 Oct 1966, Léon Rosenfeld Papers (RP hereafter), Niels Bohr Archive, Copenhagen.

had studied a few years in Rio de Janeiro and now worked in Trieste; the South African Jeffrey Bub, who had received his PhD degree under Bohm at Birkbeck College, University of London, in 1966 and now was a Research Specialist at the University of Minnesota in Minneapolis; and the French Marxist Jean-Pierre Vigier in Paris, whom Tausk had met in São Paulo in 1954 while Vigier was working with Bohm there. Tausk encountered Vigier again in Trieste, who extended an offer to Tausk to work with him and Louis de Broglie on the measurement problem at the *Institut Henri Poincaré* in Paris.

Tausk also sent a copy of his preprint to Loinger, who was now at the University of Pavia, and one also came into the hands of Rosenfeld at NORDITA in Copenhagen. In 1965 Rosenfeld had written a paper explicitly defending DLP's theory (Rosenfeld 1965), which Tausk also had criticized. Loinger and Rosenfeld now not only disagreed with Tausk's preprint, they were enraged by it.

Daneri, Loinger, and Prosperi (DLP) had considered their work to be "an indispensable completion and a natural crowning of the basic structure of present-day quantum mechanics," being "firmly convinced that further progresses in this field of research will consist essentially in refinements" of their approach (Daneri et al. 1966, p. 127). Note the rhetorical aspect of their immodest claims. Tausk, however, now declared that, contrary to DLP's claims, "no connection between ergodicity and reduction of state has been established," and he pointed to a class of measurements, Renninger's negative-result measurements, "for which ergodicity considerations are obviously irrelevant." Tausk bluntly concluded: "Recent claims by the same authors ... and L. Rosenfeld ..., which hold this attempt to be of fundamental importance, are thereby contradicted" (Tausk 1966, abstract). We comment further on Tausk's arguments and style of writing in the Appendix.

5.4 Loinger's and Rosenfeld's Attacks

Loinger was the first to react to Tausk's preprint. On September 9, 1966, he wrote an open letter to Gilberto Bernardini,[22] President of the *Società Italiana di Fisica* (SIF), requesting that it be published in the *Bollettino della S.I.F.* In it Loinger deplored the increasing number of worthless preprints that were being sent out from various institutions (implying especially the ICTP in Trieste), and were then being submitted for publication to *Il Nuovo Cimento*, the official journal of the SIF. To combat this pernicious practice, Loinger offered two suggestions: First, *Il Nuovo Cimento* should publish the title, author, and institution of all papers it rejected for publication, thus forcing irresponsible institutions to control the quantity of worthless papers they released for publication. Second, the SIF should institute an annual

[22] Angelo Loinger [in Italian] to President of the Società Italiana di Fisica, Pavia, 9 Sept 9, 1966, Klaus S. Tausk Personal Archive, São Paulo (KST, hereafter).

"antiprize" (*antipremio*) for the worst preprint written in Italy—and the preprint that should be selected for the first antiprize, lest it "escape," was Tausk's!

Loinger's attack thus was directed not only at Tausk's preprint, but also at the entire ICTP in Trieste. A couple of weeks later, on September 22, 1966, Loinger also sent an open letter to the widely circulated Italian magazine *L'Europeo*, questioning the financial support that the Italian government was providing to the ICTP, and criticizing the doubtful rigorousness of the papers emanating from it. His view here was a common one among European and American physicists. In fact, the absence of internal control over papers being written at the ICTP was intentional: Abdus Salam, its Director, wanted to maximize the publication opportunities of scientists from Third World countries.[23]

On September 20, 1966, Rosenfeld wrote a letter to Salam, calling Salam's attention to Tausk's preprint. Rosenfeld began by implicitly but clearly questioning the publication policy of the ICTP:

> From the inexhaustible flow of preprints from your Institute I picked out the other day one with the somewhat bombastic title "Relation of Measurement with Ergodicity, Macroscopic Systems, Information and Conservation Laws" by a certain K.S. Tausk.[24]

That opening sentence, coming from a leading theoretical physicist who earlier had questioned locating the ICTP in Trieste, certainly appears to have been an attempt to intimidate Salam, who was constantly striving to demonstrate that the ICTP was worthy of support on the basis of its scientific merits. But Rosenfeld went much further, declaring that Tausk's preprint

> is such incredible thrash [*sic*] that I hardly could believe my eyes when I read it. I feel that I ought to write you about it in the event that (as I hope) this masterpiece has just escaped your attention.... The author is, I suppose, very young and inexperienced; one good turn you could do him, since you presumably know him better than I do, would be to represent that before blandly assuming that the trivialities which fill his paper could have been overlooked by such people as Niels Bohr and Heisenberg, he might perhaps reflect that *he* could be the one who misses the point.[25]

Note again Rosenfeld's inference that there was a lack of control at the ICTP over the preprints that were being sent out under its banner.

Salam replied to Rosenfeld 1 week later, saying "I wish to tender to you my sincerest apologies for Mr. Tausk's paper which reached you."[26] He explained the ICTP's rules governing the distribution of preprints, and how Tausk had managed to put an ICTP cover on his internal report.

> Mr. Tausk is a special pupil of Mario Schönberg in Brazil. I have not had a chance to see him yet. He is due to leave us at the end of this month to join the Vigier group in Paris. I

---

[23] A. Loinger, "Scienza e quattrini," *L'Europeo* 39 (September 22, 1966), 3. De Greiff (2001, Chap. 6).

[24] Rosenfeld to Salam, Copenhagen, 20 Sep 20 1966, KST.

[25] Ibid.

[26] Salam to Rosenfeld,Trieste, 26 Sep 1966, RP.

> would request you that you may consider this episode as part of the old battles and in no case an expression of opinion from the Centre here.[27]

These "old battles" were the earlier battles over the interpretation of quantum mechanics. Rosenfeld was appeased, writing to Salam one week later:

> Since, however, this is clearly a case of lack of foresight with no evil intent on his [Tausk's] part, I think one ought not be too severe with him and rather dismiss the whole matter without more ado. I am glad to know (for the centre's sake) that Tausk's paper will not receive more publicity from the centre, but I have no illusions about what the Vigier group is going to do with it. However, this is another story.[28]

Rosenfeld had succeeded in neutralizing Salam. Tausk no longer would be supported by the Director of the ICTP, the institution where he had written his preprint and from which he had circulated it. Thus began Tausk's isolation from the community of theoretical physicists.

5.5 Bohm's, Jauch's, and Fonda's Defenses of Tausk

Meanwhile, Luciano Fonda, Director of the International Advanced School of Physics, a division of the ICTP, had written to two experts on the foundations of quantum mechanics, David Bohm at Birkbeck College, University of London, and Josef Maria Jauch at the University of Geneva, asking them for their opinions of Tausk's preprint. Bohm responded in a short handwritten letter on September 26, 1966, also sending copies to Salam, Tausk, and Paolo Budini, Deputy Director of the ICTP, saying: "I have read Dr. Tausk's paper, and I feel that what he writes is correct. I myself would suggest that he should publish his paper as a short article."[29] A week later Bohm also wrote a three-page typed letter to Tausk, clarifying "the confusion between the individual and the ensemble, which is contained in the argument of DLP."[30] Given Bohm's heterodox position on the foundations of quantum mechanics, however, it is not clear whether his support was helpful or unhelpful to Tausk. In any case, his opinion perhaps did not carry much weight among most quantum theorists at the time.

Jauch responded to Fonda on October 4, 1966, declaring that a "criticism of the paper by Daneri et al. is certainly most useful," and agreeing with Tausk's conclusion that "no connection between ergodic properties of the measuring apparatus and the reduction of state has been established by DLP."[31] Jauch noted, however, that certain statements in Tausk's paper were unclear and a few arguments badly

[27] Ibid.

[28] Rosenfeld to Salam, Copenhagen, 4 Oct 1966, RP.

[29] Bohm to Fonda, with copies to Salam, Budini, and Tausk, London, 26 Sep 1966, KST.

[30] Bohm to Tausk, London, 1 Oct 1966, KST.

[31] Jauch to Fonda, Geneva, 4 Oct 1966, RP; the conclusion Jauch quoted was from Tausk (1966, p. 22).

constructed, and he complained that Tausk had failed to cite earlier work, in particular Wigner's (1963) and Jauch's (1964) papers.[32] "In conclusion, I should say that a paper in this form would not be permitted to leave my institute. In [*sic*] the other hand a criticism of Daneri et al. is necessary and could be made in a more objective and dignified way on several grounds."[33]

Meanwhile, Tausk had spoken to Salam, who showed him Rosenfeld's letter of October 4, 1966. Six days later, Tausk wrote directly to Rosenfeld, assuming responsibility for having broken the ICTP's publication rules, but then adding:

> Fortunately for my reputation your opinion about my paper is not universal among those who have given serious thought to the problem of measurement: Prof. David Bohm thinks that what I wrote is correct, and he advised me to publish it. Prof. Louis de Broglie has sent me one of his books with the inscription "avec l'homage de l'auteur" in acknowledgment of this paper. A letter from Prof. G. Süssmann contains the following: "I have read your paper with great interest. What you have said about DLPI and about Rosenfeld's commentary seems to me to be completely evident."[34]

In view of Bohm's and Jauch's letters to Fonda, Daniele Amati, Paolo Budini, and Fonda wrote an open letter on behalf of the International Advanced School of Physics of the ICTP to the *Società Italiana di Fisica* (SIF), arguing that it would be a mistake for the SIF to establish an antiprize for the worst paper published in *Il Nuovo Cimento* because

> it could easily be the cause or the effect of personal issues. For example, the work of Tausk, indicated by Loinger as worthy of the year's antiprize, contains a severe criticism of a paper by Loinger himself, coauthored by Daneri and Prosperi.[35]

They then summarized Bohm's and Jauch's opinions of Tausk's preprint, which prompted an immediate and angry response from Loinger in Pavia, who wrote to the President of the SIF on October 20 regarding their "stupefying open letter," and concluded that "if Bohm and Jauch have really declared, with respect to the aforementioned masterpiece, what Amati, Budini, and Fonda claim, then they lost an excellent opportunity to remain silent."[36]

Three days earlier, on October 17, Fonda had written to Tausk:

> I have received the answer from Jauch and I see that he agrees with you on your criticism to Loinger's paper. I have agreed with professor Budini that your paper will be supported by the Advanced School of Physics; however, in that case we want you to take into account the

[32] The missing citations were Wigner (1963) and Jauch (1964).

[33] Jauch to Fonda, 4 Oct 1966, ibid.

[34] The Roman numeral I in DLPI denotes Daneri, Loinger, and Prosperi's first paper (1962) in contrast to their second paper of 1966. Tausk to Rosenfeld, 10 Oct 1966, RP. Tausk is quoting Georg Süssmann [in German] to Tausk, Frankfurt, 16 Sep 1966, KST. Süssmann's original German is: "Ihre Arbeit habe ich mit grossem Interesse gelesen.Was Sie zu DPLI und zu Rosenfelds Kommentar sagen, leuchtet mir durchaus ein." We are grateful to Ernst Hamburger for the translation of the German letters and texts for us.

[35] Daniele Amati, Paolo Budini, and Luciano Fonda in Italian to President of the Società Italiana di Fisica,Trieste, 11 Oct 1966, RP.

[36] Angelo Loinger [in Italian] to President of the Società Italiana di Fisica, Pavia, 20 Oct 1966, RP.

> suggestions and criticism of professor Jauch to your manuscript. Once you have revised your manuscript, please send it to me and I will forward it to the journal you prefer.[37]

Tausk later claimed that he never received a copy of Jauch's letter of October 4 to Fonda.[38] He never revised his manuscript and did not return it to the International Advanced School of Physics to be forwarded for publication. He claimed that he did submit an article on his work to the *American Journal of Physics*, but that its editor had received negative reports from two referees and hence had declined to publish it.

Picture 5.1 Klaus Tausk (1927–2012)

5.6 Further Developments

Tausk did not know that Daniele Amati had sent his preprint to the Northern Irish theoretical physicist John Stewart Bell at CERN[39] in Geneva, and that Bell also had received reprints of Loinger's papers. Bell commented to Loinger on his and his colleagues' work in a letter of October 26, 1966:

> It appears to me that ergodicity is relevant in showing the approximate absence of interference phenomena with macroscopically different states. But I think that nobody doubted this, and so am unable to attach fundamental importance to the formal discussion. This feature of large systems is for me about as relevant to the question of principle as is, for example, apparent macroscopic irreversibility to the question of reversibility of the fundamental Hamiltonian.... I am unable to accept all the details of Tausk as justified criticism of your paper. But I think his main points are right, and his general position sound.[40]

[37] Fonda to Tausk, Trieste, 17 Oct 1966, KST.

[38] Tausk's interviews with O. Pessoa, and Jauch to Fonda, 4 Oct 1966, RP.

[39] Conseil Européen pour la Recherche de Nucléaire

[40] Bell to Loinger, Geneva, 26 Oct 1966, RP.

Loinger replied immediately and harshly: "Dear Prof. Bell, I think that you have not understood the essence of the problem of quantal measurement. Yours sincerely, A. Loinger."[41] By this time, of course, owing to the groundbreaking papers that Bell had published two years earlier, he was becoming known as the most profound theoretical physicist working on the foundations of quantum physics.[42]

In the meantime, Jeffrey Bub also had received a letter from Loinger criticizing some aspects of a paper that he and Bohm had published in 1966 (Bohm and Bub 1966) and, in his reply to Loinger, Bub had reproduced a number of Tausk's ideas. Thus, when Bub acknowledged the receipt of Tausk's preprint on November 15, 1966, he told Tausk that it had "clarified several points which I had not understood properly before."[43] In fact, Bub was the only person ever to cite Tausk's preprint in the literature—in an article of 1968 in which he criticized DLP's theory of measurement, declaring that, "Certain aspects of the following analysis have been influenced by a critical article on the D-L-P theory by K.S. Tausk..." (Bub 1968, p. 505, n. 10).

Jauch, Wigner, and Yanase thoroughly criticized DLP's theory in a paper they submitted for publication in late November 1966 (Jauch et al. 1967). They noted that DLP did not address the problem of negative-result measurements, but in this connection they did not mention Tausk, the first theoretical physicist to make this criticism. Jauch, as noted above, had become aware of Tausk's criticism when he reviewed Tausk's preprint, and had informed Wigner of it in a letter of September 16, 1966, saying:

> I should perhaps mention that there has recently appeared an internal report from Trieste (ICTP internal Report 14/1966) written by K.S. Tausk which criticizes the paper by Daneri et al. rather severely. This paper contains some interesting points which should perhaps also be discussed in our paper.[44]

Wigner never mentioned Tausk's preprint in subsequent letters to Jauch. Further, Wigner wrote the first draft of their joint paper with Yanase, while Jauch made the final modifications to it.[45] Jauch therefore should at least have introduced a citation to Tausk's preprint, which was known to him but not to Wigner and Yanase, but he did not do so. Franco Selleri at the University of Bari, Italy (to whom we shall return below) later commented caustically: "This is a further example (I had some myself) of how some well known physicists are eager of appropriating contributions coming from authors when they judge it safe to do so."[46]

[41] Loinger to Bell, Pavia, 31 Oct 1966, RP.

[42] The papers are Bell (1964) and Bell (1966), reprinted in Bell (2004).

[43] Bub to Tausk, Minneapolis, 15 Nov 1966, KST.

[44] Jauch to Wigner, 16 Sep 1966, Eugene P. Wigner Papers, Manuscripts Division, Department of Rare Books and Special Collections, Princeton University Library, Box 71, Folder 3.

[45] Wigner to Jauch, 6 Sep 6, 5 Oct, 25 Oct 1966, Wigner Papers, ibid., Box 94, Folder 7; Wigner to Jauch, 22 Nov 1966, ibid., Box 71, Folder 3; Jauch to Wigner, 13 Oct 1966, ibid., Box 71, Folder 3; and Jauch to Wigner, 16 Sep 1966, Box 71, Folder 3, ibid.

[46] Franco Selleri, "Comments on the Thesis 'A Medida na Mecânica Quântica' by K.S. Tausk," 1972, 2 pp. KST.

It may be that Jauch and Wigner, the two senior authors of their joint paper with Yanase, failed to cite Tausk's preprint because of his vaguely unfavorable image in the eyes of some European theoretical physicists, who pictured him as an unqualified Third World physicist, a polemist who criticized the orthodox interpretation of quantum mechanics without understanding it and, moreover, sympathized with the views of the French Marxist Vigier and his group in Paris. Perhaps Jauch and Wigner also did not want to align themselves and their work with criticisms that had been advanced in one of the numerous ICTP preprints, especially in one that had been written by a virtually unknown theoretical physicist.

5.7 Return to Brazil

After leaving the ICTP in Trieste, Tausk spent some time in his hometown of Graz, Austria, and then returned to São Paulo, Brazil, to finish his PhD thesis. His advisor, Mario Schönberg, was extremely angry with him owing to the "scandal" he had precipitated in Europe. Schönberg and Rosenfeld were old friends; both were experts on cosmic-ray physics, and both were Marxists who were involved in international peace movements. Schönberg also was a close friend of the Milanese physicist Piero Caldirola, whom he had met in Rome in 1938.[47] Schönberg evidently heard what his student Tausk had done at the ICTP from Rosenfeld, Salam, Caldirola, or someone else—and was greatly distressed and embarrassed by it.

Working alone, Tausk finished a draft of his thesis in 1967, writing it in Portuguese. In addition to the material in his controversial preprint, Tausk included a chapter in which he showed (possibly for the first time) that nonlocality in correlated systems cannot be used to transmit signals. Sometime later that year, in a first discussion of his work before an advisory committee (equivalent to a qualifying examination) whose members included Schönberg and the other Brazilian physicists Antônio Piza and Yogiro Hama, Schönberg severely criticized Tausk's work. The committee concluded that Tausk's thesis could not be defended as it was; Tausk would have to rewrite certain parts of it.[48]

Tausk did, and a few months later in 1967 defended his thesis (Tausk 1967), which turned out to be another traumatic experience for him. His advisor, Schönberg, refused to attend his defense because, according to Tausk, Schönberg would not talk to him. Tausk's thesis examination board, which included the important Brazilian theoretical physicist Jorge Swieca at the University of São Paulo, who was highly critical of Tausk's work, almost flunked him. The only

[47] Caldirola (1984, p. 228) wrote in a Festschrift to Schönberg, "the author never forgot the precious advices received from Mario at the beginning of his scientific career in 1938 at Roma University." For a biographical note on Schönberg, see Fernandes et al. (2008). Schönberg's scientific papers are in Schönberg and Hamburger (2009), Schönberg and Hamburger (2013).

[48] Tausk's interviews with O. Pessoa.

Brazilian physicist who read and approved of Tausk's work, according to Tausk,[49] was the renowned experimental physicist Cesare Lattes at the new University of Campinas, who telephoned Tausk after his defense, asking Tausk to send him a copy of his thesis, which Lattes read overnight after he received it and then telephoned his approval of it. Lattes's favorable judgment, however, probably did not greatly influence the opinion of other Brazilian physicists.

Five years later, in 1972, Franco Selleri at the University of Bari, who was then deeply involved in examining the foundations of quantum theory, visited the University of São Paulo on the invitation of the theoretical physicist Henrique Fleming. While there, Selleri wrote a review of Tausk's thesis whose tone was similar to Bohm's and Jauch's: Selleri pointed out certain misunderstandings of Tausk, but overall he was sympathetic to Tausk's views. Thus, he noted that there were four weak points in Tausk's thesis but also eight original contributions in it, concluding that:

> Tausk's thesis was very interesting reading and many physicists could no doubt benefit from it, once the philosophical ambiguities are cleared up. With more self-criticism Tausk probably will be able to contribute significantly to the understanding of the structure of the physical world.[50]

5.8 Tausk's Preprint and the Rosenfeld-Wigner Dispute

Loinger's and Rosenfeld's angry reactions to Tausk's preprint cannot be fully understood unless we consider the dispute on the measurement problem in which they were involved at the time. Thus, the thermodynamic-amplification program for solving the measurement problem, which had arisen in the 1950s and early 1960s, reached its most developed form in DLP's theory. Rosenfeld supported their theory, but a few other theoretical physicists criticized it, especially Wigner, who followed von Neumann's approach, describing the measurement apparatus plus quantum object as a quantum-mechanical closed system, and suggesting that human consciousness plays an ineluctable role in the reduction of the wave packet.[51]

As we have seen in Chap. 4, in response to Wigner's and other criticisms, Daneri, Loiner, and Prosperi published a second paper in 1966, raising the temperature of the controversy by declaring that Wigner, Abner Shimony, P.A. Moldauer, Yanase, Jauch, and others had not made "new substantial contributions to the subject [the measurement problem]" (Daneri et al. 1966, p. 120). Their paper, which Tausk criticized in his preprint, also upset Wigner, who wrote to Jauch on September 6, 1966:

[49] Ibid.

[50] Selleri, "Comments on the Thesis", op. cit.

[51] Wigner (1963, p. 7). For Wigner's reaction to the DLP's papers, see also Chap. 4.

> I just finished reading the article of Daneri Loinger and Prosperi in the July issue of Nuovo Cimento and am really a bit irritated by it. First of all, it is not good taste to say about a set of articles that they do not make substantial contributions to a subject. Needless to say, I am less concerned about myself than about other people who are much younger than I am and whose future careers such statements may hurt.... I am also saddened by Rosenfeld's endorsement of the article which, after all, considers it axiomatic that macroscopic systems have only states which can be described by classical mechanics. This is, of course, in conflict with quantum mechanics...[52]

Wigner, in particular, was concerned that the future careers of Shimony and Yanase, his former doctoral students at Princeton University, might be damaged by DLP's attack. Three months later, on December 1, 1966, Jauch, Wigner, and Yanase (now at Sophia University in Tokyo) submitted their detailed response to *Il Nuovo Cimento* for publication (Jauch et al. 1967).

The Austrian-English experimental physicist Otto Robert Frisch called attention to this dispute in his opening lecture at a meeting on the foundations of quantum theory in 1968:

> I understand that at present there exists a controversy, roughly speaking between a group of people which includes Wigner as the best known person and another group centred on Milan in Italy [DLP], and that these two have different views on how this reduction [of the wave packet during a measurement] happens. (Frisch 1971, p. 14)

The alignment of Wigner on one side of the dispute and of Rosenfeld on the other reflected their different intellectual heritages on the foundation and interpretation of quantum theory, with Wigner defending von Neumann's point of view and Rosenfeld Bohr's, Wigner stressing the axiomatization of quantum mechanics and Rosenfeld a more phenomenological approach. Their dispute, however, also reflected their divergent philosophical and political commitments, Wigner being a right-wing idealist who supported the American-Soviet arms race, and Rosenfeld being a left-wing Marxist who supported nuclear disarmament.[53] This division among American and European quantum theorists was common at the time. That it affected the controversy precipitated by Tausk's preprint is clearly indicated in a letter that Frisch wrote to Hugo Tausk, who was both Frisch's cousin and Klaus's father, on September 16, 1967:

> I have occupied myself a few times with Tausk's work, but I am not a theoretician and could not follow it. The questions which he addresses (essentially the question of the reality of the external world) seems to me very interesting. The orthodox Copenhagen interpretation says that physics does not deal with things but with measurements. That tastes like idealism, and is therefore rejected by the communists. Vice versa also applies, since anyone here in the West who doubts the orthodox interpretation—even for objective reasons—is suspect of communism. All this with the complexities and meaninglessness of a religious war, complete with converts: the greatest defender of the orthodoxy is a communist [Rosenfeld], and many in the opposition are fully bourgeoise....[54]

---

[52] Wigner to Jauch, 6 Sep 1966, op. cit.

[53] On their political commitments, see Chap. 4; on Rosenfeld's beliefs, see Jacobsen (2012).

[54] Frisch [in German] to Hugo Tausk, Geneva, September 16, 1967, KST.

Klaus Tausk became embroiled in this dispute, perhaps without being fully aware of it, when he distributed his preprint in August 1966, thus aligning himself with Wigner and Jauch, the most prominent critics of Rosenfeld and of Daneri, Loinger, and Prosperi. At the same time, the Wigner-Rosenfeld dispute actually seems to have contributed to the acceptance of work on the foundations of quantum mechanics as a legitimate field of research.[55] Ironically, Tausk thus helped to legitimize a field of research in physics in which he himself could no longer participate actively as a protagonist.

Conclusions

Tausk's promising research career on the foundations of quantum mechanics was cut short. He had made a bad name for himself in this field in Europe, and its study was considered to be unimportant in Brazil. In fact, this field gained general respect in Europe and America and other developed countries only in the 1970s (Freire Jr. 2004, 2009). Tausk received no support from his thesis advisor Schönberg, and consequently was unable to revise his 1966 preprint and 1967 thesis for publication. His somewhat aggressive, arrogant, or in Jauch's words not very "dignified" style of writing contributed to his negative image, suggesting that psychological factors can be significant in the acceptance of scientific concepts.

Tausk applied for and was granted a second scholarship from the Brazilian *Conselho Nacional de Pesquisa* (CNPq) to work with Jean-Pierre Vigier in Paris in 1968,[56] but he was unable to do much work owing to the strikes and political turmoil there at the time. Returning to Brazil, he pursued an unimpressive career at the University of São Paulo, concentrating on his classes (he created a course on Groups and Tensors) and publishing very little. He became something of a folkloric figure in the Physics Institute, but did not gain much sympathy owing to his difficult personality. Further, in defense of his work, he could present only a few letters from individuals and the book that Louis de Broglie had inscribed to him. These documents, some of which were written by theoretical physicists like David Bohm who

(continued)

[55] See Freire Jr. (2004) and Chap. 4.

[56] *Conselho Nacional de Pesquisa* Process number 0208/67, Arquivos do CNPq, Museu de Astronomia, Rio de Janeiro. In justifying the award of the scholarship, Tausk gave Mario Schönberg, José Goldenberg, and Hans Joos as his references. He also attached Bub's letter to him of November 15, 1966 and a letter of invitation to him of July 28, 1966, from Vigier in Paris, where he planned to study elementary particles within Vigier's approach, specifically to "analyze the possibility of unifying the external dynamical symmetry of Elementary Particles with its internal symmetry, by introducing the De Sitter space." Tausk's request was supported favorably by José Goldenberg, who commented on Tausk's work in Trieste: "This work of his on the measurement theory in quantum mechanics attracted considerable interest and, because of it, he was invited by Prof. Vigier for a period of work in Paris."

themselves were considered to be heterodox, were insufficient to gain support for the work of a young and unknown physicist. Tausk's tragedy was not that he got involved in a significant controversy on the interpretation of quantum mechanics, but that his work was forgotten.

Tausk thus was a kind of antihero in modern physics. He had original insights that were incorporated into the emerging field of the foundations of physics, since his 1966 preprint was read by physicists who came to play significant roles in this field. But he came from a Third World country, entered physics relatively late in life, chose a field of research of low scientific prestige at the time, made a few errors in his preprint, alienated his thesis advisor, was unable to publish his work in refereed journals, and had a difficult personality. Physicists have to learn how to write papers in an appropriate format, language, and degree of physical and mathematical detail to be accepted by others working in the field. Tausk lacked this ability. Salam, Director of the ICTP in Trieste, and Tausk's thesis advisor Schönberg, aligned themselves with Rosenfeld, turning Tausk into a scientific orphan. Attacking well-known scientists can lead to professional suicide.

A vital part of a physicist's training involves the development of social skills necessary to succeed in advancing his or her arguments and career. These include taking gossip into account,[57] adopting an appropriate tone in a controversy, recognizing the right moment in which to intervene, and, most importantly, judicially choosing allies and rebuffing enemies. Tausk's career thus reveals a great deal about how competing scientists and their research programs interact, how philosophical and political commitments influence their scientific views, and how severe the difficulties are for someone doing science at the scientific periphery.

The Tausk controversy also reveals much about the kind of tacit knowledge scientists learn during their education and training. Young scientists can be wasted if they are not taught how to conduct themselves in scientific controversies, which is an art that goes well beyond reason and logic. The Tausk controversy exposes the risks and consequences of trying to participate in a scientific controversy in the absence of proper training and guidance. One value of the history of science is that it can be useful in showing young scientists the extent to which science is a social practice.

Salam's remark about the controversy that Tausk had precipitated, that one should "consider this episode as part of the old battles,"[58] displays this social dimension and suggests an analogy between scientific controversies and military warfare. Both have winners and losers, but one may lose a battle

(continued)

[57] See Traweek (1988, pp. 121–122)

[58] Salam to Rosenfeld, 26 Sep 1966, op. cit.

while winning the war. In the debates over the interpretation of quantum mechanics, some like Niels Bohr won battles and some like David Bohm lost battles, but Bohm persevered in his hidden-variables program and in the end won some battles, or at least left his mark on the battlefield. There also, however, are those who lose a battle and then surrender. That seems to have been what Tausk did.

Appendix: Summary of Tausk's Arguments

Tausk's arguments against Daneri, Loinger, and Prosperi's (DLP's) theory may be summarized as follows:

DLP's Theory Deals Only with the Statistical Case Tausk presents the reduction or projection postulate for an individual, "pure" case, and contrasts it with a statistical version, which he calls the "weak reduction postulate." He then argues that what DLP derive in their paper is not the projection postulate in the pure case, but in the statistical case (Tausk 1966, p. 4). If so, then the "measurement problem" is not solved, and DLP's theory fails. Bohm accepted this argument in his letter to Tausk of October 1, 1966 (op. cit.), and Bub developed it in his paper of 1968 (Bub 1968).

DLP's Analysis Is Circular Tausk argues that DLP's description of measurement as occurring in two stages is circular. His argument, however, seems to follow from an incorrect reading of DLP's theory, which Jauch said was one of the "many details with which I disagree."[59]

The Ergodic Hypothesis Plays No Role in DLP's Theory Tausk (1966, p. 20) suggests that the use of the ergodic hypothesis in DLP's theory plays only a "purely psychological role," a view that is based upon some sort of misunderstanding.

Negative-Result Measurements Refute DLP's Theory This argument, which we have examined above, is correct in that it shows that amplification is not necessary for state reduction. However, as we noted, contrary to what one might expect, the existence of negative-result measurements does not refute DLP's theory, which, as Loinger (1968, pp. 246–248) argued, does not explicitly mention amplification. In any case, after Tausk presents his argument, he gives an example of his not very elegant style of writing that contributed to the negative reception of his preprint, declaring that: "To our mind, this argument shows that all attempts to fulfil [sic] the program of DLPI belong to the realm of wishful thinking or, occasionally, of just wishing" (Tausk 1966, p. 23).

Tausk made three additional points in his 1966 preprint and in his 1967 doctoral thesis, as follows:

[59] Jauch to Fonda, 4 Oct 1966, op. cit.

The Conservation of Angular Momentum Paradox In Sect. 5 of his preprint and in his thesis,[60] Tausk raises an apparent paradox concerning the angular momentum of an atom that passes through a Stern-Gerlach apparatus. Assuming that before detection the component of its angular momentum along the line joining the two magnets is zero, immediately after detection it is nonzero, either "up" or "down," depending upon which of the two detectors is triggered. Tausk asks how this apparent violation of conservation of angular momentum can be explained. A few years later, however, he realized that it could be explained by assuming that angular momentum is transferred to the Stern-Gerlach magnets.[61]

Critique of Heisenberg's Epistemic Conception of Reduction In his book, *Physics and Philosophy* of 1958, Heisenberg (1958, pp. 54–55) claimed that state reduction expresses nothing more than an increase of our knowledge of a quantum-mechanical system. Tausk (1966, p. 32) criticizes this view and suggests that quantum mechanics requires a completely new foundation.

No-Signaling Theorem In his doctoral thesis, Tausk proved that an ensemble of two correlated particles, I and II, prepared in the same composite state, can never be used to transmit information at a speed greater than the speed of light (Tausk 1967, pp. 29–31). This probably is the first time that a physicist proved this rather simple result, which is known in the literature as a no-signaling theorem and is attributed to Philippe Eberhard (1978, on 416–417).

Finally, it is curious that Tausk continues by analyzing the famous Einstein-Podolsky-Rosen paper of 1935 (Einstein et al. 1935), stating that they do not make use of the reduction postulate. That is incorrect: they do make explicit use of it. This illustrates both some of the shortcomings of Tausk's work and, because this error remained in his thesis even after he defended it, shows that the Brazilian community of physicists was still not well prepared to understand and discuss such philosophical subtleties as we have noted above on the foundations of quantum mechanics.

References

Andrade, A. M. R.: Os raios cósmicos entre a ciência e as relações internacionais. Ciência, Política e relações internacionais ensaios sobre Paulo Carneiro. M. C. Maio. Rio de Janeiro, Editora Fiocruz: 215-242 (2004)

Ballentine, L.E.: Statistical interpretation of quantum mechanics. Rev. Mod. Phys. **42**(4), 358–381 (1970)

Bell, J.S.: On the Einstein Podolsky Rosen paradox. Physics **1**, 195–200 (1964)

Bell, J.S.: On problem of hidden variables in quantum mechanics. Rev. Mod. Phys. **38**(3), 447 (1966)

[60] Tausk (1966) and Tausk (1967).

[61] Tausk's interviews with Pessoa, op. cit.

Bell, J.S.: Speakable and Unspeakable in Quantum Mechanics: Collected Papers on Quantum Philosophy. With an Introduction by Alain Aspect. Cambridge University Press, Cambridge (2004)

Bohm, D.: A suggested interpretation of the quantum theory in terms of hidden variables—I & II. Phys. Rev. **85**(2), 166–179 (1952). 180–193

Bohm, D., Bub, J.: A proposed solution of measurement problem in quantum mechanics by a hidden variable theory. Rev. Mod. Phys. **38**(3), 453–469 (1966)

Bohr, N.: Atomic Physics and Human Knowledge. Wiley, New York (1958)

Bub, J.: Daneri-Loinger-Prosperi quantum theory of measurement. Nuovo Cimento B **57**(2), 503–520 (1968)

Caldirola, P.: Teoria della misurazione e teoremi ergodici nella meccanica quantistica. G. Fis. **6**, 228–237 (1965)

Caldirola, P.: A geometrical model of point electron. Revista Brasileira de Física **Volume especial - 70 anos de Mário Schönberg**, 228–260 (1984)

Collins, H.M., Pinch, T.J.: The Golem: What Everyone Should Know About Science. Cambridge University Press, Cambridge (1993)

Cushing, J.: Quantum Mechanics—Historical Contingency and the Copenhagen Hegemony. The University of Chicago Press, Chicago (1994)

Daneri, A., Prosperi, G.M., Loinger, A.: Quantum theory of measurement and ergodicity conditions. Nucl. Phys. **33**(2), 297–319 (1962). reprinted in J. A. Wheeler and W. H. Zurek, ed., Quantum Theory and Measurement, Princeton University Press, 1983, pp. 1657–1679

Daneri, A., Loinger, A., Prosperi, G.M.: Further remarks on relations between statistical mechanics and quantum theory of measurement. Nuovo Cimento B **44**(1), 119–128 (1966)

De Greiff, A.: The International Centre for Theoretical Physics, 1960–1979: Ideology and Practice in a United Nations Institution for Scientific Cooperation and Third World Development. PhD dissertation, Imperial College of Science, Technology and Medicine, London (2001)

d'Espagnat, B.: Conceptual Foundations of Quantum Mechanics, 2nd edn. Addison-Wesley, Redwood City, CA (1989). 1st ed 1971

Dicke, R.H.: Interaction-free quantum measurements—a paradox. Am. J. Phys. **49**(10), 925–930 (1981)

Eberhard, P.H.: Bell's theorem and different concepts of locality. Nuovo Cimento B **46**(2), 392–419 (1978)

Einstein, A., Podolsky, B., Rosen, N.: Can quantum-mechanical description of physical reality be considered complete? Phys. Rev. **47**, 777–780 (1935)

Everett, H.: Relative state formulation of quantum mechanics. Rev. Mod. Phys. **29**(3), 454–462 (1957). Reprinted in: J. A. Wheeler & W. H. Zurek (Eds.), Quantum theory and measurement (pp. 315–323). Princeton: Princeton University Press, 1983. Page numbers refer to the reprint

Fernandes, N.C., Cattani, M., Ventura, I., Ueta, K., Salinas, S.R.: Mário Schönberg on his 70th Birthday. Revista Brasileira de Fisica **Volume especial - 70 anos de Mário Schönberg**, V–VI (2008)

Freire Jr., O.: The historical roots of "foundations of quantum mechanics" as a field of research (1950–1970). Found. Phys. **34**(11), 1741–1760 (2004)

Freire Jr., O.: Quantum dissidents: research on the foundations of quantum mechanics circa 1970. Stud. Hist. Philos. Mod. Phys. **40**(4), 280–289 (2009)

Frisch, O.R.: The conceptual problem of quantum theory from the experimentalist's point of view. In: Bastin, T. (ed.) Quantum Theory and Beyond—Essays and Discussions Arising From a Colloquium, pp. 13–21. Cambridge University Press, London (1971)

Garuccio, A., Leone, M.: La fisica teorica tra Milano e Pavia (1945–1965). Per una storia della Fisica Italiana 1945–1965.I. Física della matéria, fisica teorica, insegnamento della Fisica G. Giuliani. La Goliardica Pavese, Pavia, 35–80 (2002)

Greiff, A.D.: The tale of two peripheries: the creation of the International Centre for Theoretical Physics in Trieste. Hist. Stud. Phys. Biol. Sci. **33**(1), 33–59 (2002)

Heisenberg, W.: Physics and Philosophy; The Revolution in Modern Science. Harper, New York (1958)
Jacobsen, A.: Léon Rosenfeld—Physics, Philosophy, and Politics in the Twentieth Century. World Scientific, Singapore (2012)
Jammer, M.: The Philosophy of Quantum Mechanics—The Interpretations of Quantum Mechanics in Historical Perspective. Wiley, New York (1974)
Jauch, J.M.: Problem of measurement in quantum mechanics. Helv. Phys. Acta **37**(4–5), 293–316 (1964)
Jauch, J.M., Wigner, E.P., Yanase, M.M.: Some comments concerning measurements in quantum mechanics. Nuovo Cimento B **48**(1), 144–151 (1967)
Jordan, P.: On the process of measurement in quantum mechanics. Philos. Sci. **16**, 269–278 (1949)
Körner, S.: Observation and interpretation—a symposium of philosophers and physicists. Butterworths, London (1957)
Loinger, A.: Comments on a recent paper concerning quantum theory of measurement. Nucl. Phys. A **A108**(2), 245–249 (1968)
McMullin, E.: Scientific controversy and its termination. In: Engelhardt, H.T., Caplan, A.L. (eds.) Scientific Controversies: Case Studies in the Resolution and Closure of Disputes in Science and Technology, pp. 49–91. Cambridge University Press, Cambridge (1987)
Renninger, M.: Messungen ohne storung des messobjekts. Zeitschrift Fur Physik **158**(4), 417–421 (1960)
Rosenfeld, L.: Measuring process in quantum mechanics. Suppl. Prog. Theor. Phys.: 222–231 (1965)
Rosenfeld, L., Cohen, R.S., Stachel, J.J.: Selected Papers of Léon Rosenfeld. D Reidel Publishing Company, Dordrecht, Holland (1979)
Schönberg, M., Hamburger, A.I.: Obra Científica de Mario Schönberg—Volume 1–1936–1948. EDUSP, São Paulo (2009)
Schönberg, M., Hamburger, A.I.: Obra Científica de Mario Schönberg—Volume 2–1949–1987. EDUSP, São Paulo (2013)
Süssmann, G.: Über den Meßvorgang. Bayerische Akademie der Wissenschaften, Mathematisch-Naturwissenschaftliche Klasse, Abhandlungen Neue Folge **88**, 3–41 (1958)
Tausk, K.S.: Relation of measurement with ergodicity, macroscopic systems, information and conservation laws. ICTP Intern. Rep. **14**, 34 (1966)
Tausk, K.S.: A Medida na Mecânica Quântica. PhD dissertation, Universidade de São Paulo (1967)
Traweek, S.: Beamtimes and Lifetimes : The World of High Energy Physicists. Harvard University Press, Cambridge, MA (1988)
van Hove, L.: The ergodic behavior of quantum many-body systems. Physica **25**, 268–276 (1959)
Von Neumann, J.: Mathematische Grundlagen der Quantenmechanik. Julius Springer, Berlin (1932)
Von Neumann, J.: Mathematical Foundations of Quantum Mechanics. Princeton University Press, Princeton, NJ (1955)
Wakita, H.: Measurement in quantum mechanics. Prog. Theor. Phys. **23**, 32–40 (1960)
Wheeler, J.A., Zurek, W.H.: Quantum Theory and Measurement. Princeton University Press, Princeton, NJ (1983)
Wigner, E.P.: Problem of measurement. Am. J. Phys. **31**(1), 6–15 (1963)

Chapter 6
"From the Streets into Academia": Political Activism and the Reconfiguration of Physics Around 1970

Abstract The political and cultural unrest of the late 1960s influenced the debate on quantum physics by helping those who wanted to push research on foundations from the margins to mainstream physics. The Italian Physical Society was at risk to split apart due to political dissensions in the universities, and the president at the time, Toraldo di Francia, thought that bringing a controversial scientific topic to the forefront might stall the split. Franco Selleri's proposal to dedicate the 1970 Varenna summer school to the foundations of quantum mechanics was accepted and Bernard d'Espagnat was invited to head it and set diplomatic rules for managing the controversy. Varenna was the Woodstock of quantum dissidents. Wigner made the keynote address and different interpretations for quantum theory were presented. Shimony and Bell spoke on experiments for testing locality and quantum mechanics. Zeh presented what would be later called the decoherence approach to the measurement problem. The quantum dissidents left political dissidence aside to concentrate on the quantum controversy. Later on, political dissidence escalated in European physics settings while on the other side of the Atlantic, the editor of *Physics Today*, under pressure from those who wanted the American Physical Society to rally against the Vietnam War, decided to feature less controversial topics. Bryce DeWitt was invited to publish a paper on the many interpretations of quantum mechanics, including Everett's many worlds interpretation. A huge debate erupted in the magazine after DeWitt's paper. In this chapter we will show how physicists exploited the political climate of the late 1960s to push for changes in the science establishment, including its research agenda.

6.1 Introduction

The 1960s and the early 1970s are landmarks in twentieth century history concerning cultural turning points and political unrest. The emergence of counter-cultures, rock music, drugs, sexual liberation, environmental concerns, feminist movements, and protests against the Vietnam War and the political establishment featured in the news at that time. The expressions "*soixante-huitards*" and "'68 generation" are used nowadays as part of the common lexicon when identifying particular social sensitiveness. The implications of the 1968 events were far

O. Freire Junior, *The Quantum Dissidents*, DOI 10.1007/978-3-662-44662-1_6

reaching. As one contemporary observer noted, "the student strike is a new phenomenon in European history. Students in the capitalist countries usually do not strike. But now, all under the heaven is great chaos." These were the words of Mao Zedong, while trying to obtain insights from those events for Chinese geopolitics (apud Kissinger 2011, p. 207). More recently, the sociologist Michael Hölscher (2012) considered those events exemplary of a "transnational social movement," writing

> Specifically, as a transnational social movement, the generation of '68 connects such diverse events as the Prague Spring, the Summer of Love in the United States, the Paris May, the international anti-Vietnam congress in Berlin, Zengakuren's attack on Tokyo, the Tlatelolco massacre in Mexico, and the protests of the Black Power movement in the United States after the assassination of Martin Luther King Jr.

As for science, it could be thought it was to a certain extent the ongoing political activism of a small but vocal group of scientists which emerged after World War II in the wake of the production and dropping of the atomic bombs by the US, the beginning of the Cold War, and the arms race between the US and the USSR. However, it was more than this. Indeed, in the late 1960s, this political activism underwent a phase transition and the physics community faced new challenges. These included the research strike at MIT against the military use of research results, senior physicists such as Hans Bethe publicly speaking out against the arms race, the American Physical Society meetings being upset by political protesters, and the Nobel Prize winner Gell-Mann being obstructed from giving a talk at the centenary Collège de France. What set this period apart from previous political activism was that now scientists were being accused by other scientists of collaborating with the military on the application of science. As recalled by Ravetz (1990, p. 902), "the complicity of American science in some of the most reprehensive dirty tricks of the dirty Vietnam war was signaled by dissident students and researchers." He continues that this was part of a wide criticism of the early vision of modern science leading to the result that "all the contradictions in the ideology of science that had been latent, through the centuries of triumph, now became manifest." In a certain sense, the social agreement around the use of physics research in military applications, so typical in the Cold War times, had been called into question.

"Although the millenarian aspirations of the 1960s, in politics and in experience, are now reduced to an object of historical study, the permanent changes achieved then should not be underestimated," according to Ravetz (1990, p. 901). However, as noted by the historian Eric Hobsbawm (1994, p. 613), "there are as yet no properly historical treatments of the social and cultural revolutions in the second half of the century," despite his chapter "Cultural Revolution" in *Age of Extremes*. Thus those times remain an open field for historical research. As for the permanent changes in science, some terrain has already been charted, and the current work aims to contribute towards fulfilling it. A short and far from comprehensive account of the issues already analyzed through scholarly works include the following: Daniel Kevles (1978) identified the 1970s as a time when there was a decline in public support for the increasing budgets spent on physics, typical of Cold War times in the 1950s. Gary Werskey (2007), while reviewing the Marxist critique of capitalist science, dedicated a section on the British scientific Left of the 1970s, which included his own influential book *The Visible College*. Paul Forman (2012) saw the early signals of the end of modernity and the dawn of postmodernity, marking the transition between the two cultural-historical epochs, in "the dramatic fall, between the early 1960s and the early 1970s, in the cultural valuation of professions and of disciplines." The sociologist Kelly Moore (2008) analyzed the political engagement of American scientists from 1945 to 1975, discussing the changes in the social authority of the scientific enterprise. Andrew Jamison (2012), writing on science and technology in postwar Europe, titled a section "From the 1960s to the 1980s: A Period of Debate and Reform" but he did not deal with the influence of such events on the very content of science. Stevens (2003) remarked that American physicists from the high energy domain changed their discourse on the importance of their field from its value in national defense and scientific competition to its cultural value. Later, David Kaiser (2012a) spotted the influence of the counter culture, hippies in particular, in the development of the field of foundations of quantum mechanics. Kaiser has shown us in a telling case that the no-cloning theorem, nowadays a central piece in quantum information, resulted from instigations from physicists gathered at Berkeley who identified themselves with the counter culture trends of the time. Kaiser also exemplified how the wide cultural environment may shape the production of textbooks, analyzing the inception of two of them in the US in the mid-1970s (Kaiser 2012b).[1] Matthew Wisnioski (2012, pp. 11–12) analyzed the "ideology of technological changes" resulting from the tension among American engineers between 1964 and 1974 related to perceived "out-of-control technology." However, other studies have noted the destructiveness of this radicalization concerning some scientific institutions, at least on the Italian scene. According to Capocci and Corbellini (2002), and Cozzoli and Capocci (2011), the political context played a role in thwarting institutions in health sciences, such as the International Laboratory of Genetics and Biophysics in Naples and the Italian Higher Institute of Health, which suffered from the accusation of being too American in style, a deadly sin in Italy at that time. However, in spite of the value of such works, the influence of the context on the practice of science, on its research agenda, and its relationship with the public at large has yet to be more extensively charted.

In a previous paper (Freire Jr. 2004) I suggested that the context of political and cultural unrest of the late 1960s might have helped to open the way for the emergence of marginal themes such as foundations of quantum physics. My purpose in this chapter is to substantiate such a suggestion. I will approach this theme through two case studies. The first one concerns the Italian Physical Society (SIF, in Italian), its summer school (Enrico Fermi school, held yearly in Varenna), which in the early 1970s included Foundations of Quantum Mechanics, History of

[1] The books analyzed were Misner, Thorne, and Wheeler's *Gravitation* and Capra's *The Tao of Physics*.

Physics in the twentieth century, and Physics and Society as some of its themes (d'Espagnat 1971; Weiner 1977).

The Italian Physical Society was created in 1935 in Bologna in the wake of the flourishing of Italian research in modern physics led by Enrico Fermi and the young physicists gathered at Via Panisperna, where the Physics Institute in Rome was located. The Fascist racial laws dispersed the group and destruction from World War II represented a strong setback for the Italian physics. However, particularly under the leadership of Edoardo Amaldi, a remnant from Fermi's team, this community was rebuilt and the Nobel Prize awarded to two of them, Fermi and Emilio Segré, contributed to a renewed self-esteem of Italian physicists. They relaunched the traditional journal *Il Nuovo Cimento*, whose creation dates from the nineteenth century; created the Enrico Fermi summer schools, in Varenna, in 1953; some of them also supported the creation of the International Centre for Theoretical Physics, in Trieste, and associated to the creation of CERN. In the mid-1960s thus the Italian physics community was recovered and its association and its summer school and journal garnered prestige inside Italy and in the world physics community (Amaldi et al. 1997, 1998; Salvini 2005).[2]

Recollections from both senior and young Italian physicists who acted in that context indicate that political motivations lay behind the decision to hold such schools. For instance, the young physicist Augusto Sabbadini recalled that "the '68 movement was still in full swing and I felt very much part of it. The sense of openness, the readiness to consider things in new and unconventional ways had to some extent spread from the streets into the academia, Varenna was partly an expression of that;" while the president of the Italian Physical Society, Toraldo di Francia, said that the inclusion of such themes in the summer schools was a way to prevent society from splitting due to the political tensions of the time.[3] The political background motivating the promotion of the school dedicated to the foundations of quantum mechanics was noted by observers external to the Italian physics milieu. According to the recollections from Bernard d'Espagnat, who was already a senior French physicist in 1970,

> I suppose there was, coming from the grassroots, some demands on calling into question the received views. It was 1970, that is, a time with much political and intellectual agitation. In Italy, like in France, there were young physicists who were activists. This might have played a role.[4]

---

[2] Amaldi reports, in Amaldi et al. (1998, pp. 244–246 and 285–286), the creation of the Varenna summer school, the reorganization of the Italian Physical Society, and the journal *Il Nuovo Cimento*. His account emphasizes the role of Giovanni Polvani, as the new president of the society, in all these initiatives and how he was inspired by the creation of the Les Houches summer school in Theoretical Physics by the French physicist Cécile Morette (later C. DeWitt-Morette). On the role played by the Varenna summer schools, in the 1950s and 1960s, see also Schweber (2014).

[3] Augusto Shantena Sabbadini to the author, e-mail, 18 Jan 2011. Toraldo di Francia, interview with Olival Freire, Florence, 01 July 2003, deposited at the Center for History of Physics – American Institute of Physics, College Park, MD.

[4] "Je suppose aussi qu'il y avait, en provenance de 'la base', une certaine demande de mise en question des idées reçues. C'était en soixante-dix, c'est-à-dire, à une époque de pas mal d'agitation politique et intellectuelle. Et, en Italie comme en France, il y avait des jeunes physiciens qui

For the second case, we move to the American scene, where reminiscences from John Clauser, the leader of the first successful experiment on Bell's theorem, suggest he changed his research from the subject of cosmic background radiation to alternative interpretations of quantum mechanics for cognitive and political reasons. While a graduate student at Columbia University, influenced by *l'air du temps* of the protests against the Vietnam War, he recalled that he wanted to shake the world, and quantum mechanics was one of the targets of this desire. According to his words, "the Vietnam War dominated the political thoughts of my generation. Being a young student living in this era of revolutionary thinking, I naturally wanted to 'shake the world.' Since I already believed that hidden variables may indeed exist, I figured that this was obviously the crucial experiment for finally revealing their existence" (Clauser 2002, p. 80). Our focus, however, will be on the role played in the US by the magazine *Physics Today* in opening the debate on the diversity of interpretations of quantum mechanics in 1970. Our point is that both the opening of such a debate and the way it was received were influenced by the political climate of the times. This influence is better documented in the Italian case, where accommodating to the political climate was influential in the decisions of the Italian Physical Society. In the American case, while such influence is very plausible, documentary evidence is not so strong.

This chapter is organized as follows. The second section covers the background to the SIF's decision to hold such schools and the sensitivities around their themes. Then I present the two schools, their results, and the political climate and the influence of this on the subjects in question. In the fourth section I briefly present the immediate continuity of political unrest and the fading of the political climate. The fifth section is dedicated to the conflicts experienced by the American Physical Society and *Physics Today* as a consequence of the political tensions of the times. In the sixth section the focus is placed on the opening of the debates on the quantum controversy in the magazine *Physics Today* and the reception of such an opening. Then I present the conclusions.

6.2 The Mesh of Science and Politics: The Varenna Summer Schools

The political unrest of the late 1960s put politically active young physicists in particular under pressure. Some of them, as we have seen in the case of Clauser, reacted to this cultural ambiance by focusing their research on foundations of physics issues. Historically, research on the foundations of physics has been a way to criticize established scientific doctrines, not unlike the case of Ernst Mach

contestaient. Cela a peut-être joué son rôle." Bernard d'Espagnat, interviewed by the author, Paris, 26 Oct 2001, Niels Bohr Library & Archives, American Institute of Physics, College Park, MD [AIP hereafter]. D'Espagnat made these comments without being asked about this kind of background.

and his criticism of mechanics in the nineteenth century. This was the case of Tito Tonietti, an Italian physicist who got his degree in 1966 and would later go on to a career in the history of physics. According to his testimony,

> I graduated with an original thesis on gauge theories, but at that time nobody cared about it. Then, in the political context of late sixties, we hoped for a deep change not only in society, but even in the way of doing physics. So, we started from the foundations of Q.M., which I learned from Messiah.[5]

Tonietti's motivation to look at the foundations of quantum physics is an example of the way of thinking that went into the decision making of the Italian Physical Society when it took the stand of dedicating one of its summer schools, held in Varenna, to the subject. Several testimonies, in addition to Tonietti's, converge on the influence of the political dissension of the times in that decision as well as to the decision to dedicate another school to the subject of history of physics and its social implications. Angelo Baracca, for instance, had graduated in 1965 and was "Professore incaricato" of Statistical Mechanics at the University of Florence. He was then "engaged in high-energy physics research and was also very sensitive to the general flux of critical considerations and practices against institutional science and its social commitments." According to his recollections about these schools,

> I was among the young physicists who participated in the Meeting and Assembly of the Italian Physical Society, in which we criticized institutional research, and expressly proposed this School in the programs of the Varenna Summer Schools, with the proposal of introducing a reflection on the foundations of Physics. In fact we had a subsequent initiative, proposing and getting a Summer School on the History of Physics, that we considered a concrete way to study the social implications of Science.[6]

It would be misleading, however, to think about these proposals as an initiative only from young physicists. Franco Selleri, who formally presented the proposal on foundations of quantum mechanics, was then a mature particle physicist, aged 36 years old. As he recalled,

> I was then a member of the directive board of the Italian Physical Society so I suggested that a summer school be organised in Varenna by the Italian Physical Society on the foundations of quantum mechanics. The idea was accepted and d'Espagnat was named director of the school and many influential people were invited.[7]

In his talk at the Varenna school Selleri presented a research program designed to reform quantum physics and in which physics, philosophical and social commitment made up equal parts. Among the arguments for adopting a realistic philosophy, he claimed,

---

[5] Tito Tonietti to the author, e-mail, 19 Jan 2011. Messiah is a reference to the classical quantum physics textbook written by the French physicist Albert Messiah (1961).

[6] Angelo Baracca, e-mail to the author, 17 Jan 2011.

[7] Franco Selleri, interview with Olival Freire, Bari, 24 June 2003, deposited at AIP.

> In this time where the social responsibility of the scientist is so strong, where the destruction or the survival of the world depends also on him, it is important to develop a science not in basic contradiction with the social reality. The foundations of physics (and in particular of quantum mechanics) should then better be based on a realistic philosophy. (Selleri 1971)

The following year, Selleri (1972) published a long paper in *Critica Marxista*, an Italian cultural and political magazine, in which the connection between his approach to quantum problems and philosophical choices were made rather more explicit. He used the thesis of the divorce between the two cultures (scientific and humanistic), a thesis developed by the British chemist-novelist Charles P. Snow (1959), to argue that contemporary physics was suffering from three weaknesses, namely absence of history of science, distance from philosophy, and abandonment of the idea of physical reality. Then Selleri developed his arguments mobilizing Marxist authors such as Marx, Engels, and Lenin to criticize what he considered to be the standard interpretation of quantum theory, that is, the complementary interpretation. Finally he established connections between physicists' political stands and their views of the interpretation of quantum physics. Selleri labeled critics of complementarity, such as Einstein and Schrödinger, progressive, and its supporters, such as Bohr, Heisenberg, Born, and Jordan, conservatives.[8] Selleri maintains this intertwinement between science, philosophy, and politics until today, as he stated: "it is our duty to build a science that can be communicated to everybody," and added "at those times I was thinking in terms of the working class, the working people. [...] Seeing that important ideas of the realist people like de Broglie and Einstein and so on were as much as possible forgotten, I understood that the scientific community does not work properly." Thus for him criticisms of the foundations of quantum theory, philosophical realism, and socialist inclinations were aspects which come together.[9]

Yet more meaningful was the fact that these proposed themes for the summer schools, expressed as they were in that context, were readily accepted by the board of the Italian Physical Society. In fact, there was a kind of political agreement on the board led by Giuliano Toraldo di Francia (1916–2011), the president of the society. A senior researcher in optics who went on to win the Max Born medal from the Optical Society of America, Toraldo di Francia led the society between 1968 and 1973. In his later memoirs, he defined himself as a left-wing man, while not a communist, very sensitive to the philosophical dimensions of science. In 2003, he

[8] In the 1970s, the Italian scholars who were Marxism-inclined and involved with the quantum controversy reproduced the same divide from the early 1950s, a divergence we examined in Chap. 2. Thus, contrasting with Selleri, the philosopher Silvano Tagliagambe, supported by Ludovico Geymonat, published a translation of the relevant Soviet papers on the quantum controversy with an analysis favorable to Fock's stand. Tagliagambe's work, however, apparently did not influence the physicists who were interested in the quantum controversy and himself, after a while, moved towards different subjects. See Tagliagambe (1972) and Freire Jr. (2011).

[9] Franco Selleri, interview with Olival Freire, *ibid.*

recalled the agreement he reached with his peers and how correct it was later considered in hindsight.

> I found myself in a way, as we say, between the hammer and anvil, between the two, because on one part I had my professors, my old teachers who were not fascists, not at all, but certainly reactionary, and the young people who were pro-Communists, but too much. I said no, you are right here and you are wrong here, and I found myself in a very different position. But later it was acknowledged that by taking that standpoint, I saved the Italian Physical Society. Because it certainly risked being split in two at that time. I said no, we cannot split, we must continue to do very good research in particle research as has been done; but also take care of our needs of the society, which particles don't do, but I could do, and later was recognized that it was a reasonable standpoint.[10]

As for the school dedicated to the history of physics in the twentieth century, traces of the political background of the times may be seen in the topics Charles Weiner chose to include on the agenda of the school. Weiner was then the head of the Center for the History of Physics at the American Institute of Physics and had been chosen by the board of the Italian society to be the director of the school, which was held in 1972. Alongside the traditional topics in the history of science, there were topics reflecting socio-political issues. Among the former, we find "Roots of Modern Physics," "Origins and Development of Quantum Theory," "History of Nuclear and Particle Physics", while among the latter there was "The Rise of Big Science," "Historical Prospectives on Physics, Technology and Society," and "The Social Role of the Professional Physicist." The "subsequent initiative" mentioned by Baracca was adopted, according to historian of physics Arturo Russo's recollections, at the 1971 SIF congress held in L'Aquila. The society had asked Giovanni Jona-Lasinio, who lectured on the history of physics and was a reputed physicist in areas such as quantum field theory and statistical mechanics, to organize the summer school on the history of physics in the twentieth century. Indeed, Charles Weiner was invited to be the director of the school and Jonas-Lasinio its organizer. Still according to Russo, in 1970, the SIF had organized a conference on "Science in the contemporary society" where critical reflections on the current trends in science were presented by, among others, Marcello Cini, Silvio Bergia, and Toraldo di Francia. This conference published its proceedings under the less neutral title "Science in the capitalistic society." We will see more on the politics around these schools later.[11]

As the Italian case is so well documented with regard to interactions between the political and cultural context and scientific content itself, or at least the agenda of

---

[10] Toraldo di Francia, *op. cit.*

[11] Russo, Arturo. (2007). Writing the history of modern physics in Italy: a personal reflection. In S. Boudia, D. Pestre, and S. Soubiran (orgs.), 'Writing the History' of the Physical Sciences after 1945: state of the art, questions, and perspectives, Strasbourg, 7–9 June 2007; unpublished papers for private circulation. I am indebted to Xavier Roqué for bringing Russo's paper to my attention. The proceedings of "Science in the capitalistic society" are in Società Italiana di Fisica (1971). Charles Weiner to Toraldo di Francia, 17 June 1971. Archives of the Italian Physical Society [ASIF hereafter], Bologna.

research, we may ask how singular was this case. The issue is more pertinent if one considers that no major Italian physicist openly criticized the orthodox view of quantum mechanics before 1970. While the question is difficult to give a clear-cut answer to, historians have given hints about how deeply rooted the Italian 1968 crisis was compared to other cases. In his comparative history of postwar Europe, the historian Tony Judt concluded his comparison between the French and the Italian 1968 events, after considering the social context, including universities, of both countries (Judt 2005, p. 416)[12]:

> Whereas French students had played with the idea that public authority might prove vulnerable to disruption from below, a caprice that Gaullism's firmly-grounded institutions allowed them to indulge with impunity, Italy's radicals had good reason to believe that they might actually succeed in rending the fabric of the post-Fascist Republic—and they were keen to try.

Indeed, to fully understand the particularities of the Italian case we should consider both the labor and social movements, which were increasing since the mid-1960s, and its deployments, which ran till the early 1980s, a period known now in Italian history as the Years of Lead. While the latter radicalization falls beyond the timeline of our case and its historical documentation is still today hugely controversial, in fact its beginning predates the 1970 Varenna summer school. As remarked by Judt, "on April 24th 1969, bombs were planted at the Milan Trade Fair and the central railway station. Eight months later, after the Pirelli conflicts had been settled and the strike movement ended, the Agricultural Bank on the Piazza Fontana in Milan was blown up," to conclude that "the 'strategy of tension' that underlay the lead years of the Seventies had begun." Furthermore, according to Judt (2005, pp. 476–477), "from 1977 to 1982 especially, the country was under the siege from random acts of extreme violence by far Left, far Right and professional criminals alike." Though it is beyond the scope of this study to consider events in those later years, the singularity and extension of the Italian upheavals have been commented on by analysts from distinct ideological perspectives. Thus the politician-turned-historian Lucio Magri, writing "a possible history of the Italian Communist Party," christened the political context in question "Italy's Long Sixty-Eight" (Magri 2011, p. 195).[13]

---

[12] Some readers may inquire about bringing together Judt and Hobsbawm as commentators of the same events, given their ideological opposite stances. I may defend my procedure citing Hobsbawm's obituary of Judt in which the book I am quoting—*Postwar*—is highly appreciated. See "After the Cold War - Eric Hobsbawm remembers Tony Judt", *London Review of Books*, 34 (8), 26 April 2012, p. 14; available at http://www.lrb.co.uk/v34/n08/eric-hobsbawm/after-the-cold-war, accessed on 10 April 2014. I am thankful to Thiago Hartz for bringing this obituary to my attention.

[13] For studies related to the upheavals in Italy, see also Cento Bull and Giorgio (2006).

6.3 The Schools and Their Results

6.3.1 1970: Foundations of Quantum Mechanics

The 1970 Varenna Summer School was successful due to the scientific and philosophical discussions it stimulated. Its 84 participants, the quality of their talks, the quick publication of its proceedings (d'Espagnat 1971), the first debates on Bell's theorem and experiments, the debates on the quantum measurement problem, and the diversity of its speakers, including John Bell, David Bohm, Eugene Wigner, Louis de Broglie, Josef-Maria Jauch, Abner Shimony, Heinz-Dieter Zeh, Franco Selleri, and Bryce DeWitt, created an air of excitement around research into foundations. In the case of this school the subject already had a small but unconnected community and the school helped network these scientists, bringing together most of the physicists who would go on to contribute to the blossoming of this research in the 1970s. For instance, John Bell, Abner Shimony, Bernard d'Espagnat, and Franco Selleri—all of them worked on Bell's theorem. In the case of Bell and Shimony they had already been working on this subject, and in the case of d'Espagnat and Selleri they turned their attention to it. However, they had been working independently before becoming part of a network (Freire Jr. 2006, p. 592). The following years, Shimony would spend a time with d'Espagnat in France and d'Espagnat would help Alain Aspect to look for support for his experiments. For Selleri it was the first opportunity to become acquainted with the theme, which would go on to occupy all his professional energy from then on, becoming the most vocal critic of loopholes in these experiments, which according to him would save local realism. As for d'Espagnat, he was then beginning to work on foundations and the success of the school led him to be regarded as one of the experts in the field. He had already published a book (d'Espagnat 1965) on the subject and would publish an influential textbook on conceptual issues in quantum physics (d'Espagnat 1989 [1st edition 1971]). It was also an important event for Zeh, who needed encouragement for the research he was beginning and which would later lead him to the decoherence effect (Freire Jr. 2009, pp. 281–282). The school was also the event at which Bryce DeWitt expressed his recent conversion to Everett's many-worlds interpretation. It conveyed the feeling of openness towards the existence of a diversity of interpretations of quantum mechanics rather than just the complementarity interpretation. The school was also helpful for some of its attendees who would later become leaders in the research on the foundations of quantum physics. This was the case with Basil Hiley and Emilio Santos. Hiley was then an assistant of David Bohm at Birkbeck College in London and would become the key protagonist in Bohm's quest for a mathematical treatment for the ideas of wholeness and implicate order. Santos would work on stochastic electrodynamics and implications for the interpretation of quantum theory and animate conferences on foundations in Spain. Finally, the school strengthened the existence of a quantum measurement problem, and Eugene Wigner, as one of its keynote speakers, presented the diverse proposals to solve it. The debate between Wigner and Giovanni Prosperi, who had clashed on the quantum measurement problem the decade before, presented in Chap. 4, was an attraction of the school. In short, the school encouraged physicists to change their research agenda to include these topics and/or encouraged those already working on them, thus contributing to the professional recognition of a research theme in serious need of attention.

E. Wigner	The subject of our discussions
J. M. Jauch	Foundations of quantum mechanics
H. Stein and A. Shimony	Limitations on measurements
M. M. Yanase	Optimal measuring apparatus
B. d'Espagnat	Mesure et non séparabilité (Revue sommaire)
G. M. Prosperi	Macroscopic physics and the problem of measurement in quantum mechanics
J. Kalckar	Measurability problems in the quantum theory of fields
J. S. Bell	Introduction to the hidden-variable question
A. Shimony	Experimental test of local hidden-variable theories
L. Kasday	Experimental test of quantum predictions for widely separated photons
B. S. DeWitt	The many-universes interpretation of quantum mechanics
H. D. Zeh	On the irreversibility of time and observation in quantum theory
G. Ludwig	The measuring process and an axiomatic foundation of quantum mechanics
F. Herbut and M. Vujičić	On a new development in the description of correlations between two quantum systems
A. Frenkel	Superselection rules and internal symmetries
K. E. Hellwig	Measuring process and additive conservation laws
L. de Broglie	L'interprétation de la mécanique ondulatoire par la théorie de la double solution
J. Andrade e Silva	Une formulation causale de la théorie quantique de la mesure
F. Selleri	Realism and the wave-function of quantum mechanics
H. Neumann	Seminar notes
D. Bohm	Quantum theory as an indication of a new order in physics
A. Shimony	Philosophical comments on quantum mechanics

Lecturers and talks at the Varenna 1970 school on Foundations of Quantum Mechanics

Its success was to a large extent a result of the abilities that d'Espagnat brought to its organization. In the mid-1960s, while pursuing a career in high energy physics at CERN, he was attracted by the increasing interest in the foundations of quantum mechanics. At CERN he became close to John Bell on recognizing that they shared a common interest in the foundations of quantum physics. A major contribution of d'Espagnat to the research on foundations required, however, not only his scientific training and philosophical inclination but also his diplomatic skills. When the council of the Italian Society of Physics decided to dedicate one of its Varenna Summer Schools to the foundations of quantum mechanics it was taking a decision to focus on a theme that was itself controversial. Since World War II the summer schools had been a privileged gathering of young promising physicists to train in research at the frontiers of physics. However, the issue arises: How should one train scientists on issues where no consensus exists? d'Espagnat dealt with this by setting the standards of behavior scientists should adhere to in case of controversy. He set out these rules in the invitation letter he sent to all participants. These diplomatic rules included (d'Espagnat 1971)[14]:

> 1) We should not take as our goals the conversion of the heretic but rather a better understanding of his standpoint. 2) We should not suggest that we consider as a stupid fool anybody in the audience (lest the stupid fools should in the end appear clearly to be ourselves!). 3) We should try to cling to facts. 4) Nevertheless, we should be prepared to hear without indignation very nonconformist views which have no immediate bearing on facts.

Considering the role played by d'Espagnat in this school and the increasing role he would play in this field, a few more words on him seem appropriate. He has been interested in both philosophy and science since his school days. Indeed, he got a French *Baccalauréat* in Mathematics and Philosophy but realized that investigations into the philosophy of science in the twentieth century required scientific training, which led him to a career in physics. In the mid-1960s, while pursuing a career in high-energy physics at CERN, he was gradually attracted by the increasing interest in the foundations of quantum mechanics. In 1965 he published *Conceptions de la physique contemporaine—Les interprétations de la mécanique quantique et de la mesure*, which would be followed 10 years later by the influential *Conceptual Foundations of Quantum Mechanics*, both part of his series of much-praised books (d'Espagnat 1965, 1989). The Varenna school and the growing interest in foundations during the following decade absorbed almost all of d'Espagnat's energy. In 1977 he visited the US as a Visiting Professor at the University of Texas, Austin, a place that John Archibald Wheeler had succeeded in establishing as a major American institution devoted to research on the foundations of quantum mechanics and general relativity. d'Espagnat arrived in the US continuing his crusade to "tolerate difference of views," as noted by Wheeler. However, discussions there with Wheeler, Everett, DeWitt, Henry Stapp, George Sudarshan, and James Hartle led d'Espagnat to change his former view on the interest of American physicists in foundational issues, as he wrote to Wheeler, "In fact, these three weeks made me discover both very attractive specific problems and also some aspects of the general trend of ideas in the United States that were novel to me, and that may perhaps correspond to a genuine evolution."[15] Eventually,

[14] d'Espagnat's diplomatic skills, however, were not enough to keep Léon Rosenfeld on the list of lecturers, as we have discussed in Chap. 4.

[15] d'Espagnat to Wheeler, 27 Apr 1977, Wheeler Papers, Series II—Box DE, folder d'Espagnat. "Tolerate difference" and references to the discussions are in Wheeler's notebook, pp. 145–149. Idem, Series V, Notebook October 1976–December 1977. Wheeler Papers, American Philosophical Society, Philadelphia, PA (*WP* hereafter). Bernard d'Espagnat, interviewed by the author, 26 Oct 2001, op. cit., AIP.

d'Espagnat abandoned high-energy physics and followed his dream to devote himself full-time to research into science and philosophy, now always related to the foundations of quantum mechanics.

The success of the 1970 school on foundations of quantum mechanics as a school was not, however, self-evident for all those present. The exceptionality of a summer school dedicated to a scientific controversy, about subjects on which there was a wide diversity of conflicting standpoints, was noted by one of its younger participants. Anders Barany was then 28 years old, a PhD student in theoretical physics at the University of Uppsala, Sweden. Barany had high expectations about the school as his doctoral research concerned the foundations of quantum mechanics, in particular the quantum measurement problem. However, for him such a school was different. According to his recollections,

> What most impressed me was that many (maybe even most) of the highly qualified scientists lecturing at the school could not cooperate to try and help the students form a coherent picture of the different problems connected with the foundations of quantum physics. Instead they immediately started arguing with each other and at some points were really fighting each other (almost physically!). In retrospect, having both attended and organized a large number of scientific meetings, I would not really call this a summer school, where "school" in some sense means that it should play an educational role, but rather a conference for mature scientists trying to bring forward their own results and messages.[16]

On his return to Sweden Barany filed the papers he had collected on the foundations of quantum mechanics and "did not touch them for many years." He went on the reflection, "If scientists such as Wigner, Jauch, Bohm, etc, could not agree on a proper direction to go, how could I even think of making a contribution?"[17]

As for the political background of that school, recollections from the attendees highlight the role played by the Italian and French students. "I remember that the Italians discussed politics a lot, but I don't remember what were the topics," are the memories from the German Michael Drieschner.[18] The German H. Dieter Zeh had more vague recollections, but they also point towards the Italians, meshing their political commitments with their critical stances about quantum physics: "Somebody told me that there were many communists (hence materialists) among the Italian organizers. Hence their problems with non-realism."[19] Zeh's and Selleri's cases should be contrasted to shed light on the development of the research on foundations. They had opposing philosophical views on quantum issues. While

---

[16] Anders Barany to the author, e-mail, 13 Jan 2011.

[17] *Idem.*

[18] Michael Drieschner to the author, e-mail, 24 Jan 2011.

[19] Dieter Zeh to the author, e-mail, 15 Jan 2011.

Selleri supported realism and criticized the usual quantum theory, Zeh had no qualms with non-realism and trusted the linear mathematical apparatus of quantum physics to obtain new quantum features (Freire Jr. 2009). As diverse as their stands were, both benefited from the creation of a professional and intellectual space for the foundations of quantum physics, which the 1970 Varenna school indeed was. Returning to politics, the reminiscences of the Swede Anders Barany are revealing[20]:

> There were violent political discussions going on, mainly emanating from the Italian (or "Latin") students. Most of the discussions were held "out-of-lectures", but as I remember it, a student "agitator" occupied the stage and read or wanted to read a revolutionary manifesto in front of the audience. This is where Wigner intervened and in my memory he managed to calm down the student and there was instead a peaceful discussion.

Barany's memories bring to surface two aspects noted by several participants. The political climate, while strong, did not interfere with the workings of the school and the singular role of the physicist Eugene Wigner. Indeed, the political episode most recalled is the incident related to July 4th involving Wigner. The incident may be described using various testimonies, including those from Basil Hiley, Emilio Santos, Andor Frenkel, Rémy Lestienne, and Giovanni Prosperi. Hiley, when asked about political incidents in the school, recalls:

> Yes, there was one classic political episode that took place. Wigner decided to throw a party and he happened to choose the 4th July. Remember that was the time of Vietnam and the Italian students, mainly Italian, tried to get the date changed. They were not going to take part in a party on 4th July. Of course Wigner refused to change the day. There was a lot of arguments and protests, a lot of hot air over this, but the party eventually went ahead on the 4th July as planned. It went off without incident.[21]

Another testimony, from the Spaniard Emilio Santos, while essentially similar to Hiley's, suggests a generational divide between those who protested and those who went to the party offered by Wigner:

> I remember very well that Eugene Wigner invited people to a party on the occasion of the National Day of USA, July 4, and there were some (mainly young) people who attempted to convince other people not to attend the party because the USA was involved in the Vietnam war at that time. Actually most senior people (myself in particular) attended the party. I remember that Wigner was very kind with the same young people who boycotted the party and there is even a photo of Wigner with them.[22]

[20] Anders Barany, *op. cit.*

[21] Basil Hiley to the author, e-mail, 22 Jan 2011.

[22] Emilio Santos to the author, e-mail, 19 Jan 2011.

Picture 6.1 Participants at the 1970 Varenna school on foundations of quantum mechanics. Wigner is seated in the middle, the only one wearing tie. On his right: Sabbadini and then Bell. Reproduced with permission from "Foundations of Quantum Mechanics", Proceedings of the International School of Physics "Enrico Fermi", course IL, edited by B. d'Espagnant (Academic Press), © SIF, 1971

The Hungarian physicist Andor Frenkel recalls that the hot political climate was not only related to the date and the Vietnam War, but also to the fact that Wigner's support for the war was well known: "Professor Wigner was a supporter of the war against Vietnam, and many participants of the School wanted the US out of Vietnam."[23] Recollections from Giovanni Prosperi describe some of those who opposed Wigner's initiative more vividly: "On Sunday there was no lesson and as it was USA independence day, Wigner wanted to give participants a party. It was the time of the Vietnam War and Selleri, who was quite left wing, was rather disappointed and tried to prevent it calling Toraldo di Francia in Bologna. However, Wigner was very determined and the party took place."[24] The party organized by Wigner was the political event most recalled among the attendees, highlighting Wigner as the main conservative pole, both for his professional prominence and his support for the Vietnam War. However, there was a kind of peaceful coexistence and competition, to use the political jargon, between the political conflicts and Wigner's role as a supporter of the debates around the foundations of quantum physics. Tonietti's reminiscences are evidence of such a coexistence:

> The intermingling of political and "technical-physical" arguments was evident, and palpable during these days in Varenna. Still keenly, I remember a direct discussion with E. Wigner at the dinner table. First we talked a little about the best formalism for Q.M.; then Wigner said that students of the '68 movement "must be mad". To that I replied: "They must be clever". Our conversation stopped. Toraldo di Francia, and Franco Selleri both sided with the Partito Comunista Italiano, I sided with the students' movement against the Soviet Union hoping for a New Left.[25]

[23] Andor Frenkel to the author, e-mail, 13 May 2011.

[24] Giovanni Prosperi to the author, e-mail, 28 Jan 2011.

[25] Tito Tonietti, *op. cit.*

Some political anxiety related to the school on the foundations of quantum mechanics survived the school and surfaced when d'Espagnat was editing the proceedings. He was afraid of the influence of the political context on the editing process. The source of d'Espagnat's worries, as expressed in his letter to the president of the Italian Physical Society, came from some of Franco Selleri's statements. "According to him, in fact, a trend would have appeared in [the Italian Physical] society whose goal would be to intimately associate scientific activities with activities from a different order."[26] D'Espagnat was then reassured by Toraldo di Francia, who wrote to him saying that,

> The directors of the Varenna's courses were, always, the final judges of what should be included in their proceedings and your case is not an exception. By the way, it is true that there is in our society the trend of not occupying itself only with technical issues, and this trend is more and more strong and well founded. However, surely, by foundations of quantum mechanics we understand the foundations of quantum mechanics, and this is the title of the proceedings to be published. Thus I hope to have dissipated your anxieties.[27]

In other words, what Toraldo di Francia was saying meant that while the political context had been influential in the choice of the subject of the school, its influence should be restricted to only this. The society's editing processes would not be influenced by it.

Before moving to the school dedicated to the history of physics, it is useful to enlarge the biographical information about Franco Selleri, because among the Italian physicists who suggested the school on foundations he would become in the future years a leader in this field of research. In 1958 Franco Selleri got his PhD in physics in Bologna where he was educated under the influence of the physicist Giampietro Puppi. In the following 10 years Selleri undertook a successful career in high-energy physics, which included original contributions such as the one-pion exchange model. These achievements assured him a position at Bologna University when he returned from a series of fellowships in Switzerland, France, and the US.[28] Disenchanted with the political and cultural climate in the Department of Physics at Bologna University and attracted by an invitation to start theoretical physics at a new university, he moved to Bari in 1968, where he remained till the end of his life. Selleri had become frustrated with the scarce amount of physical realism one could

---

[26] "D'après lui, en effet, une tendance se serait manifestée dans votre Société, ayant pour but d'associer intimement aux activités scientifiques des activités d'un autre ordre." Bernard d'Espagnat to Toraldo di Francia, 03 Sep 1970. ASIF.

[27] "Les directeurs des cours de Varenna ont été, toujours, les derniers juges de ce que doivent contenir les compte-rendus et votre cas ne fait pas d'exception. D'ailleurs, si il est vrai que dans notre Société la tendance à ne s'occuper pas seulement de questions techniques est de plus en plus forte et très bien justifiée, il est neanmoins certain que par 'Fondements de la mécanique quantique' nous entendons les fondements de la mécanique quantique. Et tel est le titre du volume de compte-rendus qui va être publié. J'espère d'avoir dissipé comme ça vos perplexities." di Francia to d'Espagnat, 16 Sep 1970; d'Espagnat to di Francia, 22 Sep 1970, emphasis in the originals; all letters in the ASIF.

[28] Tarozzi and van der Merwe (2004). The following citations and information come from Franco Selleri, interviewed by Olival Freire, 2003, *op. cit.*

lend to the approaches used in particle physics and began to see "the problems in elementary particle physics [as] due to the fact that quantum mechanics is poorly understood and anyway is a very abstract idea." He eventually became a full-time researcher working on the foundations of quantum mechanics and more recently on the theory of relativity. During this transition, his reading of d'Espagnat's 1965 *Conceptions de la physique contemporaine* was influential, as he realized how many interpretations could be accommodated by the quantum formalism.

Selleri elaborated a unique approach to foundations, combining his mistrust of the quantum formalism with an agenda to appeal for more and more experiments on these issues. Indeed, not only did he disagree with the complementarity interpretation for philosophical reasons, but he also thought that Hilbert space as a mathematical structure for quantum theory would eventually be considered erroneous and be replaced by a description in the normal space-time frame. He was among the first to suggest experiments to test de Broglie's wave plus particle picture, the "double solution," to expose loopholes in most of the first Bell's theorem experiments, and to suggest testing Bell's inequalities in particle physics. Experimental work to date has frustrated his expectations as it has confirmed quantum mechanics' predictions. However, his role as a kind of critical consciousness of experiments on Bell's theorem has probably been responsible for his high regard in this field. Furthermore, Selleri has mixed his defense of a realistic approach to quantum mechanics with what he considers wider social responsibilities.[29] In the 1980s he was responsible for building a bridge between critics of quantum mechanics and the philosopher Karl Popper, who was himself concerned with this physical theory, bringing the controversy in quantum mechanics to a wider audience. Selleri and Sexl's (1983) widely translated book, *Die Debatte um die Quantentheorie*, was part of this endeavor.

Turning his research to foundational issues did not cause any major damage to Selleri's professional career. Although his full professorship was postponed for 10 years, until 1980, explained by him as a result of his switch, he has never encountered major professional obstacles to his field of research: "I have been treated fairly. I have not been discriminated for the activity I did." Leaving a major center to work in a new center, Bari, with younger physicists also eased potential obstacles, as he acknowledged: "In Bologna it would have been more difficult." In hindsight, Selleri also considers that the Italian environment also contributed to the new stage of his career, "anyway, I have the feeling that Italy is more tolerant than other countries to the foundations of quantum mechanics," a feature he attributes to a factor not yet studied by historians: Enrico Fermi's criticisms of quantum

[29] "I always thought [...] that it is our duty to build a science that can be communicated to everybody. And at those times I was thinking in terms of the working class, the working people. That is to say, if the only way to understand what I'm doing is to study differential equations or Hilbert space, [...] there is too high a threshold. If instead I build physics in three-dimensional space and in time according to the rules of causality then I can communicate my results." Franco Selleri, interviewed by Olival Freire, *idem*.

mechanics. However, in the early 1970s he did not think of Italy in the same way. For him,

> For some time Paris was looking as the most interesting place on earth for my type of research, because de Broglie was alive. There was the de Broglie Foundation, there was Vigier, there was d'Espagnat. There were young leftists like Paty and Levy–Leblond interested in fundamental questions, so it seemed like a paradise.

Such a conducive climate did not last. According to his recalling,

> I have seen with time the paradise melting away completely, slowly, because eventually de Broglie died, the de Broglie Foundation after his death followed a path of low profile and it was not anymore a real defense of de Broglie's ideas, d'Espagnat changed completely his philosophy, and Vigier was very difficult to agree with from the beginning, because he considered himself a nonlocal realist, a position whose motivation for me is still very difficult to understand. [. . .] And then the young leftists converted to the orthodox line of thought as well, so with time nothing was left, and Paris disappeared from my horizon.

The turn of events in Paris was for him indicative of a negative feature of science at the time. He thinks that among the founding fathers of quantum mechanics there were two conflicting camps of equal sizes and it became "99 to 1" in favor of the complementarity view among those active in research. He suspects that dogmatism was the cause of the change. "How was it possible, through repression and control of positions and publications? And then also a lot of dogma exists. People do not dare to oppose important ideas."[30]

6.3.2 1972: History of Physics in the Twentieth Century

The school on the history of contemporary physics, under the direction of Charles Weiner, brought together both junior and senior historians of physics. Among them were Joan Bromberg, Robert Cohen, Gerald Holton, Marcello Cini, Yehuda Elkana, John Heilbron, Paolo Rossi, Max Jammer, Jerome Ravetz, and Martin Klein. Among the lecturers there was also a number of senior physicists such as the Nobel Prize winner Paul Dirac, Hendrik B. G. Casimir, Viktor Weisskopf, Edoardo Amaldi, and Léon Rosenfeld, who had a deep interest in the history of physics (Jacobsen 2008). Weiner had tried to bring Thomas Kuhn and Paul Forman, but was unsuccessful. The students and participants included a number of people who would later be directly or indirectly related to the history and philosophy of physics. Among them were Françoise Balibar, Angelo Baracca, Silvio Bergia, João Caraça, Penha Dias, Giulio Giorello, Sandro Petruccioli, Arcangelo Rossi, Carlo Tarsitani, John Worrall, Fritjof Capra, Manuel Doncel, Salvo d'Agostino, Noretta Koertge, Jean-Marc Lévy-Leblond, Luis Navarro, and Tito Tonietti. Other participants, such as Giancarlo Ghirardi and Constantine Philippidis, would later give meaningful contributions to the foundations of quantum physics, as we will see later.

---

[30] All quotations from Franco Selleri, interviewed by Olival Freire, *ibid.*

Compared to the school on the foundations of quantum mechanics, the school on the history of physics can be said to have been less influential in the reshaping of its own field insofar as it was already a well-defined scholarly field at the time. And yet, it was part of the process through which a number of young Italian physicists moved towards the history of physics, a process that only stabilized a few years later after facing other professional obstacles, according to analysis by Arturo Russo, who was one of the young physicists who converted to the history of science.[31] At any rate, the lectures became references for scholarly work (Weiner 1977) and it is right to say that a school dedicated to the history of physics organized by a professional society of physicists was, and is, unusual.[32]

M. J. Klein	The beginnings of the quantum theory
J. L Heilbron	Lectures on the history of atomic physics 1900–1922
P. A. M. Dirac	Recollections of an exciting era
J. Bromberg	Dirac's quantum electrodynamics and the wave-particle equivalence
H. B. G. Casimir	Development of solid-state physics
H. B. G. Casimir	Superconductivity
H. B. G. Casimir	Some recollections
P. Rossi	From Bruno to Kepler: man's position in the cosmos
Y. Elkana	The historical roots of modern physics
G. Holton	Electrons or subelectrons? Millikan, Ehrenhaft and the role of preconceptions
P. A. M. Dirac	Ehrenhaft, the subelectrons and the quark
E. Amaldi	Personal notes on neutron work in Rome in the 30s and the post-war European collaboration in high-energy physics
M. J. Sherwin	Niels Bohr and the atomic bomb: the scientific ideal and international politics, 1943–1944

(continued)

[31] The historian David Cassidy makes an analysis of the cultural unrest of the late 1960s and its influence on the history of physics with some similarities with the Italian case we are presenting. According to Cassidy (2011b, p. 141), "in the critical social environment of the day, historians and sociologists began to dismantle the apolitical, asocial, amoral ideology regarding the disinterested, value-free purity of physics," and yet, "The utilization of social perspectives, historian Paul Forman argued at the time, was essential to achieving intellectual independence from physicists' constructs and practices." See also Cassidy (2011a). On Russo, see Russo (2007), unpublished, *op. cit.*

[32] On Rosenfeld, see Jacobsen (2012). On the absence of Kuhn and Forman, see Charles Weiner, cable, 19 Feb 1972, ASIF. The following Brazilians were enrolled in this course: Amélia Império Hamburger, Enio Frota da Silveira, Penha Maria Cardoso Dias, Ernst Hamburger, and Ennio Candotti. The Hamburger couple were prevented from participating by the Brazilian military dictatorship (1964–1985) as they were then on trial for political offences. Promemoria per la SIF – Elenco dei candidati accettati per Il 3° corso di Varenna; Amélia e Ernst Hamburger, cable, 27 June 1972; Amélia and Ernst Hamburger to Toraldo di Francia, 26 July 1972; ASIF.

L. Kowarski	New forms of organization in physical research after 1945
W. Goldstein	Science, politics and international affairs
V. F. Weisskopf	Physics and physicists the way I knew them
H. B. G. Casimir	The relations between science and technology

Lecturers and talks at the Varenna 1972 school on History of Twentieth Century Physics

Some of the young attendees as well as some lecturers at the 1972 school on the history of physics were also interested in the debates on the foundations of quantum theory, the subject of the 1970 school, although they did not attend it. Some cases may illustrate this overlapping. Lévy-Leblond had been interested in the foundations of quantum physics since the mid-1960s, an interest which was not independent of his political engagement. He was immersed in the influence of Italian communism among the French communists, which had led to renewed attention to the role of science as a cultural phenomenon. The other influence came from reading the philosopher Gaston Bachelard through courses delivered by the Marxist philosopher Louis Althusser in Paris. All these influences merged in a very singular style which characterized Lévy-Leblond's contributions to the debate on the quantum foundations, contributions marked more by the critical analysis of established concepts than by new theoretical developments.[33] Max Jammer, one of the lecturers, was then writing a piece of history which would become a reference in the field of the quantum controversy (Jammer 1974). In 1985 Giancarlo Ghirardi proposed the Ghirardi–Rimini–Weber theory (GRW), which became a landmark in the quantum controversy as a systematic attempt to solve the quantum measurement problem, namely the collapse of the quantum states during measurements, through an addition of a stochastic term to Schrödinger's equation (Frigg 2009). Finally, Constantine Philippidis, a student of David Bohm and Basil Hiley, would a few years later renew the early Bohmian approach by using computers to generate the first graphic displays of the quantum potential and trajectories obtained through that approach, as we have seen in Chap. 2 of this book.

[33] For Lévy-Leblond's early interest in the foundations of physics, see the debate among Michel LeBellac, Jean-Pierre Vigier, François Lurçat, Pierre Lehmann, and himself in *Clarté* (n° 53, pp. 14–43, Janvier 1964), which was the magazine of the French Communist Student Union. Talk with the author, Nice, 5 November 2012.

Picture 6.2 Italian Physical Society, Varenna on Lake Como, Villa Monastero, 31st July–12th August 1972. Summer school on the History of Twentieth Century Physics. AIP Emilio Segre Visual Archives

When the 1972 Varenna school was held, political activism among the young scientists against the Vietnam War had escalated following the revelations of American official documents about the JASON project. This project had gathered a number of elite American scientists, most of them physicists, to advise the US in defense matters, including the Vietnam War (Moore 2008, p. 170; United States Department of Defense 1971; Finkbeiner 2006; Aaserud 1995). On 13 June 1972 protests reached a new level when French activist physicists impeded the Nobel Prize winner Murray Gell-Mann, one of the JASON scientists, from giving a talk at the prestigious Collège de France (Moore 2008, p. 172). This event motivated a number of similar incidents in Europe and in the US with JASON scientists being publicly targeted and criticized for their activities related to the military. The French physicist Jean-Marc Lévy-Leblond, one of the leaders of the protests which blocked Gell-Mann at the Collège de France, wrote to the Varenna school organizers asking, "would you be interested in a seminar on 'Radical views about science to-day?'"[34] At the school itself, apparently there was no internal conflict as there had been at the 1970 school involving Wigner. Indeed there was no politically conservative pole at the school as most of the professors, and naturally the students,

[34] Jean-Marc Lévy-Leblond to G. Jona-Lasinio, 28 June 1972, ASIF.

were left-wing inclined. As recalled by Joan Bromberg, one of the young historians to talk at the school, "My memories of that Varenna conference: that the students were busy with student protests while the faculty, a bunch of old leftists, were baffled as to how to respond."[35]

The apex of the political climate at the school was the approval of a manifesto, with 58 signatories, criticizing the involvement of scientists in the Vietnam War. Some fragments of the manifesto may give us a flavor of the discussions at the Varenna school:

> In recent weeks diplomats, journalists and responsible visitors to North Vietnam have reported the bombing of dykes [sic] by the United States Airforce. Officials of the U.S. government have acknowledged that several dykes have in fact been damaged by bombing. [...] The last tactics in the American war has been made possible by a systematic application of scientific discoveries for military purposes. [...] These new technologies have been fostered by scientists working in such projects as the Jason program of the Institute for Defense Analysis. This program has enlisted more than 30 top rank physicists, including five Nobel prize winners. [...] Our discussions have convinced us that it is no longer possible to separate our attitude on these issues from our professional activities. [...] We also call for the immediate ending of the bombing of Vietnam and the total withdrawal of American forces ...[36]

6.4 Ongoing Political Activism and Its Later Fading

Manifestations such as the expulsion of Gell-Mann from the Collège de France in Paris and the Varenna 1972 manifesto also happened in other places, among them with Gell-Mann again at CERN, in Geneva, Sidney Drell in Rome, and John Archibald Wheeler in Erice, Sicily. One of the most telling events happened in Trieste, at the International Center for Theoretical Physics during the symposium "Development of the physicist's conception of nature," to honor the 70th anniversary of the physicist Paul A. M. Dirac, held on 18–25 September 1972. The triggering events were the presence of Wheeler and Wigner, both well known to be JASON members and supporters of the American war in Vietnam. In addition to street demonstrations, including clashes with the riot police, there was the "Trieste Letter," signed by 450 scientists or students. It is noteworthy that the only public reaction from the conservative quarters came from Wigner, who carried a poster at the

[35] Joan Bromberg to the author, e-mail, 28 Jan 2011.

[36] The manifesto is reprinted in "The War Physicists", a volume with documents from the manifestations organized and published by Bruno Vitale (1976). Parts of the manifesto are translated into French in Jaubert and Lévy-Leblond (1973, pp. 186–187). These two sources are good repositories of original documents related to the political unrest of the late 1960s and early 1970s. Gell-Mann's episode at the Collège de France, Drell's at Cargese, and the events in Trieste and Varenna are recorded, under the title "European Confrontation Spoils Jason's Summer Vacation," in the American magazine *Science for the People*, 4(6), 9–14, 1972. The creation of the Science for the People movement will be presented later. I am thankful to Virgile Besson for calling my attention to this record.

symposium opening session with the words "I am flattered by your accusations. They are compliments for me."[37] The events around the Varenna summer schools were not a European singularity. Indeed, as remarked by Kelly Moore analyzing the American case, rebel scientists "disrupted many of the public rituals that had traditionally provided science with public demonstrations of unity around shared rules for social action, such as professional meetings and awards." Moore also indicates the political connection across the Atlantic, "[these] activities were paralleled in Europe in the summers of 1971 and 1972 by student activists in Italy and France. These campaigns were led by younger scientists who, like their American counterparts, wanted the United States to withdraw from Vietnam." She went on to conclude, "the war in Vietnam drew scientists into activism, both as targets and as active participants in social movements" (Moore 2008, pp. 19–20 and 171).[38]

Picture 6.3 Cover of "The War Physicists", collection of documents from European physicists protest against physicists' involvement in the Vietnam War, organized by Bruno Vitale

[37] For documents related to that demonstration, see "The War Physicists" (Vitale 1976, pp. 100–143). See also the report in the French newspaper *Le Monde* on 30 September 1972.

[38] For a review of Moore's book, see Harper (2009).

Science political activism continued in the 1970s but its strength began to fade, or at least to be socially assimilated or transmuted, and this lies beyond the scope of the present study. However, the fate of the 1972 Varenna manifesto is evidence of the disagreements among young scientists and students and senior scientists despite sharing a common stand against the Vietnam War. The manifesto was intended to have a wide circulation, particularly to be published in professional vehicles, which failed to occur. H. B. G. Casimir, then a senior physicist in Holland and one of the lecturers at the 1972 Varenna school, expressed his doubts about the content of the manifesto:

> For the time being I feel still reluctant about the publication [of the manifesto] in Europhysics News. [...] Also I must confess that I don't feel too happy about the actual text. [...] whereas a statement by physicists urging their colleagues to abstain from military work may have some effect, the statement by that same group that America should immediately withdraw his troops is somewhat ridiculous and therefore weakens the possible impact.[39]

Stronger disagreement was expressed by the physicist and Nobel Prize winner Hans Bethe, then an open critic of the arms race, in a letter to Bruno Vitale accusing him of misrepresenting the involvement of JASON scientists with the Vietnam War. According to Bethe:

> It would be unfair to the members of Jason, and to other American scientists who do some occasional consulting for the military establishment, to take the opinions of Drs. Wheeler and Wigner as typical of the Jason group or of these other scientists. In contrast to the great majority of American scientists, including those consulting for Jason, Drs. Wheeler and Wigner still support the American war in Vietnam. While it may be interesting to have a discussion with them, it would certainly not give a fair picture of the opinion of these other American scientists.[40]

The fate of the Varenna manifesto also reveals a cleavage between American and Italian physicists' sensitivities to the approach adopted in the manifesto. At the International Conference on High Energy Physics held in Batavia, Chicago, in September 1972, the Varenna statement was widely distributed and signatures requested but only 22 signatures were collected, two of them from American scientists. In contrast, at the 1972 Annual Congress of the Italian Physical Society in Cagliari, November 1972, the general meeting of the members of the society endorsed the Varenna statement (Vitale 1976, pp. 143–147). Officially, the manifesto survived, as Charles Weiner summarized at the introduction of the proceedings:

> Another important feature of the Varenna summer school was the intense and spirited discussion that engaged a large proportion of the faculty and students in informal evening sessions. The school took place during the Vietnam War, and the lectures on the social and political history of physics gave rise to concern about the role of physics in contemporary

[39] H. B. G. Casimir to J-M Lévy-Leblond, 04 Sep 1972, published in "The War Physicists" (Vitale 1976, p. 97).

[40] Hans Bethe to Bruno Vitale, letter, 12 September 1972, published in *The War*, *op. cit.* pp. 120–121.

history. These discussions culminated in a statement, drafted by some of the participants, condemning the war and the use of physics to prosecute it. The statement was approved by most of the school's participants. (Weiner 1977, p. xi)

In the mid-1970s the political activism was further fading. An evidence of this fading is telling for our story because it is related to foundations of quantum mechanics. It was the colloquium held in Strasbourg in 1974 to commemorate the fiftieth anniversary of the creation of this physical theory (Lopes and Paty 1977). The gathering was organized by the French physicist Michel Paty, then making a conversion to a career in philosophy of science, and the Brazilian physicist José Leite Lopes, who was at time exiled from his country due to the military dictatorship prevailing in Brazil. Paty and Lopes, together with the philosopher Hervé Barreau, were the main organizers of one of the research teams dedicated to history and philosophy of science created in the wake of political turmoil and reorganization of the French academic system after 1968. They ran the colloquia and the publication titled *Fundamenta Scientiae* (Berthelot et al. 2005, p. 49). Paty and Lopes invited both sides of the previous political battles for the conference: John Wheeler and Jean-Marc Lévy-Leblond, open supporter of the Vietnam War and activist against it, respectively.

In a sign of the times, differently from the Trojan war in the play by the French dramatist Jean Giraudoux (Giraudoux and Fry 1955), which eventually happened, the Strasbourg war did not take place. Interest in the foundations of quantum mechanics from both sides and the imminent resolution of the war in favor of Vietnam paved the truce. "Since you had shown some interest in my ideas at the Colloquium in Strasbourg, I indulge in sending you along various pieces of work," wrote Lévy-Leblond cordially to Wheeler. Remains of the expected battle can however be found in this correspondence. Lévy-Leblond also sent to Wheeler a copy of a letter he had sent to Paty before the colloquium, assuring him, who "was somewhat anxious about a possible clash in Strasbourg," that in spite of the divergences among them about the invitation to Wheeler, there will not be any political conflict during the event. Wheeler jotted on the letter a few words to his secretary "you may be interested in the opinions of this left wing activist!" and carefully translated several words from the French to the English in the letter Lévy-Leblond had sent to Paty, which evidences his own interest in the whole affair.[41]

The Italian case is then an evidence of how physicists who were politically active were able to open the borders of the physics discipline to include and value themes which were not usual in the discipline's agenda, namely foundations of quantum mechanics and history of physics. The activists were joined by physicists who, independent of their political stances, were interested in promoting the research on such themes or accepted them as a strategy to accommodate social

[41] J-M Lévy-Leblond to John Wheeler, 13 May 1974; Lévy-Leblond to Michel Paty, 18 Feb 1974, J. A. Wheeler Papers, Series I – Box L, folder Lévy-Leblond, American Philosophical Society, Philadelphia.

and professional tensions. In the case of foundations of quantum mechanics, this confluence allowed the creation of a professional space to boost its research.

6.5 On the Other Side of the Atlantic: The Schwartz Amendment

On the other side of the Atlantic related histories were unfolding and we need come back to the late 1960s in order to follow the events. On one hand, an enduring hostility of the American Physical Society and its magazine *Physics Today* towards the debates and manifestations against the Vietnam War. On the other hand, from 1970 on, a warm reception of *Physics Today* to debates on the interpretation of quantum theory. I would like to ask the following: Did the experience of the American physics establishment in dealing with the former have an influence on the latter? More broadly, did the very existence of a political controversy within the physics community create a more permissive atmosphere for a philosophical controversy? These are at least plausible conjectures, even though, as we shall see, we have at present no direct evidence of links.

The entire story of the protests in the United States against the escalation of the Vietnam War and the racial discriminations from the mid-1960s on has been the subject of many scholarly works.[42] While we focus here on the protests against the Vietnam War, due to its close influence on the case we are studying, the US indeed lived through the 1960s a wave of manifestations related to racial discrimination, which had its apex at riots following the assassination of Martin Luther King, Jr., on April 4, 1968 (McLaughlin 2014). As for the protests against the Vietnam War, physicists and physics students actively participated in those protests but the subject was far more controversial than in Europe insofar as a number of very influential American physicists took stands supporting the war. Eugene Wigner and John Archibald Wheeler, protagonists in our history, were among them. The American Physical Society (APS), the American Institute of Physics, and the magazine *Physics Today* were some of the scenes for conflicts among physicists but they were not the only places. The magazine has been the traditional house organ of the American physics establishment, formally linked to the American Institute of Physics, while APS is by far the largest professional organization in the American Institute of Physics. Many protests targeted research facilities installed at the university campuses where military research was taking place and later protests were directed against the scientists who were involved in the JASON project, particularly the work related to the electronic fence in Vietnam.

Divisive conflicts involving *Physics Today* began when Charles Schwartz, physics professor at the University of California at Berkeley, wrote to R. Hobart Ellis Jr., editor of the magazine, on May 28, 1967 asking him to publish a letter in

[42] See Moore (2008) and references therein.

which he suggested "that the membership of the American Physical Society or the American Institute of Physics shall be polled to yield a consensus opinion about the War in Vietnam." In the letter, Schwartz left no doubt about the stand he expected to be taken by American physics organizations: "I am one of a great many Americans who believe that the present course of this country's actions in Southeast Asia is wrong, deadly wrong." Unlike those who think that these organizations should not "offer public opinions on every question of the day," he argued, "my position regarding the Vietnam War is that this is a matter of such vital urgency that we cannot remain quiet, that we ought, as respected members of an important contributing profession in this country, to give this problem our best study and then speak out to the public of what we see." *Physics Toda*y's editor and the board of the societies procrastinated answers and the unpublished letter triggered one of the most divisive episodes in the history of American physics.[43]

To cut a long story short, as his request was not answered Schwartz came up with the idea of amending the APS constitution in order to allow its members to discuss and vote on public issues, an episode known as the Schwartz Amendment. The APS board of directors refused to submit this amendment to vote, arguing with the APS bylaw regulations, then after protests the board withdrew and the amendment was submitted to voting. However, "when the ballots were sent out in May 1968, the board included a statement of its opposition to the amendment," and the amendment was defeated by a three-to-one margin. The movement behind the Schwartz amendment ultimately led to the creation of a new organization, later known as Science for the People (SftP). The new organization was the idea of three physics professors—Martin Perl (Stanford, later he would share the 1995 Nobel prize), Schwartz (Berkeley), and Marc Ross (Michigan)—and a postdoctoral student, Michael Goldhaber. The idea was presented for consideration at the 1969 meeting of the APS.[44]

In the development of this story, the prestige and credibility of *Physics Today* and the APS were called into question. Schwartz blamed the APS for its early refusal to submit his proposal to vote accusing it of a "censorship completely alien to the principles of free discourse upon which a scientific community is built." Twenty years later he defended the same stand: "this was direct political censorship." The *Physics Today* editor had to argue with some of his peers abroad to defend the procedures of APS and *Physics Today*. Thus, he wrote to the editor of the *New Scientist* in the UK criticizing an editorial in the magazine—entitled "Physics

[43] C. Schwartz to R. Hobart Ellis Jr, 28 May 1967, Physics Today Papers, Correspondence 1948–1970, Box 20, AIP.

[44] Soon after the creation of this organization it began to publish a bi-monthly magazine with the same title, *Science for the People*. Later the organization changed its name to Science and Engineers for Social and Political Action (SESPA). The magazine was published from 1970 until the late 1980s. It provides a window to analyze how deep, wide, and entrenched the radical criticism to science was in the US in those times. On the events leading to the creation of the Science for the People organization, including quotations, see Moore (2008, pp. 133–157). A report on its first decade is Greeley and Tafler (1980).

Revolution"—for describing events in US physics in a biased manner. To the editor of the *German Physikalische Blätter*, he wrote saying that it was wrong to report that "the letter on the Vietnam question was rejected by the society's newspaper" because the APS does not have a newspaper and "the letter was rejected by PHYSICS TODAY, which is a publication of the American Institute of Physics." This was a rather defensive action—administrative definitions—because in fact both the magazine and the organizations were acting in agreement in the whole affair.[45]

Eventually, APS and *Physics Today* accepted some political changes, albeit minor ones when compared to the changes proposed by Schwartz and his colleagues. In particular the APS accepted the creation of a forum to deal with the relationship between physics and society, which eventually led in 1972 to the setting up of the Forum on Physics and Society. The proposal came from another Schwartz, this time Brian Schwartz, a theoretical physicist at MIT. "It was like APS was being squeezed from the east coast and the west coast by two Schwartzes," according to Brian Schwartz's later recollections. Charles Schwartz would later acknowledge these changes, "about that time, as you say, within APS, things happened; Physics Today certainly opened up. There was the formation of the Forum on Physics and Society."[46]

It was against this background that *Physics Today* decided to deal with softer issues rather than the hot political issues of the times. It took up the controversy over the interpretation of the quantum theory. Unlike what we had seen in the case of the Italian Physical Society and its Varenna summer school, where the connections between political unrest and the choice of themes for the schools are documented, in this case such a connection is plausible, persuasive, and contributes to making sense of the events we are presenting, but it is less documented. One of the main protagonists in this process was R. Hobart Ellis Jr., then the magazine's editor. He was an expert in nuclear engineering who had expressed some interest in conceptual issues in quantum physics. To exemplify this interest, he used the *Physics Today*'s column entitled Phimsy, usually dedicated to short notes with a comic flavor, to write a note entitled "The function is the particle." A little earlier, informed that Robert B. Lindsay would make a lecture tour for Sigma XI addressing the theme "Physics: To What Extent Is It Deterministic?", he wrote to him asking for a manuscript for *Physics Today*. About the same time, he wrote to George Trammell, from Rice University, asking for a reprint of a paper by Trammel and saying that "in reading it I see that you appear to be concerned with something that has intrigued me for some time—the real meaning of the wave function." Then he went to ask him for a good paper about this issue for the magazine. While he did not

---

[45] C. Schwartz, "censorship", in *Physics Today* 19, August, 9–10. C. Schwartz, "political censorship", in Interview of Charles Schwartz by Finn Aaserud on 15 May 1987, AIP. R. Hobart Ellis Jr. to Bernard Dixon, 30 Oct 1969, and R. Hobart Ellis Jr. to Ernst Brüche, 21 Oct 1968, Physics Today Division Records, 1948–1970, Boxes 18 and 20, respectively, AIP.

[46] Interview of Charles Schwartz, *op. cit.* Interview of Brian Schwartz by Patrick McCray on 10 Aug 2001, AIP.

get much reaction from Lindsay and Trammel, things would be different with Bryce DeWitt.[47]

6.6 *Physics Today* and the Second Life of Everett's Quantum Proposal

Bryce DeWitt was a theoretical physicist, trained at Harvard with a PhD under Julian Schwinger. From 1956 to 1971 he was at the University of North Carolina at Chapel Hill. He worked on the quantum theory of the gravitational field and since 1957 had followed the appearance of Everett's interpretation, as we have seen in Chap. 3. Ten years later, in 1967, DeWitt revived Everett's idea and certain factors were influential to the success of this revival. First, DeWitt used Everett's ideas in an intellectual context that was different from that of the dispute concerning the interpretations of quantum mechanics. Following a suggestion by John Wheeler, DeWitt arrived at a formalism that could describe the wave function of the whole universe. Its interpretation was far from evident, since there was no external observer to make a measurement in the sense of the ordinary quantum mechanics. In addition, he needed to combine general relativity and quantum theories, which was at the time, and still is, far from being a solved problem. In the early 1960s, already working on the quantum theory of gravity, he went into battle with Léon Rosenfeld about the proper interpretation of Bohr and Rosenfeld's thoughts about the measurement of quantum fields.[48] From this episode DeWitt inferred the idea of the existence of a dogmatic circle around the Copenhagen interpretation which was hampering research on the foundations of quantum physics. In a letter to Wheeler, he wrote: "The digs at the Copenhagen School were never meant to be included in the published version. [...] I must confess I made those digs, slightly maliciously, for your benefit. [...] (Re my use of the word 'rigid' in referring to the 'Copenhagen doctrine', how would you describe Rosenfeld's attitude on the subject?)"[49]

[47] *Physics Today*, August 1969, p. 21. R. Hobart Ellis Jr. to R. B. Lindsay, 01 March 1968; R. Hobart Ellis Jr. to G. T. Trammell, 26 March 1968, Physics Today Division Records, 1948-1970, Boxes 20 and 17, respectively, AIP.

[48] On this subject, see the doctoral dissertation of Thiago Hartz (2013) and the paper (Hartz and Freire Jr. 2015).

[49] DeWitt to Wheeler, 20 Apr, 1967; Wheeler Papers, Series I—Box Co-De, Folder DeWitt, *WP*. This letter refers to the paper "Quantum Theory of Gravity" (DeWitt 1967), whose manuscript he had sent to Wheeler's evaluation. In this manuscript, but not in the published version, DeWitt had written: "It is [...] clear that the quantum theory of space-time must ultimately force a deviation from the rigid Copenhagen doctrine." Wheeler circled the rigid Copenhagen doctrine and jotted "unfortunate." DeWitt also wrote "as conventionally formulated quantum mechanics comes in two packages: (1) formalism and (2) interpretation, the latter being supplied by a licensing office in Copenhagen." Wheeler jotted "bad tone." The manuscript is at the Bryce Dewitt's personal files, University of Texas at Austin, Austin, TX. I am grateful to Thiago Hartz for sharing with me the information about the manuscript. The full background of this story is analyzed by Hartz in the paper "Bryce DeWitt's road to the Many Worlds", forthcoming.

In the first of a trilogy of highly cited papers on the quantum theory of gravity (DeWitt 1967; Freitas and Freire Jr. 2003), in which what is now called the "Wheeler–DeWitt equation" appeared, DeWitt considered Everett's interpretation adequate to make sense of the tentative quantized equation of the whole universe. The reason was that this interpretation does not call for an "external observer" to perform measurements. According to DeWitt (1967, p. 1141),

> Everett's view of the world is a very natural one to adopt in the quantum theory of gravity, where one is accustomed to speak without embarrassment of the "wave function of the universe." It is possible that Everett's view is not only natural but essential.

DeWitt was not only interested in using Everett's idea in the context of cosmology but he also wanted to advertise Everett's ideas among physicists. He went on to rechristen it as the "many-worlds" interpretation, a label far from Everett's own goals, but responsible for popularizing the interpretation beyond the circle of professional physicists. In the late 1960s DeWitt found a wider audience than Everett had gotten 10 years before. The interest in Everett's ideas came not only from their implications for cosmology but also from the changing views shared by many physicists in the early 1970s about the matter of the foundations of quantum mechanics. DeWitt's paper was well received both among cosmologists and the flourishing community of foundations of quantum mechanics (Freitas and Freire Jr. 2003). He was invited to lecture on the relative states interpretation at the Varenna 1970 summer school, as we have seen earlier, and his proposal to publish a bibliographical review on the foundations of physics, which was part of the dissertation of Neil Graham, his PhD student, was well received by the *American Journal of Physics*' editors, who published it in a prestigious section of the journal, its "Resource Letters" (DeWitt and Graham 1971). He managed to publish the whole of Everett's dissertation, still then unpublished (DeWitt et al. 1973). Among his allies he would find a powerful and influential one, the editor of *Physics Today*.

As part of his crusade to favor Everett's ideas and favoring the diversity of interpretations of quantum theory, DeWitt approached R. Hobart Ellis, Jr., then editor of *Physics Today*. He asked the editor about the interest of the publication in "initiating another vigorous debate in a different area, which is also of keen interest to most physicists, namely the interpretation of quantum mechanics," and suggesting himself to write about Everett's interpretation. The reference to "another vigorous debate" comes from the fact that DeWitt had written a letter on tachyons, particles with speed greater than light [*Physics Today* 22(12), 1969]. He was surprised at the strong interest of Hobart Ellis Jr. in the subject. The editor wrote, "Your letter of 21 October strikes a very responsive chord. For a long time I personally have been dissatisfied with the apparent contradictions that physicists appear to be ready to live with in quantum mechanics and its interpretation." Hobart Ellis continued presenting the current state of the quantum interpretation issue as analogous to that which had preceded the Copernican Revolution at the dawn of modern science, which could be considered an outrage by physicists aligned with Bohr's complementarity views. Still, according to Hobart Ellis, "someone has compared the present situation with that in which cycles and epicycles could explain all the movements in the heavens and science was well satisfied with the view until the Copernican theory took over. I feel the comparison is particularly apt." Next, Hobart Ellis mentioned his previous project for publishing physicists' replies to a questionnaire on the subject, a project that failed due to the scant time available to do it. Finally, he concluded that "it seems to me that the article you propose would be a very interesting and useful contribution to Physics Today", but added that "in fact I think a general review of different interpretations of quantum mechanics without special emphasis on any one would be of interest".[50] The paper was published along the lines suggested by Hobart Ellis Jr., while keeping the author's stand favoring Everett's interpretation (DeWitt 1970). It is noteworthy that since the revival of the debate on the interpretations of quantum physics, with David Bohm's proposal in the early 1950s (see Chap. 2), this was a rare episode, one of the first times the influential *Physics Today* would open its pages to a major paper on the controversy on the foundations of quantum theory.[51]

Picture 6.4 Bryce DeWitt, circa 1970. Courtesy of North Carolina Collection, University of North Carolina at Chapel Hill Library

Some time after the publication of DeWitt's paper, "Quantum mechanics and reality," Harold L. Davies, who had succeeded Hobart Ellis Jr. in the magazine

[50] Bryce DeWitt to R. Hobart Ellis Jr, 21 Oct 1969; Hobart Ellis, Jr. to Bryce Dewitt, 24 Oct, 1969. Physics Today Division, Records, 1948–1970, AIP.

[51] Evidence about how influential that article was is the fact that nowadays it gathers 126 citations, which is a meaningful figure if one considers that *Physics Today* is not a technical physics journal. Source: Web of Science, consulted on 17 June 2013. In the 1950s there was only a paper (Margenau 1954) while in the late 1960s papers on Landé's new book on the interpretation of quantum mechanics appeared [see Shimony (1966), Landé (1967), and Born and Biem (1968)]. There was also a paper by W. E. Lamb (1969).

editorial office, undertook a typical editorial procedure, that of putting emphasis on a certain subject through calling a debate on it. *Physics Today* published in the same article, under the title "Quantum-mechanics debate," six long letters by L. E. Ballentine, Philip Pearle, Evan H. Walker, Mendel Sachs, Toyoki Koga, and Joseph Gerver, with different but critical points of view on Everett's interpretation, besides DeWitt's reply. The article was followed by several letters debating the theme, one of which remarked very incisively on the changing mood among physicists concerning the subject, a mood quite different from that in which Everett's interpretation had emerged 10 years before.[52] In fact, M. Hammerton (1971), from the Medical Research Center, Cambridge, UK, captured this changing mood, writing in a clear-cut manner,

> The very interesting contributions to the quantum mechanics debate in your April issue, and the paper by DeWitt which triggered them, exemplify the highly complex and subtle ways in which scientific opinion can change. When I was an undergraduate reading physics 20 years ago, [. . .] the Copenhagen line was "scientific," anything else was meaningless, mumbo-jumbo, or, at best, mistaken. Now the curious thing is that, as far as I am aware, there has been no major finding or theoretical insight that could be held to demolish or supersede this interpretation. Nevertheless, there is now considerable dissatisfaction with it, and a willingness to regard other points of view—for example, hidden variables—as being at least respectable.

Hammerton did not exploit the "highly complex and subtle ways in which scientific opinion can change." Surely, however, these ways included not only the standard cognitive factors, such as empirical evidence and theoretical constructions, but also social factors that may be related to the contextual setting of the production of science. In our case, opening a debate on quantum physics was a minor problem—to editors of *Physics Today*—when compared with the strong debate they had to host about the political role of the American Physical Society, a debate mainly related to the widespread idea of physics being closely related to the military efforts in the Vietnam War.

Conclusion

Our histories corroborate some general findings about the political and cultural unrest of the late 1960s and about history of physics at large. Through Europe and across the Atlantic, it was a true "transnational social movement," to use Hölscher's (2012) terminology. However, the transnational feature did not mean that topics such as foundations of quantum physics were equally supported in all the places involved. From our cases, we can see that ultimately the Italian Physical Society opened more room for this topic than its

(continued)

[52] See "Quantum-mechanics debate" (Ballentine et al. 1971). "Still more quantum mechanics," with letters by G.L. Trigg, M. Hammerton, R. Hobart Ellis Jr., R. Goldston, and H. Schmidt (Ellis 1971; Goldston 1971; Hammerton 1971; Schmidt 1971; Trigg 1971).

consort, the American Physical Society, did. While the former opened its prestige summer school to the subject of foundations of quantum mechanics, the latter opened the pages of its magazine to the same subject. While both societies opened room for the same subject it seems that the Italian experience was larger and more effective than the American one. These cases thus seem to corroborate an analogous conclusion reached by Kaiser (2012a, p. 121) comparing the policies of the *Physical Review* and *Nuovo Cimento*—the leading scientific journals of the respective societies—towards papers on the controversy over the quantum interpretation. As for Paul Forman's analysis of the rupture with disciplinarity as a sign of the appearance of postmodernism (Forman 2012), our characters confirm this assessment, but only to a certain extent. At least on the Italian scene, some of the young physicists who asked the SIF to dedicate its summer schools to subjects such as foundations and history of physics eventually left physics and did not get professionally involved with either physics or the history of physics. However, some others followed the disciplinary path to become part of the professional communities dedicated to such subjects.

The decision of the Italian Physical Society to dedicate some of its summer schools to topics such as foundations of quantum physics and history of physics in the twentieth century was neither motivated by developments in Italian physics nor physics at large. Indeed, their main instigation came from a convergence of interests among young Italian activist physicists and senior ones. The former looked for a way to do physics outside the mainstream. Senior physicists attempted to appease an upset scientific community, which included bringing topics which were at the margins of physics, such as the foundations of quantum physics, into the spotlight of the international physics community. The case concerning *Physics Today* in the turmoil of the late 1960s indicates how much easier it was for the American magazine of the American Physical Society to open its pages to a controversial topic in science than to open its pages to the heated political debate concerning the Vietnam War.

In a broader manner, the episodes we have analyzed confirm features of the workings of science that have been exploited by other historians. In fact, as noted by historians of science such as Timothy Lenoir (1997), Alexis de Greiff (2002), and Alexei Kojevnikov (2004), the sphere of politics does not necessarily hamper the practice of good science. The Italian case, in particular, reminds us of the case built by the historian Paul Forman when he argued on the influence of social factors in the direction of the research adopted in the building of the quantum theory in the early 1920s.[53] As science is a locally

(continued)

[53] Forman's paper on the dispensing of causality in quantum mechanics and the context of Weimar Republic is Forman (1971). This now classic paper is reprinted in Forman et al. (2011, pp. 85–201).

situated cultural practice, it is influenced by all the other dimensions from culture to politics and economics. According to the historian Norton Wise (2010), the cultural context, broadly conceived, may be considered a repository of resources that individuals as active protagonists may mobilize for the development of their intellectual and professional agendas (Wise 2010, pp. 430–431).[54]

This seems to have been the case of these Enrico Fermi summer schools in Varenna in the early 1970s and the debate on interpretations in the pages of *Physics Today*. Groups of physicists, young and not so young, exploited the political climate of the times to push for changes in the science establishment, suggesting and organizing schools which helped to foster the field of foundations of quantum physics as well as the field of the history of contemporary physics and publicizing the diversity of interpretations of quantum theory. Through these engagements they contributed to change the professional and intellectual environment, allowing a freer development of the research on the foundations of quantum theory.

References

Aaserud, F.: Sputnik and the Princeton 3 – the National-Security Laboratory that was not to be. Hist. Stud. Phys. Biol. Sci. **25**, 185–239 (1995)

Amaldi, E., Battimelli, G., De Maria, M.: Da via Panisperna all'America: i fisici italiani e la seconda guerra mondiale. Editori riuniti, Roma (1997)

Amaldi, E., Battimelli, G., Paoloni, G.: 20th Century Physics: Essays and Recollections – A Selection of Historical Writings by Edoardo Amaldi. World Scientific, River Edge, NJ (1998)

Ballentine, L., Pearle, P., Walker, E.H., Sachs, M., Koga, T., Gerver, J., Dewitt, B.: Quantum-mechanics debate. Phys. Today **24**(4), 36–44 (1971)

Berthelot, J.-M., Martin, O., Collinet, C.: Savoirs et savants les études sur la science en France. Presses universitaires de France, Paris (2005)

Born, M., Biem, W.: Dualism in quantum theory. Phys. Today **21**(8), 51–56 (1968)

Capocci, M., Corbellini, G.: Adriano Buzzati-Traverso and the foundation of the International Laboratory of Genetics and Biophysics in Naples (1962–1969). Stud. Hist. Philos. Biol. Biomed. Sci. **33**, 367–391 (2002)

Cassidy, D.: Paul Forman and the Environment and Practice of Quantum History. Weimar Culture and Quantum Mechanics – Selected Papers by Paul Forman and Contemporary Perspectives on the Forman Thesis C. Carson, A. Kojevnikov and H. Trischler, pp. 263–276. Imperial College Press & World Scientific, London (2011a)

Cassidy, D.: A Short History of Physics in the American Century. Harvard University Press, Cambridge, MA (2011b)

---

[54] Wise's paper was part of an ongoing debate on the historiography of physics involving Wise, Forman, and the reception of Forman's paper on the quantum physics in Weimar Germany. Unfortunately, however, most of this debate has happened through unpublished papers. On this debate, see Paul Forman, Reflections on the rejection of "Weimar Culture, Causality, and Quantum Theory" by modern and by postmodern historians of science, 2007, unpublished paper. I am thankful to Forman for bringing this paper to my attention.

Cento Bull, A., Giorgio, A.: Speaking Out and Silencing: Culture, Society and Politics in Italy in the 1970s. Legenda, London (2006)

Clauser, J.: Early history of Bell's theorem. In: Bertlmann, R.A., Zeilinger, A. (eds.) Quantum [Un] Speakables – From Bell to Quantum Information, pp. 61–98. Springer, Berlin (2002)

Cozzoli, D., Capocci, M.: Making biomedicine in twentieth-century Italy: Domenico Marotta (1886–1974) and the Italian Higher Institute of Health. Br. J. Hist. Sci. **44**(4), 549–574 (2011)

d'Espagnat, B.: Conceptions de la physique contemporaine – les interprétations de la mécanique quantique et de la mesure. Hermann, Paris (1965)

d'Espagnat, B.: Foundations of Quantum Mechanics – Proceedings of the International School of Physics "Enrico Fermi". Academic, New York (1971)

d'Espagnat, B.: Conceptual Foundations of Quantum Mechanics, 2nd edn. Addison-Wesley, Redwood City, CA (1989) [1st ed 1971]

DeWitt, B.S.: Quantum theory of gravity. I. Canonical theory. Phys. Rev. **160**(5), 1113–1148 (1967)

DeWitt, B.S.: Quantum mechanics and reality. Phys. Today **23**(9), 155–165 (1970)

DeWitt, B.S., Graham, R.N.: Resource letter IQM-1 on interpretation of quantum mechanics. Am. J. Phys. **39**(7), 724–738 (1971)

DeWitt, B.S., Everett, H., Graham, N.: The Many-Worlds Interpretation of Quantum Mechanics: A Fundamental Exposition. Princeton University Press, Princeton, NJ (1973)

Ellis, R.H.: Still more quantum mechanics. Phys. Today **24**(10), 11–13 (1971)

Finkbeiner, A.: The Jasons: The Secret History of Science's Postwar Elite. Viking, New York (2006)

Forman, P.: Weimar culture, causality, and quantum theory, 1918–1927: adaptation by German physicists and mathematicians to a hostile intellectual environment [Reprinted in P. Forman et al. (eds.) Weimar culture and quantum mechanics: selected papers by Paul Forman and contemporary perspectives on the Forman thesis, 2011]. Hist. Stud. Phys. Sci. **3**, 1–115 (1971)

Forman, P.: Production and curation: modernity entailed disciplinarity, postmodernity entails antidisciplinarity. Osiris **27**(1), 56–97 (2012)

Forman, P., Carson, C., Kojevnikov, A., Trischler, H.: Weimar Culture and Quantum Mechanics: Selected Papers by Paul Forman and Contemporary Perspectives on the Forman Thesis. Imperial College Press & World Scientific, London (2011)

Freire Jr., O.: The historical roots of "foundations of quantum mechanics" as a field of research (1950–1970). Found. Phys. **34**(11), 1741–1760 (2004)

Freire Jr., O.: Philosophy enters the optics laboratory: Bell's theorem and its first experimental tests (1965-1982). Stud. Hist. Philos. Mod. Phys. **37**, 577–616 (2006)

Freire Jr., O.: Quantum dissidents: research on the foundations of quantum mechanics circa 1970. Stud. Hist. Philos. Mod. Phys. **40**(4), 280–289 (2009)

Freire Jr., O.: On the connections between the dialectical materialism and the controversy on the quanta. Yearbk. Eur. Cult. Sci. (Jahrbuch Für Europäihce Wissenschaftskultur) **6**, 195–210 (2011)

Freitas, F.H.A., Freire Jr., O.: Sobre o uso da Web of Science como fonte para a história da ciência. Revista da SBHC **1**(2), 129–147 (2003)

Frigg, R.: GRW theory (Ghirardi, Rimini, Weber model of quantum mechanics). In: Greenberger, D., Hentschel, K., Weinert, F. (eds.) Compendium of Quantum Physics Concepts, Experiments, History and Philosophy, pp. 266–270. Springer, Berlin (2009)

Giraudoux, J., Fry, C.: Jean Giraudoux. Tiger at the gates. (La Guerre de Troie n'aura pas lieu.) Translated by Christopher Fry. [London, Appolo Theatre, 2 June 1955.]. Methuen, London (1955)

Goldston, R.: Still more quantum mechanics. Phys. Today **24**(10), 11–13 (1971)

Greeley, K., Tafler, S.: History of science for the people: a ten year perspective. In: Arditti, R., Brennan, P., Cavrak, S. (eds.) Science and Liberation, pp. 369–383. South End Press, Boston, MA (1980)

Greiff, A.D.: The tale of two peripheries: the creation of the International Centre for Theoretical Physics in Trieste. Hist. Stud. Phys. Biol. Sci. **33**(1), 33–59 (2002)

Hammerton, M.: Still more quantum mechanics. Phys. Today **24**(10), 11–13 (1971)

Harper, K.C.: Disrupting science: social movements, American scientists, and the politics of the military, 1945-1975. Isis **100**(1), 197–199 (2009)

Hartz, T.: As heterodoxias quânticas e o olhar do historiador: uma história dos usos dos argumentos de Niels Bohr acerca da medição de campos quânticos (1930-1970). PhD dissertation, Universidade Federal da Bahia and Universidade Estadual de Feira de Santana (2013)

Hartz, T., Freire, O. Jr.: Uses and appropriations of Niels Bohr's ideas about quantum field measurement (1930-1965). Scientia Danica, Series M – Mathematica et Physica: Forthcoming (2015)

Hobsbawm, E.J.: The Age of Extremes. The short twentieth century 1914-1991. Penguin, London (1994)

Hölscher, M.: 68 Generation. In: Anheier, H.K., Juergensmeyer, M. (eds.) Encyclopedia of Global Studies, pp. 1552–1553. Sage, London (2012)

Jacobsen, A.S.: The complementarity between the collective and the individual – Rosenfeld and Cold War History of Science. Minerva **46**(2), 195–214 (2008)

Jacobsen, A.: Léon Rosenfeld – Physics, Philosophy, and Politics in the Twentieth Century. World Scientific, Singapore (2012)

Jamison, A.: Science and technology in postwar Europe. In: Stone, D. (ed.) The Oxford Handbook of Postwar European History, pp. 630–648. Oxford University Press, Oxford (2012)

Jammer, M.: The Philosophy of Quantum Mechanics – The Interpretations of Quantum Mechanics in Historical Perspective. Wiley, New York (1974)

Jaubert, A., Lévy-Leblond, J.-M.: Autocritique de la science. Éditions du Seuil, Paris (1973)

Judt, T.: Postwar – A History of Europe Since 1945. Penguin, New York (2005)

Kaiser, D.: How the Hippies Saved Physics: Science, Counterculture, and the Quantum Revival. W. W. Norton, New York (2012a)

Kaiser, D.: A tale of two textbooks experiments in genre. Isis **103**(1), 126–138 (2012b)

Kevles, D.J.: The Physicists: The History of a Scientific Community in Modern America. Knopf, New York (1978)

Kissinger, H.: On China. Penguin, New York (2011)

Kojevnikov, A.: Stalin's Great Science: The Times and Adventures of Soviet Physicists. Imperial College Press, London (2004)

Lamb, W.E.: An operational interpretation of nonrelativistic quantum mechanics. Phys. Today **22** (4), 23–28 (1969)

Landé, A.: New foundations for quantum physics. Phys. Today **20**(2), 55–58 (1967)

Lenoir, T.: Instituting Science: The Cultural Production of Scientific Disciplines. Stanford University Press, Stanford, CA (1997)

Lopes, J.L., Paty, M.: Quantum Mechanics: A Half Century Later. Holland/D. Reidel, Dordrecht/Boston, MA (1977)

Magri, L.: The Tailor of Ulm: Communism in the Twentieth Century. Verso, New York (2011)

Margenau, H.: Advantages and disadvantages of various interpretations of the quantum theory. Phys. Today **7**(10), 6–13 (1954)

McLaughlin, M.: The Long, Hot Summer of 1967: Urban Rebellion in America. Palgrave Macmillan, New York (2014)

Messiah, A.: Quantum Mechanics. North Holland, Amsterdam (1961)

Moore, K.: Disrupting Science: Social Movements, American Scientists, and the Politics of the Military, 1945-1975. Princeton University Press, Princeton, NJ (2008)

Ravetz, J.: Orthodoxies, critiques and alternatives. In: Olby, R.C., et al. (eds.) Companion to the History of Modern Science, pp. 898–908. Routledge, London (1990)

Salvini, G.: Book review: Edoardo Amaldi. Da via Panisperna all'America: I fisici italiani e la seconda guerra mondial. Isis **96**(2), 295 (2005)

Schmidt, H.: Still more quantum mechanics. Phys. Today **24**(10), 11–13 (1971)

Schweber, S.S.: Writing the biography of Hans Bethe: contextual history and Paul Forman. Phys. Perspect. **6**, 179–217 (2014)
Selleri, F.: Realism and the wave-function of quantum mechanics. In: d'Espagnat, B. (ed.) Foundations of Quantum Mechanics. Proceedings of the International School of Physics "Enrico Fermi", pp. 398–406. Academic, New York (1971)
Selleri, F.: Sull'ideologia nella fisica contemporanea. Critica marxista Quaderni **6**, 120–150 (1972)
Selleri, F., Sexl, R.U.: Die Debatte um die Quantentheorie. F. Vieweg, Braunschweig (1983)
Shimony, A.: Review of Alfred Landé, new foundations of quantum mechanics. Phys. Today **19**, 85–90 (1966)
Snow, C.P.: The Two Cultures and the Scientific Revolution. Cambridge University Press, New York (1959)
Società Italiana di Fisica: La scienza nella società capitalistica. Donato, Bari (1971)
Stevens, H.: Fundamental physics and its justifications, 1945-1993. Hist. Stud. Phys. Biol. Sci. **34**, 151–197 (2003)
Tagliagambe, S. (ed.): L'interpretazione materialistica della meccanica quantistica – fisica e filosofia in URSS. Feltrinelli, Milan (1972)
Tarozzi, G., van der Merwe, A.: For Franco Selleri on his seventieth birthday. Found. Phys. **34**(1), 1613–1615 (2004)
Trigg, G.L.: Still more quantum mechanics. Phys. Today **24**(10), 11–13 (1971)
United States Department of Defense: The Pentagon Papers; the Defense Department History of United States Decision Making on Vietnam. Beacon, Boston, MA (1971)
Vitale, B.: The War Physicists [Documents About the European Protest Against the Physicists Working for the American Military Through the JASON Division of the Institute for Defence Analysis, IDA]. B. Vitale, Napoli (1976)
Weiner, C.: History of Twentieth Century Physics. Academic, New York (1977)
Werskey, G.: The Marxist critique of capitalist science: a history in three movements? Sci. Cult. **16** (4), 397–461 (2007)
Wise, M.N.: Forman reformed, again. In: Carson, C., Kojevnikov, A., Trischler, H. (eds.) Weimar Culture and Quantum Mechanics: Selected Papers by Paul Forman and Contemporary Perspectives on the Forman Thesis, pp. 415–431. Imperial College Press & World Scientific Press, London (2010)
Wisnioski, M.: Engineers for Change: Competing Visions of Technology in 1960s America. MIT Press, Cambridge, MA (2012)

Chapter 7
Philosophy Enters the Optics Laboratory: Bell's Theorem and Its First Experimental Tests (1965–1982)

Abstract This chapter deals with the ways that the issue of completing quantum mechanics was brought into laboratories and became a topic in mainstream quantum optics. It focuses on the period between 1965, when Bell published what we now call Bell's theorem, and 1982, when Aspect published the results of his experiments. Discussing some of those past contexts and practices, I show that factors in addition to theoretical innovations, experiments, and techniques were necessary for the flourishing of this subject, and that the experimental implications of Bell's theorem were neither suddenly recognized nor quickly highly regarded by physicists. Indeed, I will argue that what was considered good physics after Aspect's 1982 experiments was once considered by many a philosophical matter instead of a scientific one, and that the path from philosophy to physics required a change in the physics community's attitude about the status of the foundations of quantum mechanics.

7.1 Introduction[1]

Quantum non-locality, or entanglement, that is the quantum correlations between systems (photons, electrons, etc.) that are spatially separated, is the key physical effect in the burgeoning and highly funded search for quantum cryptography and computation. This effect emerged in relation to the investigation of the possibility of completing quantum theory with supplementary variables, an issue once considered very marginal in physics research. This paper deals with the ways that the issue of completing quantum mechanics, especially completing it according to the criterion of locality, was brought into laboratories and, later on, became a topic in mainstream quantum optics. Discussing some of the past contexts and practices

[1] This chapter is a modified version of the work: Freire Jr., O. Philosophy Enters the Optics Laboratory: Bell's Theorem and its First Experimental Tests (1965–1982), *Studies In History and Philosophy of Modern Physics*, v. 37, p. 577–616, 2006. Additions from later archival research were introduced, references were updated, and stylistic rules were adapted to this book. Acknowledgements are in the original paper.

O. Freire Junior, *The Quantum Dissidents*, DOI 10.1007/978-3-662-44662-1_7

related to Bell's theorem, I hope to show that factors in addition to theoretical innovations, experiments, and techniques were necessary for the flourishing of research on this issue, and that the experimental implications of Bell's theorem were neither suddenly recognized nor quickly highly regarded by physicists. Indeed, I will argue that what was considered good physics after Alain Aspect's 1982 experiments was once considered by many a philosophical matter instead of a scientific one, and that the path from philosophy to physics required a change in the physics community's attitude about the intellectual and professional status of the foundations of quantum mechanics. I have argued elsewhere (Freire Jr. 2004) that a new attitude toward the foundations of quantum mechanics matured around 1970 related to subjects like the measurement problem and alternative interpretations of quantum mechanics, which were related neither to Bell's theorem nor to experimental tests. In the present chapter, I argue that even concerning Bell's theorem and its tests a similar new attitude was required. On these events and periods there are already a number of testimonies, popular science books, and science studies works.[2] This chapter, however, differs in that it attempts a historically oriented study about how what was considered a philosophical quarrel became a genuine topic of physics research.

Horne et al. (1990) have produced a historical account of the concept of entanglement. These authors showed that as early as 1926 Erwin Schrödinger realized that this concept is a consequence of the mathematical structure of quantum mechanics, and that in the same year Werner Heisenberg explained the energy structure of the helium atom using states that are entangled.[3] However, they also showed that in none of the first quantum mechanical treatments of many-body systems "was entanglement exhibited for a pair of particles which are spatially well separated over macroscopic distances" and that only with the Einstein–Podolsky–Rosen *Gedanken* experiment, proposed in 1935 (Einstein et al. 1935), was this feature of the mathematical structure of quantum mechanics explicitly discussed. Finally, they showed that Schrödinger not only reacted to this ideal experiment by introducing the term "entanglement", but also asked himself if this quantum feature would be confirmed by experiments, or not. The authors continued sketching the historical record, passing through the appearance of Bell's theorem and its first tests until the appearance of down-conversion pairs of photons, in the late 1980s, led to improved tests of Bell's theorem. This paper has a narrower timeline. I focus on the period between 1965, when John Bell published what we now call Bell's theorem,

---

[2] See, for instance, Aczel (2002), Bernstein (1991), Gilder (2008), Clauser (1992, 2002, 2003), and Wick (1995). Studies with a sociological or historical approach are Harvey (1980), Harvey (1981), Pinch (1977), and Bispo et al. (2013). The latter is a study of the techniques and instruments used in Clauser's first experiment. The authors argue that this experiment could not have been carried out earlier as it used phototubes called "quanticons" which had just arrived on the market. For this information, see Gilder (2008, p. 266).

[3] "If two separated bodies, each by itself known maximally, enter a situation in which they influence each other, and separate again, then there occurs regularly that which I have just called *entanglement* of our knowledge of the two bodies" (Schrödinger 1983, p. 161).

and 1982, when Alain Aspect published the results of his experiments violating Bell's inequalities and supporting quantum mechanics.[4] I leave aside Albert Einstein's and Niels Bohr's previous works and the debates on the interpretation and foundations of quantum mechanics in the 1930s, the debates on hidden variables triggered by the appearance of David Bohm's causal interpretation in the 1950s, which was the subject of Chap. 2 in this book, and the ongoing series of new experiments on Bell's inequalities, since the late 1980s, which will be discussed in the Chap. 8. Of these excluded topics, we only need to consider the context of the debates around Bohm's interpretation, since it strongly influenced the production and the initial reception of Bell's work. Indeed, Bell's decision to approach the hidden variable issue came from the very existence of Bohm's interpretation. Bell and his associates also inherited from the 1950s what, retrospectively Clauser (2003, p. 20) named the "stigma [...] against any associated discussion of the notion of hidden variables in quantum mechanics."

The period in focus also allows us to discuss why optics became the privileged bench for experimental tests of Bell's inequalities. John Bell himself did not think it would be so at the beginning. It were those who first pushed these inequalities into the laboratories, such as Abner Shimony and John Clauser, who realized the conceptual advantages of optical tests when compared to tests with positronium annihilation, proton scattering, and other experiments. However, in addition to these advantages, other reasons operated in favor of optics. Training in optics was an asset of many who were willing to work on Bell's theorem and, for this reason, they could do their best and achieve telling results. By 1969 there was a balanced distribution of scientific skills, Clauser and Richard Holt being the optics experimenters and Shimony and Michael Horne the theoreticians without training in optics. Later, a meaningful number of protagonists were trained in optics. One would like to inquire further into the connection between quantum optics and foundations of quantum mechanics. This question is not central to this book, but I remark that Joan Bromberg (2006), who is working on the history of quantum optics in the US, suggested that "device physics was pursued in tandem with fundamental physics, and even with research into the foundations of quantum mechanics," and that device research led to fundamental physics problems, and the latter in turn inspired new devices. Indeed, technical improvements made available while the experiments were being carried out, such as the tunable laser, dramatically improved the accuracy of the experimental results.

While following the theoretical, experimental, and technical issues related to Bell's theorem and its tests, I will pay attention to the biographical sketches of a few physicists involved in this story, addressing questions like: what factors led them to choose issues from the foundations of quantum mechanics as research themes? What issues did each one come to grips with? What were the favorable factors, and

---

[4] Bell's theorem (Bell 1964) was indeed published in 1965. In addition, for the sake of chronology, it should be noted that Bell's papers, (Bell 1964; Bell 1966), were written in the inverse order of their publication.

the obstacles, to their activities? To what extent did they succeed in their endeavor? Comparing their biographies, and thereby drawing resources from the method of prosopography in history and history of science,[5] I can draw a rough collective biography of these figures. As a whole, it is a story of success since they pushed a subject from the margins of physics to its mainstream; but it also included failures and hopes not fulfilled. It suggests to us that foundations of quantum mechanics was, at least in the period under analysis, still a job for quantum dissidents, fighting against the dominant attitude among physicists according to which foundational issues in quantum mechanics were already solved by the founding fathers of the discipline. However, the common ground of the quantum dissidents was minimal and focused just on relevance of the research in the foundations of quantum mechanics, since these scientists supported different interpretations of this physical theory and chose different approaches and issues in their research. The different features of these biographies are as enlightening as their common traits are, and I will address both the contrasts and common features in this paper.

The second section of this chapter deals with the context in which Bell's theorem was produced, its content, and its initial and uncomprehending reception; it covers the period between 1965 and 1969. Next I move on to analyze John Clauser and Abner Shimony's reactions, their proposal to submit the theorem to a viable experimental test, the involvement of American teams with these experiments, and the conflicting results among the first two experiments with optical photons. These events took place roughly between 1969 and 1974. The fourth section analyzes how the physicists involved settled the experimental tie with two new experiments carried out by Clauser and Edward Fry. This section is focused on the period between 1975 and 1976. It also treats the socialization of the physicists involved with research on Bell's theorem as well as the professional recognition of such themes of research. It pays attention to the cases of the journal *Epistemological Letters* and of the Erice 1976 meeting, which was seen by some physicists as the turning point in the acceptance that quantum nonlocality was indeed a new physical effect. The fifth section analyzes the road toward what was considered at the time the new challenge, an experiment changing the analyzers while the photons are in flight. It roughly deals with the period between 1976 and 1982, when Alain Aspect announced the results of his last experiments. Aspect's experiments resonated with the shifting attitude among physicists in the direction of a wide recognition of the importance of Bell's theorem. I conclude the chapter by drawing conclusions on the way physicists changed their views on such subjects of research and by drafting a collective biographical sketch of the physicists who brought hidden variables to the optics laboratories.

[5] For methods in prosopography, see Stone (1971) and Kragh (1987, pp. 174–181).

7.2 Bell's Theorem, the Context of Its Production, and Its Initial Reception

The title of Bell's first paper, "On the problem of hidden variables in quantum mechanics," suggests a strong relation with the research program aiming to introduce hidden variables into quantum theory that was conducted by David Bohm, Louis de Broglie, and Jean-Pierre Vigier in the 1950s. However, the title cannot be taken at face value. Although they were historically intertwined, Bell's contributions did not give a new breath to this program. Rather, Bell led the hidden variable issue in a completely new direction. Indeed, Bell's approach and main achievements in quantum mechanics are of a very different strain when compared to Bohm's (1952) hidden variable interpretation. While Bohm built models that would first mimic quantum mechanics and later on lead to distinctive results, Bell was interested in the critical analysis of the assumptions behind mathematical proofs and *Gedanken* experiments. This way, Stöltzner (2002) has argued that despite Bell's enduring criticisms of von Neumann's proof of the impossibility of hidden variable in quantum mechanics, "a mathematically minded view on the relation between the theorems of von Neumann and Bell" should consider Bell's theorem to be "a generalization of von Neumann's." Stöltzner's point is that "if one considers [. . .] Hilbert's axiomatic method as a critical enterprise, Bell's theorem improves von Neumann's by defining a more appropriate notion of 'hidden variable' that permits one to include Bohm's interpretation which recovers the predictive content of quantum mechanics." However, Bell's work has a close historical connection with Bohm's work on a hidden variable interpretation. He was directly motivated by the very existence of Bohm's proposal and by its reception among physicists. Bell's (1982, 1987) statements—"In 1952 I saw the impossible done," and "Bohm's 1952 papers on quantum mechanics were for me a revelation"—hide more truth than is usually recognized.

Picture 7.1 John Bell (1928-1990)—on the board, drawing of Aspect's 1982 experiment with two-channel polarizers. Courtesy: Nature

Born in 1928, in Belfast, Bell had no scientific or educational family background; indeed, he was the first of his family to go to high school. He went to Queen's University in Belfast, where he earned a BSc in Physics and formed the conviction that he would be a theoretical physicist. A job at the Atomic Energy Research Establishment at Harwell permitted him a leave of absence to begin his doctorate in Birmingham under Rudolf Peierls. Bell built his reputation working on high-energy particle physics theory and the design of particle accelerators, and from 1960, he and Mary Bell, his wife, worked at CERN in Geneva. His early concerns about quantum mechanics can be traced back to his undergraduate courses in Belfast, where he quarreled with Richard Sloane, his teacher, because Sloane was not able to afford him a plausible explanation of Heisenberg's uncertainty principle (Whitaker 2002, pp. 14–17). Later on, he avowed: "When I was a student I had much difficulty with quantum mechanics. It was comforting to find that even Einstein had had such difficulties for a long time" (Bell 1982, p. 989). Since then he began to think about the transition between quantum and classical descriptions of the world.[6]

---

[6] We can reconstruct this account due to Bell (1982) and Jeremy Bernstein (1991), who wrote his *Quantum Profiles* based on extensive talks with Bell and John Wheeler. Besides Bernstein, biographical information on Bell can also be collected from Shimony (2002), Whitaker (2002), and from papers by Bernard d'Espagnat, Michael Horne, and others, gathered in Bertlmann and Zeilinger (2002). For a comprehensive evaluation of Bell's scientific contributions, see Jackiw and Shimony (2002), Jackiw and Shimony (2008). His selected papers are in Bell et al. (1995) and his papers on the foundations of quantum physics are collected in Bell (2004b).

"Smitten by Bohm's papers,"[7] the Irish physicist attempted to determine what was wrong with von Neumann's proof, since it did not allow for hidden variables in quantum mechanics. Bell knew von Neumann's proof only indirectly, from his reading of Max Born's *Natural Philosophy of Cause and Chance*, but he could not read von Neumann's book because at that time there was no English edition of it. The solution was to ask Franz Mandl, his colleague at Harwell, about the content of the book. "Franz was of German origin, so he told me something of what von Neumann was saying. I already felt that I saw what von Neumann's unreasonable axiom was." He wrote to Wolfgang Pauli (Pauli and Meyenn 1999, p. 28) asking for reprints of his paper on Bohm's proposal, but he probably did not like the views expressed there since Pauli (1953, p. 33) had considered Bohm's hidden variables as "artificial metaphysics." He went to Birmingham in 1953, including hidden variables as one of the possibilities for his studies. Asked by Peierls to give a talk about what he was working on, "Bell gave Peierls a choice of two topics: the foundations of quantum theory or accelerators." Peierls chose the latter, which was the end of the first stage of Bell's involvement with hidden variables. The intermezzo lasted 10 years; he only resumed this work at Stanford, during a leave of absence from CERN.

However, as we will see, there were intermediate events which prompted Bell to resume his early reflections. In the first of the two articles on foundations of quantum mechanics he published while in the US, Bell (1966) recorded in the acknowledgments both the very origin of his investigation and earlier and later influences: "The first ideas of this paper were conceived in 1952. I warmly thank Dr. F. Mandl for intensive discussion at that time. I am indebted to many others since then, and latterly, and very especially, to Professor J. M. Jauch." In fact, while in Geneva in the early 1960s, Bell attended some seminars of the group led by Jauch at the University of Geneva. Both Jauch and Constantin Piron, his doctoral student, had converted Geneva in one of the world centers working in the axiomatic formulation of quantum physics, an approach, as we had seen in Chap. 4, remounting to von Neumann's works. Discussions with Jauch had represented a true challenge for Bell, since Jauch "was actually trying to strengthen von Neumann's infamous theorem." It was, of course, von Neumann's proof against the existence of hidden variables compatible with quantum mechanics. According to Bell: "For me, that was like a red light to a bull. So I wanted to show that Jauch was wrong." In fact, Bell (1966) opened his first paper addressing it directly to Josef Jauch: "The present paper [...] is addressed to those who [...] believe that 'the question concerning the existence of such hidden variables received an early and rather decisive answer in the form of von Neumann's proof on the mathematical impossibility of such variables in quantum theory'" (Bell 1966, p. 447).

Bell's work can therefore be placed in the crossroad between the tradition related to the reinforcing of proofs against hidden variables and the tradition of building of viable models with such variables. If the possibility of introducing hidden variables

[7] Quotations are, unless indicated, from Bernstein (1991, pp. 65–68).

in quantum mechanics was the motivation, Bell's approach was far from Bohm's, as we have remarked. Indeed, he was not interested in building viable models mimicking quantum mechanics. Instead, his works focused on the critical analysis of the assumptions behind von Neumann's proofs and its reformulations, and later on the assumptions behind the Einstein-Podolsky-Rosen *Gedanken* experiment.

From the 1950s Bell received a heritage with a double meaning. In addition to the motivation derived from the existence of Bohm's work, he needed to face a widely shared idea that the matter of hidden variables was just a philosophical controversy and not a job for professional physicists. I have analyzed in the Chap. 2 in this book the context of the disputes about Bohm's interpretation in the 1950s. Here I just need to summarize it, pointing out that the label of "philosophical controversy" resulted from the overlapping of several factors. It originated with the physicists closely associated with the Copenhagen interpretation such as Pauli (1953, p. 33), Rosenfeld (1953, p. 56), and Heisenberg (1958, p. 131). "Artificial metaphysics," "debate [in] the field of epistemology," and "ideological superstructure," were respectively the words used to dismiss Bohm's hidden variable interpretation. These terms were used in the context that Jammer (1974, p. 250) named "the almost unchallenged monocracy of the Copenhagen school in the philosophy of quantum mechanics," describing the early 1950s. Bohm and collaborators unintentionally reinforced this label because the results they were able to obtain did not conflict with the predictions of quantum mechanics; neither did they present a heuristic advantage over that theory. In addition, Bohm, Vigier, and de Broglie emphasized what they considered to be epistemological advantages of their approach when compared to the complementarity interpretation. That the idea this was a philosophical controversy was largely shared can be seen from statements written by physicists who tried impartially to represent the controversy. Albert Messiah (1961, p. 48), in his very influential textbook, published originally in 1958, wrote, "the controversy has finally reached a point where it can no longer be decided by any further experimental observations; it henceforth belongs to the philosophy of science rather than to the domain of physical science proper." A similar example is Fritz Bopp's statement, during a conference dedicated in 1957 to foundational problems in quantum mechanics (Körner 1957, p. 51): "...what we have done today was predicting the possible development of physics—we were not doing physics but metaphysics." Bopp's declaration becomes still more meaningful if one considers that he was working on another alternative interpretation, the so-called "stochastic interpretation." Finally, these widely shared views were reinforced by a pre-existent belief that foundational problems were already solved by the founding fathers of the quantum theory. In the middle of the 1960s the spirit of the "old battles" from the 1950s was still alive.[8]

[8] "Old battles" was the term used by Abdul Salam in 1966 to explain to Rosenfeld an affair that happened at the International Centre for Theoretical Physics, in Trieste, Italy, related to a paper criticizing some tenets of the Copenhagen interpretation that had been written by the physicist Klaus Tausk. This episode is analyzed in Chap. 5 in this book. Abdul Salam to Léon Rosenfeld, 26 Sep 1966, Rosenfeld Papers, Niels Bohr Archive, Copenhagen.

Bell had a fine sensitivity for the prejudices nurtured in the old battles. He knew how strong the shared belief that he needed to face was. Just when he had published his papers on the conflict between quantum mechanics and certain hidden-variable theories, he received a not encouraging letter, to say the least, from Léon Rosenfeld: "I need not tell you that I regard your hunting hidden parameters as a waste of your talent; I don't know, either, whether you should be glad or sorry for that."[9] Thus in the very period when he published his two seminal papers, he also published a third on the measurement problem in quantum mechanics, co-authored with Michael Nauenberg, a physicist from the University of California, Santa Cruz, in which they criticized the view shared by the *majority* of the physicists:

> We emphasize not only that our view [that quantum mechanics is, at the best, incomplete] is that of a minority but also that current interest in such questions is small. The typical physicist feels that they have long been answered, and that he will fully understand just how if ever he can spare 20 minutes to think about it. (Bell and Nauenberg 1966)

As the paper was dedicated to Victor Weisskopf, they rhetorically appealed for Weisskopf scientific authority to support their own research: "It is a pleasure for us to dedicate the paper to Professor Weisskopf, for whom intense interest in the latest developments of detail has not dulled concerns with fundamentals." Ten years later, even with experiments on his inequalities underway, Bell kept the same sensitivity. When Alain Aspect formulated his proposal of new experiments on Bell's inequalities, he met Bell to discuss them. Bell's first question was, "Have you a permanent position?" After Aspect's positive answer, he warmly encouraged and urged him to publish the idea, but warned him that all this was considered by a majority of physicists as a subject for crackpots.[10]

Let us now focus on the results that Bell achieved in the middle of the 1960s. In his first paper, Bell (1966) isolated the assumption of von Neumann's proof that seemed to him to be untenable, and showed that while quantum mechanics satisfies this assumption it is not reasonable to require that any alternative theory have the same property. This assumption was that "any real linear combination of any two Hermitian operators represents an observable, and the same linear combination of expectation values is the expectation value of the combination." Next, he moved to analyze the new version of the proof that had been suggested by Jauch and Piron (1963), and made a similar objection. These authors had drawn an analogy between the structure of quantum mechanics and the calculus of propositions in ordinary logic, and they introduced a generalized probability function $w(a)$. They assumed as an axiom that "if a and b are two propositions, such that for a certain state $w(a) = w(b) = 1$, then this means that a measurement of a and b will give with certainty the values 1." Bell pointed out that this property, which is valid in ordinary logic and satisfied by ordinary quantum mechanics, should not be required of theories that were supposed to be alternatives to quantum mechanics. Bell also criticized

[9] Léon Rosenfeld to John Bell, 2 Dec 1966. Rosenfeld Papers, Niels Bohr Archive, Copenhagen.

[10] Aspect (2002, p. 119). Michael Nauenberg (pers. comm., 16 April 2005) also heard this anecdote from Bell.

Gleason's (1957) work on similar grounds for his use of quantum mechanical properties that were not reasonable to require of alternative theories. Bell remarked, however, that Gleason's work did not intend to reinforce proofs against hidden variables but rather to reduce the axiomatic basis of quantum mechanics.[11]

After showing that previous proofs against hidden variables included assumptions that were not reasonable, Bell (1966) considered whether some features should be required from models with hidden variables, if these models were to be physically interesting. "The hidden variables should surely have some spatial significance and should evolve in time according to prescribed laws." He recognized that "these are prejudices," but added "it is just this possibility of interpolating some (preferably causal) space-time picture, between preparation of and measurements on states, that makes the quest for hidden variables interesting to the unsophisticated." As the ideas of space, time, and causality had not been relevant in the assumptions hitherto considered, he attempted to determine what implications follow from hidden variables related to such ideas. After recalling that Bohm's (1952) proposal was "the most successful attempt in that direction," he wrote the wave function of a hidden-variable model for the case of a system with two spin 1/2 particles. Then, he showed that this wave function is in general not factorable and presents a grossly non-local character, since "in this theory an explicit causal mechanism exists whereby the disposition of one piece of apparatus affects the results obtained with a distant piece." As the state of two spin 1/2 particles could represent a system similar to that suggested by Einstein, Podolsky, and Rosen, Bell concluded, "in fact the Einstein–Podolsky–Rosen paradox is resolved in the way which Einstein would have liked least." Bell asked himself if non-locality is the price to be paid for the existence of hidden-variable theories, and admitted that there was no proof of this. Indeed, he was already looking for such a proof while writing the paper.

To obtain such a proof, Bell took the next logical step: to isolate what reasonable assumption was behind Einstein's argument and check the compatibility between this assumption and quantum mechanics. For Bell (1964), the "vital assumption" when dealing with a two-particle system is that what is being measured on one of them does not affect the other. He recalled Einstein's dictum, according to which, "on one supposition we should, in my opinion, absolutely hold fast: the real factual situation of the system S_2 is independent of what is done with the system S_1, which is spatially separated from the former." As Bell knew that Bohm's hidden-variable theory did not satisfy this dictum, he went to build a simple model of a hidden-variable theory obeying such a supposition and showed that its results conflict with quantum mechanical predictions in very special cases. This is Bell's theorem: no local hidden-variable theory can recover all quantum mechanical predictions. In a very rough description, Bell's theorem can be derived when one considers a hidden variable model of a system with two spin $\frac{1}{2}$ particles in the singlet state moving in

[11] It was Jauch who had called Bell's attention to Gleason's paper, which is an additional piece of evidence of how influential Jauch was in Bell's resuming of his own work on hidden variables.

opposite directions, a system that is analogous to the system suggested in the Einstein–Podolsky–Rosen argument. Bell built a function that is the expectation value of the product of spin components of each particle, and using different spin components derived an inequality with this function. The theorem is demonstrated when one uses quantum mechanical predictions in such inequality, since some quantum mechanical predictions violate this inequality. Since then, many other analogous inequalities have been obtained, adopting somewhat different premises, as we will see along this chapter; thus today it is usual to speak of Bell's inequalities as the quantitative measurement of Bell's theorem. In spite of its simplicity, Bell's theorem has been considered by many one of the most important results in quantum physics since its creation in the middle of the 1920s, but awareness that the issue at stake was locality and not just hidden variables spread through a slow process. Even considering those who were already involved with foundational issues, not all of them quickly grasped the real meaning of Bell's theorem; which will be illustrated with David Bohm's and Louis de Broglie's cases. Research on the connections between quantum non-locality and relativity theory only began in the late 1970s, when the balance between experiments suggested confirmation of quantum mechanical predictions and violations of Bell's inequalities. How important as the cognitive obstacles were, however, they were not independent of attitudinal obstacles, which were mainly related to the intellectual and professional status physicists attributed to subjects such as hidden variables and foundations of quantum theory.

The simplicity of Bell's theorem has given rise to the following question, suggested by Shimony (2002): why was it Bell who arrived at such an "elegant but not very difficult" result? Reviewing Bell's steps in his career as a physicist, Shimony suggested that "Bell's moral character is primarily responsible for his discovery of Bell's Theorem," relating this discovery to his independence and tenacity in pushing critical analysis to its last consequences. As we will see, Bell actively participated in the endeavor to bring his theorem to laboratories, in the 1970s, and followed closely the wide recognition of his contribution in the 1980s. He died prematurely in 1990. In the next decade, Bell's theorem was the key concept behind the search for technological applications of quantum effects in quantum computation. Since then, his fame has only increased, and we may even be witnessing the birth of a new founder myth. According to physicist Daniel Greenberger (2002, p. 281),

> John Bell's status in our field has the same [like Isaac Newton, James Watson, and Linus Pauling] mythic quality. Before him, there was nothing, only the philosophical disputes between famous old men. He showed that the field contained physics, experimental physics, and nothing has been the same since.

Let us now come back to the reception of Bell's theorem.[12] It opened the possibility of using experimental physics in order to reject some theories and

[12] Bell's theorem paper was cited more than 4,000 times. I include information related to the number of citations of some of the main papers concerning the tests of Bell's theorem as evidence

preserve others; however, physicists did not see the subject in this way so promptly. Ballentine (1987) remarked, "the awareness of its significance was slow to develop," quoting a graph with the number of citations of Bell's (1964) paper. Indeed, consulting the *Web of Science* database one can check that before Clauser's (1969) note at the American Physical Society, and the Clauser et al. (1969) paper, only three papers cited Bell's paper: (Bell 1966), himself, and two letters that missed the point as concerns Bell's theorem.[13] Another evidence of the poor initial reaction to Bell's theorem can be found in the colloquium "Quantum Theory and Beyond" held at Cambridge in July 1969, which "intended to provide opportunity [...] to discuss some possible alternative theories to see what a real change might involve." The colloquium was organized by Edward Bastin and David Bohm, chaired by Otto Frisch, and gathered physicists interested in foundations of quantum mechanics, such as Yakir Aharonov, Jeffrey Bub, Mario Bunge, H. J. Groenewold, Basil Hiley, Aage Petersen, G. M. Prosperi, and C. F. von Weizsäcker. None of them cited Bell's works.[14] Before going to Clauser's and Shimony's reactions, however, it is interesting to see that no reaction to Bell's theorem came from exactly where it would be expected, i.e. the partisans of hidden-variable approaches such as David Bohm and Louis de Broglie.

---

of their resonance among physicists. The data were updated on 15 February 2014. I am aware of the limits of such kind of information (Freitas and Freire Jr 2003), but it could help us just as one more piece of information. According to Podlubny (2005), there is no reasonable criterion in the available literature for comparisons between "scientists working in different fields of science on the basis of their citation numbers." However, just as a guess, I took data from the number of US article output and citations of US articles, in the field of physics, for the years 1997, 1999, and 2000, available at http://www.nsf.org, and I obtained an average of 7.1 citations by article. Redner (2005), considering only citations in *Physical Review* of papers published in this journal, concluded that "nearly 70 % of all PR articles have been cited fewer than 10 times" and that "the average number of citations is 8.8." These numbers match the physicists' shared tacit perception that an article should receive more than ten citations to be known. *Spires*, Stanford's database for high energy physics preprints (http://www.slac.stanford.edu/spires/), suggests the following classification: "unknown papers (0); Less known papers (1–9); Known papers (10–49), Well-known papers (50–99), Famous papers (100–499 cites), and Renowned papers (500+ cites)." Comparison with research on the foundations of quantum theory should be taken with a grain of salt due the huge difference in the number of active physicists in such fields. The source of data is the *Web of Science*.

[13] Sachs (1969) cites Bell's paper just incidentally. Clark and Turner (1968) realized that Bell's theorem predicts a conflict between quantum mechanics and hidden variables, but they exploited neither the nature of this conflict nor viable tests to reveal this conflict.

[14] For the proceedings of the conference, see Bastin (1971). Henry Stapp claimed that a paper by himself, "widely circulated in 1968," was the first recognition of the importance of Bell's theorem. This paper "was to appear in the proceedings of Bastin's conference on Quantum Theory and Beyond, which occurred in the summer of 1968." Stapp to Clauser, 5 Feb 1975, John Clauser Papers. These proceedings did not list Stapp among the attendance of the colloquium. The preprint paper was later reprinted, with the information it had been distributed in 1968. It is H. P. Stapp, "Correlation Experiments and the Nonvalidity of Ordinary Ideas about the Physical World," LBL 5333, 9 July 1976. I am thankful to Gustavo Rocha for obtaining a copy of this paper. The first paper published by Stapp, in which Bell's theorem is explicitly considered is Stapp (1971).

It is certain that David Bohm read Bell's (1966) paper, which contains references to the other paper in which the theorem was demonstrated.[15] Bohm could have exploited the full implications of Bell's theorem. Instead, he, with his former student Jeffrey Bub, reacted by building another type of hidden-variable theory, an explicitly nonlocal one. This time they used some ideas implicit in the "differential-space" theory of Norbert Wiener and Armand Siegel, and suggested a threshold, a relaxation time of the order of 10^{-13} s, below which there would appear conflicts with quantum mechanical predictions. These ideas had been used by Bub in his doctoral dissertation.[16] They presented this theory as a candidate for solving the so-called measurement problem of quantum mechanics, which was the focus of Bub's dissertation. As far as I know this was the only time Bohm suggested a figure to contrast hidden variables with quantum mechanics. In the 1950s, he had just begun to speculate that changes in his model could produce different predictions in the domain of the size of an atomic nucleus, but did not carry out the promised changes. Immediately after Bohm and Bub's proposal, the Harvard experimentalist Costas Papaliolios tested it. Papaliolios (1967) successively measured linear polarization of photons emitted from a tungsten-ribbon filament lamp. The measurements were carried out within time intervals lesser than the threshold suggested by Bohm and Bub, and he found their theory untenable. Bohm was notified of the result before its publication and tried to reduce the reach of this experiment, "I regard our 'theory' largely as something that is useful for refuting von Neumann's proof that there are no hidden variables. I would not regard it as a definitive theory, on which predictions of experimental results could be made." In addition, he admitted that "the time, $\tau \approx \frac{\hbar}{kT} \approx 10^{-13}$ s is just a guess," and not a consequence of their theory.[17] Papaliolios conceded that "the primary purpose of

[15] Bohm and Bub (1966) cites Bell (1966), which cites the paper where the theorem is demonstrated (Bell 1964). In addition, according to Jeffrey Bub, Bell's papers and his theorem were discussed by Bohm and him. Bub also recalls that only later, while in Minnesota, he fully realized the implications of Bell's theorem. Talk with Jeffrey Bub, 22 May 2002, American Institute of Physics, College Park, MD.

[16] Wiener and Siegel's ideas are in Wiener and Siegel (1953), Wiener and Siegel (1955), Siegel and Wiener (1956). Previously, Wiener had been influenced by Bohm's works: "I have been tremendously influenced in my thinking by my conversations and correspondence with Mr. Gabor and Mr. Rothstein, and by reading a sequence of two papers [...] which appeared this January under the authorship of David Bohm." In addition, in a talk given at the MIT-Harvard physics seminars, in 1956, Siegel cogitated of experimental predictions different from the standard ones, and N. F. Ramsey (Harvard) and Martin Deutsch (MIT), who attended the talk, "were quite willing to discuss the question as a serious and legitimate claim." "Paper to be presented on May 3 [1952] before the American Physical Society by Norbert Wiener," [7pp, unpublished, Box 29C, folder 678] and Armand Siegel to Norbert Wiener, 18 May 1956, [Box 15, folder 217], Norbert Wiener Papers, Institute Archive, MIT, Cambridge, MA. Wiener's interest in the foundations of quantum mechanics has not been analyzed yet, as far as I know, in the historical and philosophical literature on this subject.

[17] Costas Papaliolios to David Bohm, 17 February 1967; Bohm to Papaliolios, 1 March 1967, 2 March 1967, 11 May 1967. Papaliolios Papers, Accession 14811, Harvard University Archives, Boxes 23, folder "Hidden variables," and 10, folder "Bohm letters," respectively (*CPP* hereafter).

[Bohm-Bub's] paper was to demonstrate, by means of an explicit theory, how one can circumvent Von Neumann's proof," but emphasized the role that experimental predictions and real experiments, such as the one he carried out, should play in the choice between theories in the foundations of quantum physics. Papaliolios did not profess an empiricist view on choice of theories, he simply noted the advantages when experimental results are available. His reply to Bohm was a premonition of the coming times in the foundations of quantum mechanics:

> It is to your credit that you make your theory testable by stipulating a definite relaxation time [...]. If [it] had been left unspecified then you could always hide behind a suitably short relaxation time thereby making the hidden variable theory experimentally indistinguishable from the usual quantum mechanics. This latter approach is not entirely without merit, but one would have to use a non-experimental criterion such as elegance, simplicity, etc., in order to choose between the two theories.[18]

While the Papaliolios' experiment did not deal with Bell's theorem, it is an interesting case for our study since it was the first time ever that an experiment was devised to test hidden-variable theories. Bohm and Aharonov (1957) had compared the ideal experiment suggested by Einstein, Podolsky, and Rosen with real data, but using results from a previous experiment that had not been designed with this goal. The context of Papaliolios' experiment indicates the shifting mentality concerning experiments and foundations of quantum physics among the physicists. It exhibits both old and new behaviors. Papaliolios was a Bell's theorem experimentalist *avant la lettre*, since for him Bohm-Bub's theory simply triggered what he had already been reflecting on, that is, he had been wondering about "hidden variable experimental possibilities," including a number of different possible approaches. For instance, he asked himself if "is there some effect which $=0$ for quantum mechanics that $\neq 0$ for hidden-variable theories?"[19] Aware of the old behavior of bias against the hidden variable subject, the U.S. Naval Ordnance Lab physicist Thomas Phipps praised the experiment and encouraged him to pursue this kind of experiments, noting, "even though this may not represent the most fashionable mode of research of the day."[20] Papaliolios received a similar and stronger encouragement from the *Physical Review Letters*' referee who evaluated his paper. "This paper should be published, but only if the author includes a discussion of the feasibility of these improvements and indicates plans to pursue the matter further." Papaliolios changed his paper according to the referee's requirement —"an experiment is now in progress to set even lower, upper bound on τ by using a thinner polarizer"—and in fact elaborated plans to work out an improved experiment, but nothing came out of these attempts.[21] Yet, Papaliolios's optical experiment exhibits the interaction

---

[18] Papaliolios to Bohm, 20 March 1967, CPP, box 23, folder "Hidden variables," ibid.

[19] "V. Experimental Possibilities?", minute by Costas Papaliolios, [w/d], CPP, box 23, folder "Hidden variables," ibid.

[20] Thomas Phipps to Papaliolios, 19 Apr 1967, Papaliolios Papers, ibid.

[21] George Trigg [editor of *Physical Review Letters*] to Papaliolios, with the referee's report enclosed, 28 Feb 1967; Papaliolios to Trigg, 7 March 1967, CPP, ibid.

between scientific experiments and new technical devices that would be a driving force in Bell's theorem experiments. After his experiment, Papaliolios approached R. Clark Jones, Director of Research of Polaroid Corporation, showing how the polarizers produced by this corporation had been useful in a foundational experiment, and asking if Polaroid could supply thinner polarizers for new experiments.[22]

Louis de Broglie, the other main proponent of hidden variables, only reacted when experiments on Bell's theorem were already being carried out, and he did not grasp their full implications. de Broglie's argument, based on an analogy between quantum states and light wave packets, was that the wave function of a two-particle system would factorize after the particles fly a certain distance, which is a conjecture due to Wendell Furry. He did not realize that experimental tests of Bell's theorem could check this hypothesis. In addition, de Broglie (1974, p. 722) considered that quantum correlations between two electrons spatially separated would imply an instantaneous exchange of information, thus violating the relativity theory; an issue not yet elucidated. A harsh controversy followed between 1974 and 1978 that pitted Bell and Abner Shimony against de Broglie and George Lochak, a former student of de Broglie.[23] Lochak (1978) still maintained, "Bell's attempt, as interesting as it is, does not say, *and cannot say*, anything decisive about the existence of hidden variables, local or nonlocal." By that time, Bohm had realized the full implications of Bell's theorem, and stated that non-locality was the most important quantum property.[24] In fact, Bell's theorem survived and was fully exploited in the hands of a new generation with very different approaches from those of the older one, as we will see now.

7.3 Philosophy Enters the Labs: The First Experiments

The philosopher and physicist Abner Shimony and the physicist John Clauser were the key figures in the move to bring Bell's theorem to the laboratory. Before analyzing what they did let us see who they were at the time. Shimony, born in 1928, had been interested in science, mainly physics and mathematics, and philosophy since his undergraduate studies at Yale, where he was a joint philosophy and mathematics major.[25] When he was finishing his doctorate in Philosophy at Yale,

---

[22] Papaliolios to R. Clark Jones, 7 March 1967, CPP, ibid.

[23] See Broglie (1974), Bell (1975), Lochak (1975). Bell's paper is followed by a discussion between Bell and Lochak. Lochak's stance provoked Shimony's irony: "In view of the extreme implausibility of such behavior [to admit a conspiratorial behavior of the detectors for explaining the experimental results violating Bell's inequalities], the local hidden-variable theories are very hard to defend, and their advocates should remember the sermon of Donne 'And therefore never send to know for whom the bell tolls; it tolls for thee'" (Shimony 1976).

[24] See Bohm (1971), and especially Bohm and Hiley (1975).

[25] Shimony's biographical sketch is based on Wick (1995, pp. 106–109), Aczel (2002, pp. 149–155), and Joan L. Bromberg, "The rise of 'experimental metaphysics' in late twentieth century

working on probability, in the winter of 1952–1953, a reading of Max Born's *Natural Philosophy of Cause and Chance* revived his interests in physics, especially in classical statistical mechanics and quantum mechanics, and prompted him to take a second doctorate in physics under Eugene Wigner at Princeton. There he worked with statistical mechanics. In the early 1960s, even before finishing his physics doctorate, Wigner's interests in the measurement problem of quantum mechanics excited him.[26] Ever after, Shimony focused on the foundations of quantum mechanics, publishing his first paper on the subject in 1963. I have argued in the Chap. 4 in this book that Wigner played a unique role in elevating the status of such issues. He was also very supportive of his students and colleagues who worked on this subject. His dispute with Leon Rosenfeld and the Italian physicists Daneri, Loinger, and Prosperi, in the second half of the 1960s, which contributed to breaking down the monocracy around the Copenhagen school, was primarily motivated by his defense of young physicists like Shimony who had been strongly and unfairly criticized. In the letter to Jauch suggesting him a joint reply to the Italian physicists, he wrote, "Needless to say, I am less concerned about myself than about other people who are much younger than I am and whose future careers such statements may hurt."[27] The lasting interaction with Wigner was fruitful for both of them. Shimony had Wigner's authoritative support for his entry into the field of foundations of quantum mechanics, and Wigner met in Shimony an informal assistant for philosophical matters.

Shimony's background in philosophy facilitated his physical research. Influence from Peirce and Whitehead facilitated his acceptance of quantum mechanics, and the remaining conflict, related to his commitment to realism, has been a major factor in his lasting "search for a world view that will accommodate our knowledge of microphysics." His double training also facilitated his professional career. He was hired by MIT's philosophy department in 1959, and gave courses on foundations of quantum mechanics there in the early 1960s. After a while, he moved to Boston University, in 1968, for a double affiliation in philosophy and physics. So he never depended exclusively on his physics training and his achievements in foundations of quantum mechanics for his professional career. His double training permitted him, however, to be considered a physicist with a philosophical culture among the physicists, and a philosopher with physics training among the philosophers. His achievements in the foundations of quantum physics have carried him to

physics," unpublished manuscript, 2004. See also Abner Shimony, Interviewed by Joan Lisa Bromberg, 2002, Niels Bohr Library, American Institute of Physics, College Park, MD.

[26] "I found your paper on the mind–body problem extremely stimulating. It is one of the few treatments of the problem which considers the mind–body relationship to be a legitimate subject for scientific investigation, without achieving this scientific status for the problem by reducing it to behavioristic or materialistic considerations." Abner Shimony to Eugene Wigner, 1 May 1961. Wigner Papers, box 94, folder 1, Manuscripts Division, Department of Rare Books and Special Collections, Princeton University Library (WigP hereafter). Shimony's first paper on the measurement problem is Shimony (1963).

[27] Eugene Wigner to Josef M. Jauch, 6 September 1966. Wigner Papers, box 94, folder 7, WigP.

a key position in this field. With Shimony, foundations of quantum mechanics entered the optics laboratory, but did not lose its philosophical implications.[28]

Picture 7.2 Left to right: Daniel Greenberger, unidentified man, Abner Shimony, and Lev Vaidman at a dinner in Baltimore, Maryland. 10 Aug 1993. AIP Emilio Segre Visual Archives

Picture 7.3 John F. Clauser. AIP Emilio Segre Visual Archives, gift of John Clauser

[28] For a sample of this approach, see Shimony (1993).

John Clauser, born in 1942, has been uneasy about quantum mechanics since his undergraduate studies at Berkeley and graduate studies at Columbia, an uneasiness that he may have inherited from his family background. Francis Clauser, his father, was an aeronautical engineer and researcher who worked with Theodore von Karman on the physics of fluids. According to the younger Clauser, "he always was trying to understand physics, and there were very strong similarities between the mathematics of fluid flow and the mathematics of quantum mechanics, and he didn't understand quantum mechanics. And he kind of pre-programmed me as the guy who might help try to solve the problem that he couldn't solve." His father was also a strong influence on Clauser's skepticism, which was very instrumental for his discovery that no previous experimental data were adequate to test Bell's theorem. "Son, look at the data. People will have lots of fancy theories, but always go back to the original data and see if you come to the same conclusions."[29]

Clauser was finishing his thesis on measurement of the cosmic microwave background under the direction of Patrick Thaddeus at Columbia University when he became interested in Bell's theorem. Clauser's doctoral training, including the measurement of microwaves, enabled him to foresee and design quantum optics experiments to test Bell's inequalities. However, the appeal coming from the discovery of an interesting—but not yet done—experiment was not the only motivation behind his quick shift to a subject related to the foundations of quantum mechanics. Pedagogical and political factors were also influential factors. Clauser's approach to physics demands visualization and construction of physical models, not just abstract mathematics. So, he collided with the traditional way in which quantum mechanics has been taught.[30] Additionally, he (Clauser 2002) read EPR's paper and Bohm's and de Broglie' works, and "while [he] had difficulty understanding the Copenhagen interpretation, the arguments by its critics seemed far more reasonable to [him] at that time." To his awareness that he had discovered that a good experiment had not yet been done, was added a political influence, "l'air du temps." As he remembers, "the Vietnam war dominated the political thoughts of my generation. Being a young student living in this era of revolutionary thinking, I naturally wanted to 'shake the world.' Since I already believed that hidden variables may indeed exist, I figured that this was obviously the crucial experiment for finally revealing their existence" (Clauser 2002).[31] To shake quantum mechanics was then the target of his desire.

[29] Clauser's biographical sketch is based on Clauser (2002), Wick (1995, pp. 103–106), and Aczel (2002, pp. 155–159). Quoted fragments are from John F. Clauser, interviewed by Joan Lisa Bromberg, 2002, pp. 3 and 19, Niels Bohr Library, American Institute of Physics, College Park, MD (*AIP* hereafter).

[30] For the teaching of quantum mechanics, see Kaiser (2007) and Greca and Freire Jr (2014). For the role of pedagogy in the production of physics, see Kaiser (2005).

[31] For this context in American physics, see Kevles (1978, pp. 393–409) and Chap. 6 in this book.

Since 1968, Shimony and Clauser, independently and without mutual knowledge, had been working on Bell's theorem. Indeed, Bell's paper early claimed Shimony's attention. He wrote to Eugene Wigner on New Year's day, 1967.[32]

> There is a paper by J. S. Bell [. . .] which I found very impressive as evidence against hidden variable theories. He shows that in an Einstein–Podolsky–Rosen type of experiment the supposition of hidden variables, with any statistical distribution whatever, is certain to disagree with some of the predictions of quantum mechanics unless there is a kind of action at a distance.

In the summer of 1968, just before beginning to teach physics and philosophy at Boston University, Shimony enlisted the physics graduate student Michael Horne for designing a "realizable Bohm-type EPR experiment" as a dissertation subject. Horne was the right choice as he was attracted to physics for the intellectual endeavor it embodied. He was captivated by physics while reading I. B. Cohen's *The birth of a new physics* in high school, and the reading of E. Mach's *The science of mechanics*, while at the University of Mississippi, led him to decide that he wanted to do research on the conceptual foundations of physics. In the early 1969, however, Shimony was surprised by Clauser's abstract in the *Bulletin of the American Physical Society* suggesting an experiment to test Bell's theorem. "We were scooped," told Shimony to Horne.[33] After consulting Wigner, Shimony decided to call Clauser and suggest collaboration, which was accepted by Clauser.[34] By the time they began to collaborate they had already independently realized that no previously available experimental results were able to test Bell's theorem and that the most adequate and viable test would be to repeat in slightly different conditions an optical experiment done by Carl Kocher as a doctoral student of Eugene Commins in 1967. The CHSH paper, which will be analyzed later, was the first result of this collaboration.[35]

The first news Bell had from the American reaction to his work did not come, however, through Shimony. According to Wick (1995, p. 106), "The letter from the American student was the first serious reaction [Bell] got to his paper—after a lag time of five years." The American student was, surely, Clauser. It is worth noting the letters exchanged between Clauser and Bell, not only for what they correctly predicted but also for their unfulfilled hopes. Clauser rightly assured Bell that the results from the Wu-Shaknov experiment with the annihilation of positronium were not adequate to test Bell's inequalities. He suggested, instead, a modified extension

[32] Shimony to Wigner, 1 Jan 1967. Wigner Papers, Box 83, folder 7, *WigP*.

[33] Horne (2002) and Horne (pers. comm., 8 June 2005). Clauser's abstract is Clauser (1969).

[34] "It was a pleasure talking to you on the telephone on Thursday. Mike Horne and I had been through a bad day after we found you had done an analysis that sounded very much like ours." Shimony to Clauser, 20 Apr 1969, Abner Shimony Papers, Box 2, Folder 9, (Early Work on Hidden Variables, 1969), Archives of Scientific Philosophy, Special Collections Department, University of Pittsburgh (*ASP* hereafter).

[35] CHSH paper is Clauser et al. (1969). Kocher's experiment is reported in Kocher and Commins (1967). For the roads of Clauser and Shimony to Bell's theorem, and for their meeting, see Wick (1995, pp. 103–113) and Aczel (2002, pp. 149–169).

of the Kocher and Commins experiment with the polarization correlation of photons from an atomic decay cascade. He also promised that "it might also be possible to 'rotate' the polarizers by means of magneto-optic effects while the photons are in flight to rule out all local hidden-variable theories;" a promise that would wait more than 10 years to be fulfilled, and then not by Clauser but by Alain Aspect with a different method for obtaining time-varying analyzers.[36] Bell rightly anticipated the results of the experiments, holding slight hope for a breakthrough[37]:

> In view of the general success of quantum mechanics, it is very hard for me to doubt the outcome of such experiments. However, I would prefer these experiments, in which the crucial concepts are very directly tested, to have been done and the results on record. Moreover, there is always the slim chance of an unexpected result, which would shake the world.

Bell also revealed that he expected to be performing experiments in particle physics: "experiments have been proposed involving neutral kaons [...] and will become practical in the course of time." Particle physics, though, never became the main bench for experiments with Bell's theorem.

It is interesting that among the three quantum dissenters, Clauser and Bell were more optimistic about the possibility of obtaining results violating quantum mechanics than Shimony, and that Clauser was by far the most optimistic among them. Before they met each other, Shimony wrote to Clauser: "Incidentally, I am amazed at your estimate of the probabilities of the possible outcomes of the experiment. I would estimate a million to one in favor of the quantum mechanical correlation function. Needless to say, I hope I am wrong in this." "Do keep imagining that it will come out *against* quantum theory; that makes it very interesting!" were words from Horne to Clauser. In 1972, when Clauser notified the results he had eventually obtained, confirming quantum mechanics predictions and violating Bell's inequalities, Shimony remarked—in his distinctive literary vein—how Clauser's early expectations were frustrated: "Your paper finally arrived today. It is a classic, but unfortunately a classic tragedy, since the hero dies, and dies nobly."[38] Shimony kept these memories of Clauser's hopes, "... he was absolutely convinced that the experiment was going to come out for the local hidden variable theory and against quantum mechanics, and it was going to be an epoch-making experiment."[39]

The CHSH (1969) paper is a fine piece in physics literature both for its concision and breadth.[40] It is an acronym for Clauser, Horne, Shimony, and Richard Holt, a Harvard University student of Francis Marion Pipkin. The authors fulfilled three

[36] Clauser to Bell, 14 Feb 1969, Clauser Papers (*JCP* hereafter).

[37] Bell to Clauser, 5 March 1969. Idem.

[38] Shimony to Clauser, 14 Jan 1972, Abner Shimony Papers, Box 1, Folder 4 (Clauser, John F.—Correspondence, 1971–1972), *ASP*.

[39] Shimony to Clauser, 20 Apr 1969; Horne to Clauser, 18 Apr 1969; both in *JCP*. Abner Shimony, interviewed by Joan Lisa Bromberg, 2002, on p. 71, *AIP*.

[40] Clauser et al. (1969). This paper has 2,286 citations.

goals. They modified Bell's theorem in order to make it usable for real, not idealized experiments, they showed that data available from previous experiments did not produce evidence against local hidden-variable theories, and they suggested a viable test with optical photons suggesting optics as the privileged bench for tests of Bell's theorem. They showed that the experiment by Madame Chien-Shiung Wu and I. Shaknov, performed in 1950 using the annihilation of positronium for measuring polarization correlation of γ photons, was not adequate to test Bell's theorem because "the direction of Compton scattering of a photon is a statistically weak index of its linear polarization," and there are no good polarizers for high-energy photons.[41] They also showed that the Kocher and Commins experiment measuring polarization correlation of photons emitted in an atomic decay cascade of calcium, performed in 1967, was in principle adequate for testing Bell's theorem but that these experimentalists had only measured angles—0° and 90°—in which there is no conflict between quantum mechanical predictions and local hidden-variable theories. For this reason the proposed experiment was essentially a revision of the Kocher and Commins experiment with angles—22.5° and 67.5°—in which there is a maximum conflict of predictions. It is curious to remark that Kocher and Commins (1967, p. 575) explicitly presented their experiment "as an example of a well-known problem in the quantum theory of measurement, first described by Einstein, Podolsky, and Rosen and elucidated by Bohr," but they were not aware of Bell's work.

In order to make Bell's theorem testable, CHSH authors introduced additional assumptions that were determined by the type of experiments they wanted to do. The three additional assumptions were related to: avoiding Bell's assumption of perfect correlation between the pair of particles; considering rates of emergence or not of photons from the filters instead of their detection (a change due to the small efficiencies of the available photoelectric detectors); and taking the probability of joint detection as being independent of the orientation of the polarization filters. They considered that the latter assumptions, which can be called a fair sampling assumption, "could be challenged by an advocate of hidden-variable theories in case the outcome of the proposed favors quantum mechanics;" but they did not assess them a flaw in the proposed experiment because "highly pathological detectors are required to convert hidden-variable emergence rates into quantum mechanical counting rates" (Clauser et al. 1969). As necessary as these assumptions were for the time we are analyzing, to perform experiments

[41] The suggestion of a test with optical photons faced competition to be established. The competition pitted the teams involved with atomic cascade and those with positronium annihilation. When both experiments were already done, Shimony wrote to Clauser (19 May 1972, Clauser Papers): "Freedman told me about the difficulties raised by Wu, Ullman, and Kasday. What has been the upshot of that? I think the only way to handle it is to continue to state, politely but firmly, that their experiment is a fine one, but much less decisive than yours, because of their additional assumption." Wu's early experiment was Wu and Shaknov (1950). Later, Kasday led the repetition of this experiment (Kasday 1971; Kasday et al. 1975).

relaxing them has been the holly grail for physicists involved with foundations of quantum mechanics.

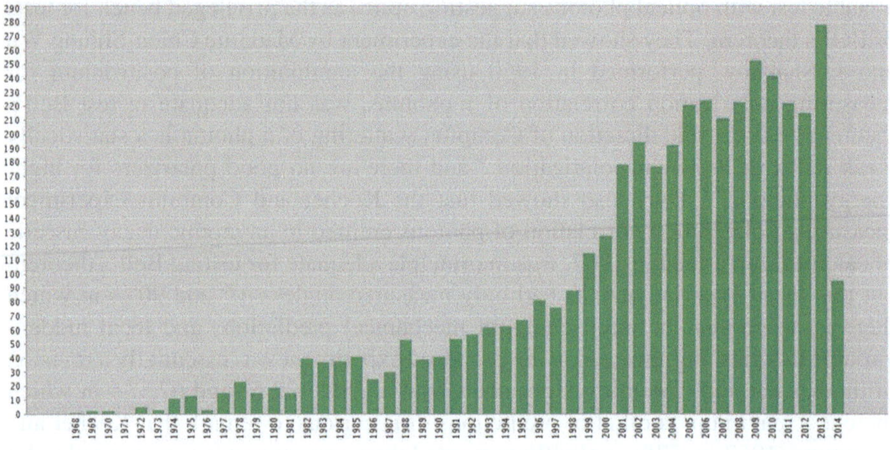

Picture 7.4 Citations of Bell's theorem paper, from 1965 to 2014. Source of the data: Web of Science

The CHSH paper awakened interest among Europeans physicists, such as Bell, Bernard d'Espagnat, de Broglie and Franco Selleri, who were already involved with hidden-variables theories. As a consequence of this paper, Shimony was invited by d'Espagnat to lecture at the Varenna school on foundations of quantum mechanics, in 1970. The Varenna summer school, that is, the "International School of Physics 'Enrico Fermi'," had been organized by the Italian Physics Society starting in 1953, and had become a traditional gathering for training European physicists in novel themes of research. The 1970 school was the first to be dedicated to the theme of the foundations of quantum mechanics, an episode we have discussed in the previous chapter. There, Shimony gave three lectures, made acquaintance with and began a lasting friendship with Bell and d'Espagnat, and became still more involved with issues related both to Bell's inequalities and measurement problem. We can conclude that by the early 1970s Shimony had received recognition in Europe and the United States for his work on foundations of quantum mechanics. It is also worth remarking that the Varenna school was an important meeting point for developing research on Bell's theorem as it brought together a number of physicists who were already working on the subject but had never met each other previously.[42]

[42] Varenna's school had 84 participants. For its proceedings, see d'Espagnat (1971). For its stimulus on the research on foundations of quantum mechanics, see Freire Jr. (2004) and the previous chapter. Shimony thinks he was invited to Varenna for his previous work on measurement problem (Shimony, interviewed by Joan Lisa Bromberg, 2002, *op. cit.*, on pp. 75–82). However, d'Espagnat remembers that it was Bell who suggested he invite Clauser and Shimony (Bernard d'Espagnat, interviewed by Olival Freire, 2001, *AIP*).

From the CHSH authors emerged the two teams that conducted the first experiments with optical photons. At Berkeley there was Clauser, who moved there as a postdoc of Charles Townes, and Stuart Freedman, a doctoral student of Commins. At Harvard, there was Holt, a student of Pipkin. Holt's work involved not only Pipkin, his adviser, but also Costas Papaliolios, who two years before had performed the experiment on the Bohm–Bub theory. At Berkeley, Clauser had the support of Charles Townes for the experiment. "Townes was the guy who actually twisted Commins' arm to put Stu Freedman on the experiment and to steer Atomic Beam Group funds into doing the experiment." However, Clauser did not receive a wide support from the Berkeley faculty. Clauser remembers that even after the experiment was performed "most of the physics faculty at Berkeley all said [this was junk], because as 'you got exactly what you expected, what was the point?" For Clauser, "they did not understand Bell's Theorem." Commins, who had performed the experiment that Clauser was repeating with some modifications, was no exception; "what a pointless waste of time all of that was" are Clauser's (2002, pp. 12–13) memories of Commins's remarks.[43] In fact, as early as February 1969, Commins had "dismissed the idea [repetition of his experiment] as worthless," and stated that "there are 'thousands' of experiments which already prove" what Clauser was looking for, which led Clauser to appeal for Townes' mediation.[44] Harvard's environment was very different. In 1969, even before his meeting with Clauser, Shimony was enthusiastic about the involvement of Harvard experimentalists. "There is a great deal of interest in the experiment at Boston University and at Harvard, though I think it could be done quickly only at Harvard. [...] The man most interested at Harvard is Costas Papaliolios. [...] However, he is very busy now, and therefore suggested it to two men whom I have not met yet. One is Nussbaum, who did his doctoral works at Harvard under Pipkin, looking at photon polarization correlation in a mercury cascade. The other is Dick Holt, another student of Pipkin who inherited Nussbaum's apparatus."[45] In the late 1969, the possibility of three tests of quantum mechanics caught the attention of popular science magazines. "Acid test for quantum theory," advertised *Scientific Research*, explaining that "three versions of the same experiment, just getting under way at three separate laboratories, may supply a definitive answer as to whether there are indeed 'hidden variables' that provide a more deterministic view of reality than is

[43] John Clauser, interviewed by Joan Lisa Bromberg, 2002, pp. 12–13, *AIP*.

[44] Clauser to Eugene Commins, 18 Feb 1969, [cc: C. Townes], Clauser to Townes 18 February 1969, Clauser Papers.

[45] Shimony to Clauser, 20 April 1969. Clauser Papers. Papaliolios recorded his first meeting with Shimony in the following way: "Hidden variables—March 18, 1969. Talked with Shimony today. (has student–Mike Horne). He pointed out 2 good references (1) Kocher & Commins, Phys Rev Lett 18, 575 (1967) (2) Bell, Physics 1, 195 (1964). It may be poss. to do an Einstein–Rosen–Podolsky exper. with Nussbaum apparatus. Shimony also left me two of his references [on the measurement problem]. See Bull of APS (for 1969 Wash. Meeting). Clauser has independently come up with the same experimental test." Papaliolios Papers, box 24, folder "EPR Experimen (Shimony-Clauser)", *CPP*.

possible with quantum mechanics."[46] Gilbert Nussbaum, then at the Bell Laboratories, did not, in fact, comply, but his place was occupied later on by the Texas A&M University physicist Edward Fry, who would be the leader of the third team to carry out experiments with optical photons.

Fry had been trained in atomic physics and spectroscopy, while doing his thesis at the University of Michigan under Bill Williams. He was appointed as Assistant Professor at Texas A&M, in 1969. There he was introduced by James McGuire, a theorist in atomic collision physics, to the CHSH paper, in "evening philosophical society discussion in late fall 1969." Fry was "immediately intrigued," figured out an experimental scheme and applied to National Science Foundation for funding. It was not a good experience for his first ever application. "The reviews did not argue about the physics, their theme was basically that NSF should not waste money on fruitless pursuits." Discouraged by the result and by technical troubles with the available devices, he did not actively pursue the intended experiments until after the Clauser and Holt results. Meanwhile, his bets for the results of such experiments were against quantum mechanics predictions. He considers that he "had the same bug as Clauser and may have even, in part contracted it from him." As he recalls, "I was really hoping for a major breakthrough in our understanding of the quantum world. Clearly, a violation of the Bell inequalities does improve our understanding; but it does not provide the dramatic overthrow of existing thought that I had anticipated."[47]

The first round of experiments ended in a tie. The experiment conducted by Freedman and Clauser (1972) confirmed the quantum mechanical predictions and violated Bell's inequalities, while the experiment held by Holt (1973) under the supervision of Pipkin produced the opposite results.[48] Freedman and Clauser observed pairs of photons emitted by transitions in calcium. They used "pile-of-plates" polarization analyzers, and the experiment ran for 200 h. For a certain

[46] *Scientific Research*, 4(23), 10 November 1969, p. 19. Information about the plans for the third experiment is independently confirmed: "There is yet a third experiment on another atom, at Bell Labs. So all in all we ought to get some firm results." Richard Holt to Frederick Belinfante, 6 Jan 1970, Shimony Papers, Box 2, Folder 9B, *ASP*. Belinfante, who was writing a book on hidden variables (Belinfante 1973), had criticized CHSH calculations and this triggered a huge correspondence among them. Most of these letters are at Box 2, Folder 9B, *ASP*. Belinfante eventually acknowledged his criticisms were unfounded.

[47] Edward Fry (pers. comm., 5 Aug 2005). James McGuire to Clauser, 3 Jan 1972, and 24 Feb 1972, Clauser Papers. The collaboration with James McGuire led to a derivation of Bell's theorem and comparison with data from the experiment by Freedman and Clauser (McGuire and Fry 1973). For the technical troubles, see Harvey (1980, p. 154). Fry considers the NSF's reviews to be a "clear window" in the physics culture at the time: "there was a real culture at the time that thought that this wasn't something you should do, in spite of it seemed things were more reasonable at Harvard and maybe even at Berkley." In addition, the reviewers wrote: "there was a specific reference to the time and money already being wasted at Berkley and Harvard and that NSF shouldn't waste any more money on this." Fry's lecture at the symposium "Optics and the Second 'Magic Decade' of Quantum Mechanics," meeting of the Optical Society in San Jose, CA, 2007. I am thankful to Joan Bromberg for sharing this manuscript with me.

[48] Freedman and Clauser's paper has 612 citations.

variable that resulted from count rates of photons they obtained 0.300 ± 0.008. For the sake of simplification, let us name this magnitude S. Bell's inequality for this variable and this experimental setup was $\leq \frac{1}{4}$, and the quantum mechanical prediction for the same variable was 0.301 ± 0.007. Holt and Pipkin observed photon pairs from transitions in the isotope of mercury ^{198}Hg. They used calcite prisms as polarization analyzers, and the experiment lasted 154.5 h. For this case, the quantum mechanical prediction for the same variable was 0.266, and they obtained 0.216 ± 0.013. As we will see later, the main effect on the physicists was the desire for new experiments. Clauser repeated Holt's experiment, Fry used new techniques for getting better results, and Alain Aspect devised what he intended to be an experiment able to settle the controversy.

In addition to these experiments with polarization correlation of optical photons there appeared experiments in fields other than optics. The experimentalist L. Kasday, a doctoral student at Columbia, the Italian group of G. Faraci in Catania and the group led by A. Wilson at Birkbeck College, in London, repeated the Wu and Shaknov experiment with γ photons, with further assumptions, and arrived at conflicting results. Kasday's and Wilson's results confirmed quantum mechanics while Faraci's results confirmed Bell's inequalities. At Saclay, France, M. Lamehi-Rachti and W. Mittig performed an experiment with spin correlation in proton-proton scattering, confirming quantum mechanics.[49] In the 10 years between Freedman and Clauser's result in 1972 and the three experimental results Aspect and his collaborators would publish in 1981 and 1982, we had 11 experiments and 8 teams involved with tests of Bell's theorem.

Since experiments with optical photons were considered by Shimony, Clauser, and Bell the most adequate to test local hidden variables, and the resonance of results from atomic cascade experiments evidences that this stance was widely shared by the physicists involved with Bell's theorem, it is interesting to focus on the period between 1973 and 1974, when the only available results from such experiments were in conflict with each other, and see how the physicists reacted to this disagreement. The physicists involved in these issues did not consider that there was a true tie. Harvard physicists did not trust their own results but suspected them of systematic errors, which, however, they could not identify. They decided not to publish the results, but Holt's work was recognized and he received his PhD degree. Behind this decision was their trust in quantum mechanics. Pipkin concisely recorded that "the measurements on the polarization correlation disagree with quantum mechanics and agree with the predictions of hidden variable theory. So far efforts to explain this discrepancy have not been successful."[50] He did not

[49] Kasday (1971), Kasday et al. (1975), Faraci et al. (1974), Wilson et al. (1976), Lamehi-Rachti and Mittig (1976). After 1978 many authors cited the review by Clauser and Shimony (1978), which has 783 citations, instead of each experimental result. This may partially explain the lesser number of citations of these papers.

[50] Pipkin, F. M. "Atomic Physics Experiments Using Fast Atomic Beams". NSF Grant proposal. [To begin on 01 June 1974]. Pipkin Papers [Accession 12802], box 25, folder "NSF Proposal 1974–1975," Harvard University Archives (*PP* hereafter).

ponder the possibility that their results had exposed a deficiency of quantum theory. Holt, in his thesis, dared more, writing, "the polarization correlation results present a far more puzzling (and perhaps) exciting prospect. The statistical accuracy is certainly great enough to allow us to say that a discrepancy with quantum mechanics exists. The arguments of the preceding section show that all the obvious sources of systematic errors have been examined." Next he attenuated his claim, "on the other side are two arguments for disbelieving the results," the first one being the results already published by Freedman and Clauser. He did not discuss why this previous result should be considered more reliable than his own results. Next, he stated the main reason for his caution, "the second argument against our results is the enormous success that has been enjoyed by quantum mechanics in the correct prediction of experimental results" Holt (1973, pp. V–27). Holt's awareness that this argument could not be definitive led him to conclude, "this, however, is not a telling argument because it is quite conceivable that a deterministic theory substructure could yield the same ensemble average as quantum mechanics in all other experimental situations save this one, in which the least intuitive features of quantum mechanics are strikingly displayed." Holt's back and forth reasoning did not surprise Harvey (1980, p. 143), who has written sociological studies on these experiments. For him, "despite its possible philosophical significance, [local hidden variables] could hardly be described as a highly *plausible* theory." Yet, he showed how the idea of plausibility, widely used by the physicists involved in the debate on Bell's theorem, was strongly dependent on "their immersion in the *culture* of physics" and could not be reduced to logical reasoning or data evaluation (Harvey 1981, p. 105). Trust in quantum mechanics was, and is, an essential part of the culture of physics.

We have good evidence of this trust in quantum theory in a review of the results written in 1974 by the French physicist Michel Paty. He was at the time an experimentalist in particle physics, very interested in epistemological issues, preparing himself for a conversion to a philosophical career by taking a second doctorate, this time in philosophy. Together with the Brazilian physicist José Leite Lopes, he ran a series of seminars and a journal under the title *Fundamenta Scientiae*. In 1974, they organized in Strasbourg a colloquium dedicated to the 50th anniversary of quantum mechanics in which Paty presented a review on "the recent attempts to verify quantum mechanics." After concluding that "the present balance sheet of the experiments designed to test Bell's inequalities is therefore as follows: three agree with quantum mechanics, and two disagree;" he asked, "has quantum mechanics now revealed its limitations, or more exactly, the limits of its field of application?" He conjectured that "This would not be unthinkable a priori, [...]. This would also be the case for a theory as powerful as quantum mechanics, which itself is highly powerful, but at the same time probably has a frail basis." Next, however, he did not accept his own conjecture, and reaffirmed a trust in quantum mechanics: "However, it may seem doubtful that such an established theory might be questioned in such simple experiments. And in fact quantum mechanics may only *appear* to be frail; its hold on our conceptions is paradoxically shown in this recent questioning: it is not quantum mechanics which is put into doubt, so much as the basis of these very experiments or at least their interpretation." Anyway, Paty was careful about this conclusion, and urged for more refined experiments.[51] This kind of discussion shows the tacitly shared view that foundational problems in quantum mechanics were already solved by its founding fathers being reinforced by the increasing practical success of this theory. Even physicists like Pipkin, Holt, and Paty, who spent time doing or reviewing such experiments and for this reason cannot be counted among those who developed prejudices against hidden-variable theories as a subject, embodied this trust in quantum mechanical predictions.

Now I want to take the case of the Harvard experimenters for close consideration. Compared to Berkeley, Harvard seems to have been a friendly intellectual and professional environment for those who were interested in Bell's theorem. We can infer this both from the involvement of its experimenters with tests of hidden variables and from the testimony of Holt. Papaliolios was a key figure in the creation of this intellectual and professional environment, since he was openly involved with tests of hidden variables since 1967 and throughout the 1970s. Acknowledgements in Holt's (1973) and Horne's (1970) dissertations, and in the CHSH (1969) paper testify to his role in the first tests of Bell's theorem. He published with Freedman and Holt a review on the status of hidden variable experiments, and in 1975 gave a course at Harvard on hidden variables.[52] Papaliolios earned a PhD from Harvard in 1965, and was a professor of physics there from that time until his retirement in 2001, while being also a physicist at the Harvard-Smithsonian Center for Astrophysics. His main research interests were related to astrophysics, but he included hidden variables as one of his topics of research in Harvard's reports to its visiting committees even before he was tenured as a Professor in 1971.[53] It was Papaliolios who invited Pipkin to undertake an experiment on Bell's theorem. Pipkin went to Harvard after his PhD in Physics at Princeton and there became a member of the faculty in 1957. Reputed as a good experimentalist both in low-energy atomic physics and high-energy particle physics, Pipkin had begun in the middle of the 1960s a line of research on precision measurement of the atomic fine and hyperfine structure, which was continuously funded by the National Science Foundation. This approach included the study of correlation of photons from atomic cascade in order to calculate lifetime of atomic states. Papaliolios' proposal was easily accommodated in Pipkin's project through the thesis work of Holt, his doctoral student, but Pipkin's interest in foundations of quantum mechanics were never as strong as Paliolios' were. Holt's history at Harvard presents nuances that show us distinctive features in the social and

[51] Paty (1977) included in his count not only optical photons experiments, but also proton–proton scattering and positronium annihilation experiments.

[52] Freedman et al. (1976). The readings for this course are in the Papaliolios Papers, box 16, folder "Fall 75 Hidden Variables—Reading Course—(P351)", *CPP*.

[53] Reports of Visiting committees are in Costas Papaliolios Papers, box 5, folder "Visiting Committee, 1970–1973," *idem*.

intellectual recognition of the importance of Bell's theorem.[54] In general, he did not encounter prejudices or disdain but merely a lack of interest. Edwin Purcell was on the committee examining his thesis and considered the subject "a worthwhile endeavor," but in his introductory quantum mechanics course, which Holt attended, quantum measurement was handled via Schwinger's measurement operator algebra, in which the full measurement problem is not explicit.[55] As a different example, he has the vivid image of the graduate quantum mechanics course given by Paul Martin, in which not only measurement was carefully presented via Kurt Gottfried's textbook, but also Bell's theorem was introduced.[56] Holt read Bell's paper but did not become interested himself, and considers that even if he was not yet a physicist he shared the attitude of the time. Thus, Holt does not speak of stigma associated with the subject but rather of little interest in the subject, and even this little interest was not unanimous. As a sign of the times, Holt opened his thesis with an approach that in other times and places would have been considered heretical. He analyzed the measurement problem in quantum mechanics and presented the "Copenhagen interpretation" as one of its possible solutions, the others being the "Everett-Wheeler interpretation," the "Wigner's idea," and the "hidden variables" (Holt 1973, pp. I–6–14). Incidentally, I remark that, 15 years before, regarding the Copenhagen interpretation as only one of the possible interpretations of quantum mechanics was removed from a dissertation before its publication. It happened with Hugh Everett's dissertation at Princeton under John Wheeler.[57]

As favorable as Harvard was for the experimental tests of Bell's theorem, its participation in our account is not a story of success. Holt and Pipkin did not trust their result, but they were not able to identify the source of errors. They circulated it as a preprint that was never published, meaning that in fact they did not claim to have found a violation of quantum mechanics prediction. In his sociological studies Harvey discussed how their cautious stance was conditioned by local and cultural circumstances such as Holt's status as doctoral student, trust in quantum theory, and a previous failure of Pipkin's, who had a violation of quantum electrodynamics that was not eventually confirmed.[58] As important as these factors could have been, and were, I think one should add a fact that was not determined by such factors. Pipkin and Holt gave up the subject, they did not pursue it to its ultimate consequences, by repeating the same experiment or by planning a new one, in spite of the interest their unpublished result awakened even beyond the group of physicists already

[54] Richard Holt (pers. comm., 21 March 2005). See also Wick (1995, p. 108).

[55] The composition of the committee who examined Holt's thesis was Pipkin, Papaliolios, and Purcell.

[56] Gottfried (1966). The whole Section IV is dedicated to "The measurement problem and the statistical interpretation of quantum mechanics." Bell's (1966) paper is suggested for reading.

[57] On this case, see Chap. 3 in this book.

[58] Harvey (1980). Pipkin's claim of a violation of quantum electrodynamics was Blumenthal et al. (1965).

involved with Bell's theorem.[59] Thus, the University of Southern California physicist Marc Levenson wrote to Pipkin, "I have obtained a preprint of your paper with Holt which casts doubt upon the validity of quantum mechanics. This result distresses me somewhat as I am expected to introduce our juniors to this subject next semester," and Levenson continued discussing possible sources of error.[60] One can conjecture that neither Pipkin nor Holt were able to foresee the importance that this subject would acquire in physics. Pipkin only took it up again in 1976, when Clauser and Fry were announcing the new results we discuss below. Pipkin then stated, "a careful study was made of systematic effects which could account for the deviation from the quantum mechanical prediction but no candidates were found. In view of the result reported by Freedman and Clauser [. . .] it was concluded that the experiment should be repeated with a somewhat different configuration of the apparatus."[61] However, they had not tried and they did not try any repetition; it was up to Clauser, Fry, and later Aspect to make new experiments on Bell's theorem. Two years later, Pipkin made his "closing arguments," while reviewing the "atomic physics tests of the basic concepts of quantum mechanics." After repeating that they had "recommended that the experiment be repeated by someone else with a different configuration of apparatus," he cited Clauser's new results and concluded, "this experiment thus indicated that the Holt–Pipkin experiment was incorrect although it did not localize the source of the error in the earlier experiment."[62] Their failure to analyze what was wrong with their experiment was probably responsible for the deletion of their role in the current story of success associated with Bell's theorem. So, Aczel (2002), in his popular science book *Entanglement*, dedicated one chapter to "the dream of Clauser, Horne, and Shimony," and another to the "Alain Aspect." Holt, Pipkin, and even Edward Fry's roles were sent to the backstage of the history. Symptomatically, Pipkin's Harvard colleagues, while writing his official obituary, did not include experiments in foundations of quantum mechanics among his achievements.[63]

---

[59] When Clauser was repeating Holt's experiment, the latter wrote to the former, "every time that Stu Freedman asks me when we're going to publish, I tell him I'm waiting for your results." Holt to Clauser, 31 Aug 1975, Clauser Papers.

[60] Marc Levenson to Pipkin, 03 December 1974. Pipkin Papers [Accession 12802], box 12, folder "NSF proposal 1974–1975," *PP*.

[61] Holt and Pipkin (1976). This report was presented at the Erice workshop by Pipkin. Similar words appeared in Freedman et al. (1976).

[62] Pipkin (1978, pp. 317–319). Until today, Holt holds the same opinion, "I think it is still accurate to say that the source of the error remains unknown." Richard Holt (pers. Comm., 21 March 2005). Clauser and Shimony's conjecture is that "stresses in the walls of the Pyrex bulb used to contain the electron gun and mercury vapor" made the glass optically active, and this systematic error was not adequately compensated. A similar problem appeared while Clauser was repeating the experiment. After the stresses were removed, "the experiment was re-performed, and excellent agreement with quantum mechanics was then obtained. On the other hand, Holt and Pipkin did not repeat their experiment when they discovered the stresses in their bulb" (Clauser and Shimony 1978, p. 1910).

[63] Gary Feldman, Paul Horowitz, Costas Papaliolios, Richard Wilson, and Robert Pound (Chairman), "F. M. Pipkin – Memorial Minute," *Harvard Gazette*, 26 Nov 1993, p. 15. At

Describing Holt and Pipkin's case as a failure risks anachronism.[64] After all, experiments of Bell's theorem became mainstream physics in the 1980s, and one needs to consider the importance of the result for them, at the time they performed the experiment. Indeed, neither for Pipkin nor for Holt did this experiment have the importance that we attribute to it today. Pipkin was developing a line of research on precision measurement of the atomic fine and hyperfine structure, from which Nussbaum's thesis on lifetime of atomic states obtained through atomic cascade experiments was one of the first results (Nussbaum and Pipkin 1967). Holt's experiment on Bell's theorem was a small extension of this method.[65] Indeed, in Pipkin's 1972 NSF proposal the completion of Holt's experiment is listed as #6 out of a list of seven goals, while it also included a precision measurement of the lifetime of the 7^3S_1 state of atomic mercury, a result that was indeed published.[66] Apparently, Pipkin's reputation as experimenter was not damaged by the unsolved problem with the test of Bell's theorem. In 1990, the physicist who evaluated his NSF grant extension in the same domain stated, "Professor Pipkin is a well-established leader in the field of high-precision measurements of fundamental atomic systems," and rated the proposal as "excellent."[67] Holt made a successful career working with precision measurement in atomic physics at the University of Western Ontario, Canada, but never returned to the subject of Bell's theorem.

The 1970s was the decade when hidden variables, an issue once considered a question of philosophical taste, entered the lab. These experimental activities did not mean, however, a decline in theoretical work on hidden variables. There was a flow of new derivations of Bell's inequalities. In 1978, reviewing the subject, Clauser and Shimony analyzed at least 11 different derivations, by Bell himself, Wigner, Frederik Belinfante, Holt, Clauser and Horne, Henry Stapp, d'Espagnat, D. Gutkowski and G. Masotto, Selleri, and L. Schiavulli, in addition to a new and

Harvard, this kind of obituary is commissioned. See Jeremy Knowles to Papaliolios, 10 Apr 1992. Papaliolios Papers, box 26, folder "Frank," *CPP*.

64 Harvey (1980, p. 158) spoke of "Holt's virtual capitulation." I think that he singled out too much Holt's profile as a graduate student. Indeed, as we have seen, it would be more reasonable to describe the case as a story of failure of the Harvard experimentalists involved, which was responsible for the deletion of their participation in the present story of success of Bell's theorem.

65 In his "Proposal: Atomic Physics Experiments Using Photon Coincidence Techniques," [1969], Pipkin listed as goal #2 "To continue the present coincidence measurements of the 4358–2537A photon cascade in mercury [...]," and listed Holt as doctoral student, but no reference was made to the hidden variable test, which would only appear in the proposal of the next year. See F. M. Pipkin, "Atomic Physics Experiments Using Fast Atomic Beams and Photon Coincidence Techniques," NSF Grant Proposal [GP22787], [1970]. He was funded $54,200.00 for 2 years, cf. Rolf Sinclair [NSF] to Pipkin, 26 May 1970. Pipkin Papers, box 23, folder "NSF Atomic 1973," *PP*.

66 F. M. Pipkin, "Proposal for a Grant from the NSF to continue atomic physics experiments using fast atomic beams and photon coincidence techniques." He was funded $112,380.00 for 2 years. Rolf Sinclair [NSF] to Pipkin 25 May 1972. Pipkin Papers, box 23, folder "NSF Atomic 1973," *PP* (Holt and Pipkin 1974).

67 Referee report on "Atomic Physics Experiments Using Lasers and Fast Atomic Beams", NSF proposal PHY-9016886, enclosed with Marcel Bardon [NSF] to F. M. Pipkin, 13 Dec 1990. Pipkin Papers, box 21, folder "NSF awards," *PP*.

different derivation by Bell, which was criticized by Shimony, Horne, and Clauser.[68] Two new books appeared, one by Belinfante (1973), entirely dedicated to the hidden-variable issue, and another by d'Espagnat (1989) [first appeared in 1971], dealing with the foundations of quantum mechanics. The derivations made by Bell and Clauser & Horne helped to focus what the issues at stake were in these experiments. Bell (1971) presented a proof that his theorem was not restricted to deterministic theories, and Clauser and Horne (1974) further developed this derivation of Bell's theorem. They showed that the available experimental data also falsified stochastic local theories, a conclusion, however, that depended on a supplementary "no-enhancement" assumption; which is weaker than the fair sampling assumption adopted by Clauser in the CHSH paper.[69] In any way, these results evidenced that it was locality and not determinism that was at stake in Bell's theorem.[70] Since then, to speak of the tests of Bell's theorem as tests of determinism, as we have seen in the notice published in *Scientific Research*, is less than accurate.

7.4 Settling the Tie and Turning the Page

Let us now consider the period between 1975 and 1976. Indeed, it was a time of new and impressive experimental results as well of new experimental challenges concerning Bell's theorem. Two new experiments, by Fry and by Clauser, promise of a new one, by Aspect, new social settings for gathering the physicists involved in the debate, and the feeling, by certain physicists, of a turning point in this story were the main features of the time.

As we have seen, Edward Fry, from Texas A&M University, became interested in experiments on Bell's theorem in the early 1970s. Having learnt from his first unsuccessful application that many physicists had disdain for these experiments, Fry changed his strategy. Now, he tried funding from Research Corporation, and added two new assets. He attached two letters supporting the application, one from Eugene Wigner, who wrote "a strong supportive letter," and the other from his

[68] Clauser and Shimony (1978, pp. 1886–1900). The debate between Bell, on one hand, and Shimony, Horne, and Clauser, on the other hand, was published in *Epistemological Letters* and reprinted in Bell et al. (1985). Clauser and Shimony's paper became the canonical review on the experiments of Bell's theorem. Wigner's (1970) paper included as a footnote a historical remark about what was von Neumann's main reason for stating the inadequacy of hidden-variable theories. This footnote stirred up a strong Clauser's (1971a, b) criticism, and the whole affair demanded the intervention of Shimony's diplomacy. Wigner to Shimony, 5 October 1970, Wigner Papers, box 72, folder 1, *WigP*, and Wigner (1971).

[69] Clauser and Horne's (1974) paper has 725 citations.

[70] The philosopher Karl Popper is an example of somebody who conjectured that the real conflict concerned determinism and not locality. For a criticism of this stance, see Bell (1972) and Clauser and Horne (1974, p. 526).

former adviser Bill Williams. And yet, he "tried to justify the pursuit of this subject in terms of a plethora of other more conventional capabilities that it also offered."[71] As a consequence of this strategy, Fry (1973) wrote that he was mainly interested in coincidence observation of optical photons from an atomic cascade as an experimental technique with variegated applications; and that experimental tests of local hidden variables were just one of the uses of such a technique, others being the determination of excited-state lifetimes and g values, branching ratios, absolute quantum efficiencies and source strengths.[72] The strategy adopted by Fry calls our attention to the kind of technical devices used in tests of Bell's theorem.

The experimental tests of Bell's theorem used technical devices that can be framed in what Peter Galison calls the "logic tradition" by contrast to those in the "image tradition." Both, according to Galison, formed two distinct traditions in the material culture of high-energy particle physics. In the logic tradition statistics played the key role and there was no room for the picture of the "golden" and unique event that was important in the image tradition. The very counterintuitive nature of nonlocality prevents us from visualizing the phenomenon at stake. In the experiments with correlation of photons coming from atomic cascades, the most sensitive pieces of the apparatus are the photodetectors and the electronics to count them if there is coincidence, that is, if the pair of detected photons comes from the same cascade decay. Indeed, Galison's (1997, p. 464) logic tradition came "out of the established electronic logic tradition of counters," which was roughly available in the 1930s but was dramatically improved during the war.[73] As we have seen, Fry was particularly interested in improved detectors. Experiments with Bell's theorem also depended on good polarizers and optical filters in addition to efficient methods to excite the atomic samples to the right level in order to obtain the intended cascade decay. In contrast to Papaliolios, who used the polarizers produced by Polaroid Corporation, experimenters dealing with Bell's theorem used more traditional polarizers, like calcite prisms and "pile-of-plates" polarizers, a cluster of glass plates arranged in certain angles, due to their large efficiency in observing linear polarization. To excite the atomic sample, the experimenters used resonance absorption with radiation emitted by lamps and passed through interference filters or electron bombardment, methods that had the undesirable effect of exciting many levels and not only those which were intended. Among all these technical devices, the technical innovation that changed the scene of these experiments in the 1970s was a new technique to excite the atomic samples, the tunable dye laser. Peter Sorokin and John Lankard invented dye lasers in 1965, but "it took several years before tunability emerged as preeminent among its properties."[74] It quickly became

[71] Edward Fry (pers. comm., 5 August 2005).

[72] Fry's style was noted by Harvey (1980, p. 156) in the following terms: "... a major part of Fry's strategy was to develop experimental techniques per se, and then apply them to a number of quite different empirical problems."

[73] For the development of such techniques during the war, see Galison (1997, pp. 239–311).

[74] Bromberg (1991, p. 184). Still according to Bromberg, "the dye laser was also discovered independently by Mary L. Spaeth and D. P. Bortfield at Hughes and by Fritz P. Schaefer and coworkers in Germany. Both of these groups published later."

a revolutionary technique for spectroscopy insofar as it yielded—within a certain range—the precise optical wavelengths one required to excite the atomic levels needed by experimenters.

Indeed, when Fry, helped by his graduate student Randall Thompson, went to carry out a new test of Bell's theorem he could take advantage of this new technique, the tunable dye laser. He used it for exciting exactly the atomic cascade of interest, and this permitted an improved speed of accumulating data. A number can summarize the improvements; while Clauser and Holt needed about 200 h for collecting data, Fry and Thompson performed the experiment in 80 min. Their results strongly violated Bell's inequalities and matched quantum mechanical predictions. The Bell's inequality under consideration was $\delta \leq 0$, the quantum mechanical prediction was $\delta_{qm} = +0.044 \pm 0.007$, and they obtained $\delta_{exp} = +0.046 \pm 0.014$.[75]

Meanwhile, Clauser repeated at Berkeley, in slightly different conditions, the experiment carried out before by Holt and Pipkin at Harvard, with the aim of breaking the previous experimental tie. The main difference was due to economic and practical reasons. He used "pile-of-plates" polarizers instead of calcite prisms. He obtained results confirming quantum mechanical predictions and violating Bell's inequalities. The experiment ran for 412 h. The Bell's inequality under consideration was $\delta \leq 0$, the quantum mechanical prediction was $\delta_{qm} = 0.0348$, and he obtained $\delta_{exp} = +0.0385 \pm 0.0093$.[76] For the second time, Clauser obtained experimental results that contradicted his hopes. He reported them to Wheeler in the following terms: "Dr. Henry Stapp here at LBL has told me that you were interested in the latest results from my experiments. These were attempting to reproduce the results observed by Holt and Pipkin at Harvard. Unfortunately, I have failed to do so, and obtained more or less good agreement with the quantum mechanical predictions."[77] Freedman and Holt also thought of repeating Holt and Pipkin's experiment, but this project did not happen.[78]

Clauser's and Holt's hopes and actual results deserve a remark on the literature in sociology of science. Harvey (1980, p. 157) convincingly argued that "the particular social, historical and cultural context in which the LHV [local hidden variable] experiments took place had a major effect on many features of these experiments." As examples of this effect, he cited location, physicists who carried them out, "the way in which they were presented, and the response to anomalous results." Holt's experiment fitted well in his claims. However, Harvey's claims were stronger, reflecting the then new trends in the sociology of science. For him (Harvey

[75] Fry and Thompson (1976). This paper received 220 citations.

[76] Clauser (1976). This paper received 159 citations.

[77] Clauser to John Wheeler, 27 Oct 1975. Clauser Papers.

[78] "You might be interested that I have decided to repeat Dick's mercury experiment here with pile-of-plates polarizers. [...] I hear you and Dick are considering collaborating on a similar repeat. Have you made a final decision on that?" Clauser to Freedman, 25 Jan 1974. Clauser Papers.

1981, p. 106), "the social and cultural, as well as the technical, context in which a scientist finds himself will influence not only the style, timing and presentation of his work but also (at least in principle) its content." This strong claim about the content of the experimental results does not meet evidence when one compares the prospects nurtured by both Holt and Clauser. As Shimony remarked, after making the caveat that he "dislike[s] the idea that experimental results are theory laden, that somehow experimenters see what they want to see," the two physicists obtained results opposed to their expectations.[79] Evidence about Clauser's hopes on violating quantum mechanics was already available at the time Harvey conducted his research. Harvey made reference to them, but did not extract the full consequences of the contrast between Clauser and Holt's cases. Nowadays, with more archival evidence available, it is harder to accept that local contexts determined the content of their experimental results.

Since the early 1970s there already had been a small community of physicists who were interested in Bell's theorem. In the middle of the 1970s they looked to reinforce their links and to create opportunities for discussions and gathering. In addition to the usual trips and leaves of absence physicists use as regular means for circulating professional information, our protagonists used two others: the Erice Thinkshops on Physics, and the journal *Epistemological Letters*. The latter was an unusual vehicle for scientific debates. It was conceived as a permanent written symposium on "Hidden Variables and Quantum Uncertainty," and defined itself in this way: "*Epistemological Letters* are not a scientific journal in the ordinary sense. They want to create a basis for an open and informal discussion allowing confrontation and ripening of ideas before publishing in some adequate journal."[80] Indeed, the journal was more than this. It published short letters, kept open debates for several issues, announced news of interest, republished some papers, and even kept a list of the recipients of the journal. It was, in a certain sense, a predecessor of the contemporary Internet discussion lists. Instead of circulating via the electronic web it was mimeographed and sent to its recipients. Thirty-six issues were published from November 1973 to October 1984. About 60 authors wrote in the journal, and Shimony, Bell, d'Espagnat, Lochak, Costa de Beauregard, P. A. Moldauer, F. Bonsack, J. L. Destouches, and M. Mugur-Schaechter wrote at least five pieces each. Many papers were indeed published elsewhere but some debates which were not well documented elsewhere, such as the refusal of de Broglie and his collaborators to accept the full implications of Bell's theorem, are uniquely recorded there. It was published in Switzerland by the "Association F. Gonseth—Institut de la mèthode," under the editorial responsibility of the philosopher of science François Bonsack, who was the secretary of this association. Shimony also acted informally as an editor, publishing short reviews on the subject and actively intervening in the

[79] Abner Shimony, interviewed by Joan Bromberg, 2002, p. 74, *op. cit.*

[80] From the back cover of all *Epistemological Letters* issues. The University of Pittsburgh has a complete collection of this journal, a gift of Abner Shimony.

debates.[81] After the publication was over, he wrote a very favorable review of its existence:

> The variety of the contributions and the vigor of the debates showed that the purpose was very well accomplished. Because of the brief time interval between issues and the absence of customary refereeing procedures, it was possible to carry on a debate more rapidly than in standard journals, and speculative ideas could be more easily made public. It is remarkable that in spite of the informality of *Epistemological Letters*, the typing of the articles, including mathematical formulae, was very accurate. The reputation of the written symposium spread rapidly, and many people throughout the world wrote to be added to the list of recipients. (Shimony 1985).

Erice, in Sicily, has been a favored destination for physics gatherings due to a conjunction of the natural and cultural appeal of the town and the restless initiatives of the Italian physicist Antonio Zichichi, head of its "Ettore Majorana Centre." Bell was close to Zichichi in their activities in high-energy physics at CERN, and this relationship allowed him and d'Espagnat to organize the meeting, which took place in April 1976, and gathered together 36 participants from 25 laboratories and 9 countries.[82] This meeting had a double importance for the history of Bell's theorem. It was an environment for the socialization of some of the physicists involved in this research, and it was the stage for presenting and discussing Clauser's and Fry's yet unpublished results. It was instrumental for physicists who were established in the field, like Clauser, or entrants, like Aspect, to network. Among the attendees there was Anton Zeilinger who would play an important role in research on entanglement from the late 1980s on, as we will see in Chap. 8. In Erice he presented a report on experiments confirming a counter-intuitive quantum prediction. This prediction states that a neutron quantum state changes its signal after a 2π rotation, only recovering the original signal after a 4π rotation. This resulted from the research on neutron wave featured in Vienna by Helmut Rauch, who was Zeilinger's doctoral supervisor (Rauch et al. 1975). Zeilinger went to Erice unaware of entanglement but came back fascinated by the subject.[83]

Clauser (1992, p. 172) described the Erice meeting in these words, "the sociology of the conference was as interesting as was its physics. The quantum subculture finally had come 'out of the closet' and the participants included a wide range of

[81] See, for instance, Horne and Shimony (1973) and Shimony (1980).

[82] A report of the conference, written by John Bell [Testing Quantum Mechanics], and the abstracts of the papers were published in *Progress in Scientific Culture—The Interdisciplinary Journal of the Ettore Majorana Centre,* 1/4, 439–460, 1976. I am grateful to Alain Aspect for sending me a copy of it.

[83] See *Progress in Scientific Culture*, ibid, pp. 443, 458–460. Zeilinger's intellectual style is marked by a deep curiosity, which was directed towards science during his undergraduate studies and favored by the flexible curriculum at University of Vienna at that time. In addition, he benefitted from Rauch's support to research on foundations of quantum mechanics and from the intellectual climate of physics in Vienna—with its mix of science and philosophy—a legacy coming from the late nineteenth and twentieth centuries. Anton Zeilinger, interviewed by Olival Freire, 30 June 2014, AIP. For Rauch's research on neutron interferometry and its relation to foundational issues, see the review Rauch (2012).

eminent theorists and experimentalists." We shall see its importance for Aspect later. Socialization and professional recognition were important issues in the Bell's theorem saga, since recognition many times did not come in due time. We have much evidence concerning the lack of professional recognition even after the first experimental tests of Bell's inequalities.[84] It was not by chance that in this very meeting, Bell, highly sensitive to this issue as we have already seen, felt the need to criticize such an attitude and to present a rationale for pursuing the experiments. "The great success of quantum mechanics in accounting for natural phenomena in general will incline most people to expect it to remain successful here. Many people will even be intolerant of the idea of actually performing such experiments. [...] We do a service to future generations by replacing gedanken with real experiments" (Bell 1976, p. 440). If the first sentence concerns trust in quantum mechanics, the second is an open criticism of the still existing prejudices. The French physicist Franck Laloe (pers. comm., 28 March 2005), a newcomer to the subject at that time, recalls "being interested in the foundation of quantum mechanics was still considered sort of bad taste by most main stream physicists." However, none of this evidence is as telling as Clauser's case.

Clauser's achievements in the 1970s were remarkable. He realized the full implication of Bell's theorem; carried out two key experiments on it, one of them being the first ever experimental result; and enhanced our understanding of the subject. In addition, he used the knowledge required for experiments with Bell's inequalities to contribute to the debate between supporters of semi-classical radiation theories, notably Edwin Jaynes, and supporters of a full quantum treatment of radiation. He showed that experimental data from the polarization correlation of photons emitted in atomic cascades were incompatible with the semi classical theories and fitted well with quantum treatments (Clauser 1972, 1974). However, in spite of his achievements, Clauser faced hindrances achieving a professional career in physics based on experiments related to foundations of quantum mechanics. Indeed, he did not get a job.[85]

The early 1970s were also the times of restrictions in the public funding of American science and this conjuncture brought consequences for the employment of the new physicists.[86] In addition to this background, to analyze why a researcher

[84] My account strongly contrasts with Wick's stance (1995, p. 244). He asked some of these protagonists "if their participation in testing quantum mechanics adversely affected their standing among their peers, their ability to obtain research funding, or their job prospects, each replied simply 'no'."

[85] Although highly considered among both quantum opticians and physicists involved with foundations of quantum mechanics, Clauser eventually shifted his interests to other issues such as nuclear fusion, X-ray imaging, and recently Talbot-vonLau interferometry. One can conjecture about the role played in his decision by his feelings of lack of recognition of the subject among physicists, and the impact of such a lack on his own career.

[86] According to Kevles (1978, pp. 421–423), this period represented "a degree of disestablishment" in American physics. It had begun in the middle of the 1960s, signaling the end of the "post-Hiroshima honeymoon," but by the early 1970s "the cutbacks [...] created an employment squeeze reminiscent of the 1930s depression."

did not get a job in a number of institutions would require a more comprehensive study. However, the available documentation shows that some physicists who decided the issue were influenced by the prejudice that experiments on hidden variables were not "real physics." Clauser faced, both at Columbia and California, the same hostile environment inherited from the 1950s. His former adviser, P. Thaddeus, wrote recommendation letters warning people not to hire Clauser if it is for doing quantum mechanics experiments, since it is "junk science." In fact, some of his possible employers thought the same, in spite of the many letters of recommendation he had.[87] Shimony reported to him that "when I saw d'Espagnat last week he had a letter from the Dep't Chairman at San Jose, inquiring whether what you have been doing is real physics. Needless to say, he'll write a strong letter answering the question in your favor. I'm sorry, from that evidence, to find that your job situation is still unsettled."[88] Terms such as "real" or "junk" science may have jeopardized Clauser's professional career, particularly when "what counted as real science" was a matter of dispute as part of the "pseudoscience wars" (Gordin 2012, p. 3). Retrospectively, Clauser admits that he was not smart enough not to give talks on Bell's inequalities while looking for a job. "I was sort of young, naïve, and oblivious to all of this. I thought it was interesting physics. I had yet to recognize just how much of stigma there was, and I just chose to ignore it. I was just having fun, and I thought it was interesting physics. I was just trying to understand what was going on."[89] Shimony, also retrospectively, suggests a sociological explanation for this fact; "certainly there was a lot of interest [in Bell's theorem], but that doesn't mean there was enough to get majority votes in physics departments to bring somebody whose main credentials were an experiment concerning hidden

[87] John Clauser, interviewed by Joan Bromberg, (2002), p. 12), *op. cit.* "What is the situation regarding employment next year? If any more letters should be written, let me know;" Abner Shimony to John Clauser, 19 May 1972. "In reply to your letter of January 9, I am happy to write in support of Dr. John Clauser as a candidate for a faculty position at UCSC. I believe he shows promise of becoming one of the most important experimentalists of the next decade. [...] I say these things in spite of the fact that Clauser's results spell trouble for my own pet theory." Edwin T. Jaynes to Peter L. Scott [Chairman, Board of Studies in Physics, University of California at Santa Cruz], 31 Jan 1973. Clauser Papers.

[88] Shimony to Clauser, 8 Aug 1972, Clauser Papers. Shimony had made his own attempt, unsuccessfully, at Boston University: "... I have an appointment with [the dean ... at Boston University] for this coming Thursday, and shall try to argue for a special appointment for you, on the grounds that this is a great opportunity for B.U. At least one of the astronomers (Hegyi) knows your work in astrophysics, and therefore I can make a case that you'll serve two Departments. The new president of B.U. has been making a splash with distinguished senior appointments, outside normal departmental budgets, and I shall argue here is an opportunity for an equally distinguished junior appointment. I wish I could count on their being imaginative enough to see their opportunity, but unfortunately I am not, and don't want you (or me) to become too hopeful." Shimony to Clauser, 20 July 1971, Shimony Papers, Box 1, Folder 4, "Clauser, John F.—Correspondence, 1971–1972," *ASP*.

[89] John Clauser, interviewed by Joan Bromberg, 2002, p. 18, *op. cit.*

variables."[90] Independent of his admitted naiveté and the job crisis in American physics, it is interesting to record how some leading physicists reacted to the physics he was doing. Clauser (2002, p. 71) reports that while he "was actually performing the first experimental test of the CHSH-Bell predictions as a postdoc at UC-Berkeley, [... he] made an appointment with Prof. Richard Feynman to discuss these same questions. Feynman was very impatient with [him]." Feynman's stance was: "Well, when you have found an error in quantum-theory's experimental predictions, come back then, and we can discuss your problem with it."[91]

The second important feature of the Erice meeting is related to the fact that it was considered a turning point towards the recognition of quantum nonlocality as a physical effect. An evidence of the high expectation the meeting awakened is the fact that Bell had especially invited the reputed University of Chicago's experimentalist, Valentin Telegdi, to give his opinion on the recent experiments performed by Clauser and Fry. Telegdi had not had any previous involvement with experiments in foundations of quantum mechanics. Bell and d'Espagnat also invited Pipkin, responsible for the only diverging result, in order to bring together the main protagonists of this story. "It will be good if our meeting can contribute to getting to the bottom of the differences between the various experiments. In any case it will be a great pleasure to see you in Erice," wrote Bell to Pipkin.[92] In his concluding remarks, Bell (1976, p. 442) was sober, "such atomic cascade experiments were reported by Clauser, Pipkin, and Fry (a very elegant new experiment). Three of the four experiments are in excellent agreement with quantum mechanics. But that of Holt and Pipkin is in serious disagreement. [...] After discussions at this meeting it remains unknown what, if anything, went wrong in the Holt-Pipkin experiment." Bell's sobriety was not widely shared. The French physicist Olivier Costa de Beauregard reported this feeling in *Epistemological Letters*: "The latest experimental results, not yet published [...] are explicitly in favor of quantum mechanics, thus confirming the reality of the paradox."[93] Indeed, the audience sensed that one page in the hidden variable story was being turned. The impression

[90] This conjecture, however, does not attenuate Shimony's criticisms: "I think he was treated very shabbily." "He's a brilliant man, a very good experimenter, and really a good theoretician also." Abner Shimony, interviewed by Joan Bromberg, 2002, pp. 82–83, *op. cit.*

[91] Feynman's opinion on Bell's theorem deserves further research to track its evolution. His views after Aspect's experiments, in 1984, will be commented upon later. After the talk reported by Clauser, while visiting Texas A&M at Austin, in 1974, Feynman was approached by Edward Fry and James McGuire to discuss their planned experiment, and reacted positively. Edward Fry (pers. comm., 5 August 2005).

[92] John Bell to Francis Pipkin, 22 Dec 1975, Pipkin Papers, box 2, folder "Correspondence January–April 1976," *PP*.

[93] "Les tout derniers résultats expérimentaux, non encore publiés [...] sont explicitement en faveur de la Mécanique Quantique, et confirment donc la réalité du paradoxe". O. Costa de Beauregard—Nouvelles du colloque sur le Paradoxe EPR au Centre Ettore Majorana, a Erice, 18–23 avril 76, *Epistemological Letters*, 10th issue, p. 26, May 1976.

was that from now on one could speak of a new quantum physical effect, quantum non-locality, as Lalöe testifies:

> This meeting coincided with an important turn in physics. Until about that time it had been possible to believe that the Bell inequalities were obeyed by Nature, since they relied on very general assumptions (very much in the spirit of relativity). [...] For some time, in particular in view of the experiments performed by Pipkin at Harvard University, some doubt remained indeed possible. But when John Clauser and his group gave their results, and then even more when Ed Fry came with another series of even more precise experiments, the agreement between the results and quantum mechanics was so impressive that no-one could anymore still think seriously that the Bell inequalities were obeyed by physics. (Lalöe, pers. comm., 28 March 2005).

Bell, Costa de Beauregard, and Lalöe's words acquire more vivid colors when compared with the following review of experiments on hidden variables written two years before by three experimentalists who were involved with the subject, Freedman, Papaliolios, and Holt.[94] After revising the results obtained by Freedman and Clauser, by Holt, and by experiments on annihilation of positrons, they concluded:

> We note, in conclusion, that the problem of the validity of local hidden variable theories rests with the experimentalists. New experiments are in progress or are being planned by several groups and we can hope for a solution in the near future. It is fair to say that the existing evidence still favors quantum mechanics; nevertheless, the question is of fundamental importance and there is too much at stake to allow any experimental discrepancy to remain unexplained. (Freedman et al. 1976, p. 57).

However, despite of Fry and Clauser's results favoring quantum mechanics, Bell did not consider the history over, and in his final report in fact he set a new agenda with a new experimental challenge. He resumed an idea due to Bohm (1951, p. 622), "that [in a EPR experiment] while the atoms are still in flight, one can rotate the apparatus into an arbitrary direction." As in the real experiments the analyzers are set before the experiment one cannot discard that the existence of an unknown subluminal interaction between the analyzers is responsible for the correlations. If experiments confirm the existence of such an interaction, Bell's condition of locality, on which Bell's theorem depends, would lose its validity. According to Bell (1976, p. 442), "now it can be maintained that the experiments so far described have nothing to do with Einstein locality. It is therefore of the very highest interest that an atomic cascade is now under way, presented here by Aspect, in which the polarization analyzers are in effect re-set *while the photons are in flight*."

---

[94] The paper was written in 1974, by initiative of Freedman and Holt, for a volume in honor of Louis de Broglie. Freedman to Papaliolios, 18 Mar 1974, Holt to Papaliolios, 1 Apr 1974, Papaliolios Papers, box 23, folder "Hidden Variables—Paper in honor of de Broglie," *CPP*.

7.5 New Challenges: "While the Photons Are in Flight"

Alain Aspect's road to experiments in the foundations of quantum mechanics reflects a transitional time more than Shimony's and Clauser's roads. It was a transitional time from the consideration of a subject as a philosophical quarrel to its recognition as an interesting field of physical research. His biographical profile not only reflects that time but also evidences his active contribution to awarding these issues the respectability they deserved. Born in 1947, Aspect did his undergraduate studies at École normale supérieure de l'enseignement technique (ENSET), in Cachan, south of Paris, while taking physics courses at Orsay. Next, he did the French *doctorat de troisième cycle*, on holography, also at Orsay. Soon after his work on holography, he became disappointed with physical research, and planned to become a teacher, having successfully ranked second in the French national contest named "aggregation." Following the French tradition of replacing military duties with civil service, he went to Cameroon, teaching there between 1971 and 1974. However, while in Africa, his plans changed once more. He realized that just teaching would become boring in the medium term, and he studied the new textbook on quantum mechanics written by Cohen-Tannoudji et al. (1973), which revived his interest in physical research, especially in optics.[95] Coming back to France, Aspect got a tenured position at ENSET, his former school; a job which gave him freedom to choose his themes of research. In the fall of 1974, interested in resuming research but on subjects in which the "quantum weirdness" appears, he headed to the Institut d'Optique at Orsay, to talk with Christian Imbert, a young professor who had done experiments related to the photon self-interference, looking for his French *doctorat d'état*. Imbert, who was in touch with Bernard d'Espagnat and Olivier Costa de Beauregard, handed him a bibliography related to Bell's inequalities, "to look into that for a possible subject."

Aspect became fascinated reading Bell's paper, realized the conflict between the experimental results of Clauser and Freedman, on one hand, and Holt and Pipkin on the other hand, and found the right subject for his dissertation: "an experiment in which the polarizers would be rotated while the photons were in flight." Aspect's project would resonate with the interest of some of our protagonists. Andre Marechal, then the director of the Institut d'Optique, asked d'Espagnat for a good subject for the dissertation of a good researcher who had come back from Africa, i.e. Alain Aspect. d'Espagnat discussed the subject with Bell and they agreed to propose to Aspect that he settle the conflicting results obtained in the previous experiments, which at the time were in a tie between Clauser's first experiment and

[95] Biographical information about Alain Aspect comes from (Aspect, pers. comm., 28 Feb 2005) and from Alain Aspect, interviewed by Olival Freire and Indianara Silva, 2010 and 2011, *AIP*. "I can say that my previous studies in quantum physics had been totally disappointing: it was just solving partial differential equations about 'rigid rotator' and so, not physics according to my view of physics. The textbook of CCT et al. totally changed my view on that," Aspect, 2005, *ibid.*

the one by Holt and Pipkin.[96] Aspect also realized that this experiment should be a long term project due to its technical challenges and the need of getting acquainted with the subject, because he had no previous training in modern experimental atomic and laser physics. However, if the intrinsically long *doctorat d'état* and his tenured position at ENSET could be accommodated by this project, he needed advisers and funding. At this point, a series of conversations involving Bell, d'Espagnat, Imbert, Costa de Beauregard, in addition to letters from Arthur Wightman and Alfred Kastler supporting the funding of such project, sealed favorably the fate of the project.[97] However, it should be noted that the French leaders in atomic and laser physics were not initially involved with Aspect's experiment, since that leadership was shared by Jean Brossel, who was the head of the laboratory at Ecole Normale Supérieure in which Kastler, Cohen-Tannoudji, and Laloë were working, and Pierre Jacquinot, head of the Laboratory Aimé Cotton.

Picture 7.5 The author and Alain Aspect, 2011

[96] Bernard d'Espagnat, interviewed by Olival Freire, 2001, *AIP*.

[97] Aspect (pers. comm., 28 Feb 2005). The acknowledgments in Aspect's (1976) proposal of experiments evidence the patronage for them: "the author gratefully acknowledges Professor C. Imbert and Dr. O. Costa de Beauregard for having suggested this study and for many fruitful discussions. He especially thanks Dr. J. S. Bell for his encouragement, and Professor B. d'Espagnat for his thorough consideration and discussion of the theoretical aspects of our scheme". The choice of recommendation letters from Wightman and Kastler were not casual. The Princeton mathematical physicist Arthur S. Wightman championed the axiomatic quantum field approach, a subject with some overlapping with foundations of quantum mechanics. For a presentation of philosophical issues in quantum field theories, see (Brown and Harré 1988), particularly the chapters by Michael Redhead and by James T. Cushing. Alfred Kastler had been awarded the 1966 Physics Nobel Prize for his works on optical pumping and was the leader figure in French atomic physics.

At the beginning, Aspect worked alone. He learned about the measurement of "photon coincidence at the electronic shop of the CEA at Saclay," borrowed equipment from them, and talked to experimentalists at Ecole Normale Supérieure and Laboratory Aimé Cotton, who were using tunable lasers and atomic beams. He also learned how to speak on Bell's inequalities to people who were not a priori interested in foundations of quantum mechanics. According to him, "I discovered that if I presented things in a very simple and naïve way, just as I had understood Bell's paper the first time I had read it, most of the public a priori skeptic (not to say more) would become interested and sympathetic."[98] Nowadays, people who attended Aspect's lectures consider he became a charismatic lecturer. Things evolved positively. In the 1976 Erice meeting, in addition to meeting physicists already involved with Bell's theorem, he met Laloë,[99] who convinced Cohen-Tannoudji that Bell's inequalities were an interesting physics topic, and introduced Aspect to him. According to Aspect, the interaction with Cohen-Tannoudji produced a "phase transition" in the way he was regarded by his colleagues around 1978–1979, when collaboration with Cohen-Tannoudji's team on a side subject, different from Bell's inequalities, allowed him to establish close intellectual links with this highly respected French physicist.

In the late 1970s, signs appeared suggesting that Bell's theorem was gaining wider recognition. In July 1979, Bell was one of the invited speakers at the Conference of the European Group for Atomic Spectroscopy, to talk on "Atomic-cascade photons and quantum-mechanical nonlocality" (Bell 1980). In June 1980, the Collège de France, under the patronage of Cohen-Tannoudji, who was then at the apex of French physics, with a chair in this prestigious French institution, organized an international colloquium on the conceptual implications of quantum physics in which Bell's inequalities and Aspect's ongoing experiments were presented.[100] The intense experimental and theoretical scientific activities on Bell's theorem awakened interest beyond physics. d'Espagnat (1979) published in *Scientific American*, the renowned American popular science magazine, a non-technical account of Bell's theorem and its experimental tests. In the late 1970s, the new trend in the sociology of science, interested as it was in the study of scientific controversies, produced a number of papers dedicated to the debate on hidden variables in quantum mechanics.[101]

Aspects' experimental results eventually came out, between 1981 and 1982. The initial project of a single experiment had become three. In collaboration with his

[98] Aspect (ibid.). As examples of his approach, see Aspect (1983, 2002).

[99] "Alain is right when he mentions our many friendly discussions at that time and later, during his thesis in Orsay. [...] Sometimes he was explaining to me things that I had not understood, sometimes it went the other way. For instance I remember that one day in Paris I suggested to him the two channel experiment (with birefringent filters), which turned out to be a good idea, but certainly not for the reasons I was proposing, which were incorrect!" Laloë (ibid.).

[100] The lecturers at the College de France were Laloë, Bell, Aspect, Shimony, and d'Espagnat, and the conference brought together 25 people, *Journal de physique*, Tome 42, Colloque C2, 1981.

[101] Pinch (1977), Brush (1980), Harvey (1980), and Harvey (1981).

undergraduate student Philippe Grangier and the research engineer Gérard Roger, he once again performed the test of Bell's inequalities, as Clauser and Fry had done, but used two tunable lasers to excite the sample, providing him a source with higher efficiency. The counting lasted 100 s. In addition, as the Paris group separated source and polarizer by 6.5 m, they could rule out Furry's conjecture, which suggested that the quantum non-locality would vanish after the photons travel a distance of the order of the coherence length of their associated wave packets; which meant a distance of 1.5 m in this experiment. In mathematical terms, a pure state would evolve towards a mixture of factorizing states. With the same collaborators Aspect used two-channel polarizers, which allowed a straightforward transposition of EPR *gedanken* experiment. In previous experiments when one of the detectors is not triggered, one could not know whether it was a result of the low efficient detectors or whether the polarizer has blocked the photon, which would be a real measurement. For this reason auxiliary experiments with the polarizers removed were needed to circumvent the intrinsic deficiency of the setup. Finally, with Jean Dalibard and Gérard Roger, he produced the first test of Bell's inequalities with time-varying analyzers. Aspect's ingenuity was to use a switch to redirect the incident photons to two different polarizers. This device works through an acousto-optical interaction with an ultrasonic standing wave in water.[102] All the experimental results violated Bell's inequalities and strongly confirmed quantum mechanics' predictions. In the first experiment the Bell inequality was $\delta \leq 0$, the quantum mechanical prediction was $\delta_{QM} = 5.8 \times 10^{-2} \pm 0.2 \times 10^{-2}$, and the experimental result was $\delta_{exp} = 5.72 \times 10^{-2} \pm 0.2 \times 10^{-2}$, which violated the Bell inequality by more than 13 standard deviations. For the second experiment, the Bell inequality at stake was $-2 \leq S \leq 2$, $S_{QM} = 2.70 \pm 0.05$, and the result was $S_{exp} = 2.697 \pm 0.015$, to that date the strongest violation of Bell's inequalities ever reported. In these experiments each run lasted 100 s. The third experiment is telling by what it was measuring, a Bell's inequality using time-varying analyzers, and due to this reason it was the result that most resonated in the physics community, but its accuracy was less than the previous ones. It tested Bell's inequality $S \leq 0$, $S_{QM} = 0.112$, and the experimental result was $S_{exp} = 0.101 \pm 0.020$, violating Bell's inequality by five standard deviations in runs which lasted 200 min.

Aspect's experiments made his professional reputation, and his ongoing research on new fundamental phenomena, like experiments with just one photon, with laser cooling below the one photon recoil, and, more recently, on the Bose–Einstein condensate, carried him to a position of leadership in quantum optics and atom

[102] Aspect et al. (1981, 1982a, b). The first paper received 764 citations, the second 1,555, and the third 1,039 citations. The choice for this switch, instead of using Kerr or Pockels cells as first thought by Clauser, was determined by the consideration that with these cells "only very narrow beams could be transmitted, yielding very low coincidence rates; as these cells heat up, and then become inoperative, long runs would be prohibited." In addition, the calibration of the system would be "exceedingly difficult" due to the need of monitoring the change of the polarizer orientations (Aspect 1976, p. 1945).

optics on the world stage, while in France his prestige can be evaluated by his 2001 election to the French *Académie des sciences* and his 2005 *Médaille d'or du CNRS*.

In 1982 Alain Aspect was one of the invited speakers at the Eighth International Conference on Atomic Physics, held in Sweden, to report on his experiments on Bell's inequalities. The American physicist Arthur Schawlow, Physics Nobel Prize winner in 1981, was requested to make the final report of the conference. He chose Bell's theorem and its experiments as the main topic of his speech:

> Physical metaphors, such as the dual concepts of particles and waves in dealing with the light and atoms, are more than just conveniences, but rather are practical necessities. [...] But the experiments on Bell's inequalities are making it difficult for us to continue using some of our familiar physical metaphors in the old ways. We are used to thinking that light waves are produced at an atom with definite polarizations and are subsequently detected by remote detectors. However, the experiments show that if anything is propagated, it seems to convey more polarization information than a transverse wave. [...] As an experimentalist, I like to think that there is something there that we call an atom, and that we can make good measurements on it if we are careful not to disturb it too much. But the experiments on polarization of correlated photons don't bear out these expectations. (Schawlow 1983)

Two years later, Feynman, who once refused to talk about hidden variables with Clauser while the first experiment was being carried out, attended a seminar given by Aspect at Caltech on the tests of Bell's theorem and wrote to him, "once again let me say, your talk was excellent." At this seminar, Aspect finished his talk by quoting a certain paper whose author derived results similar to Bell's inequalities and went on to discuss whether it was "a real problem." According to Aspect, this author gave an answer that was so unclear that he "had found it amusing to quote it as a kind of joke to conclude this presentation." Only at this point, did Aspect reveal the name of the author, Richard Feynman. According to Aspect, nobody in the audience laughed until Feynman laughed. Feynman checked the quotation and wrote to Aspect conceding he was right.[103] After this emblematic anecdote, we present a short epilogue concerning Aspect's professional recognition.

Aspect experienced the transition of the field of foundations, considered a suspicious topic, to an acclaimed one. After these experiments he considered that there was no foreseeable groundbreaking achievement to be obtained with new experiments on Bell's theorem and moved to other subjects. Initially he worked on quantum optics and later he helped shape atomic optics. Acknowledgment of his contributions to our current understanding of quantum mechanics has been growing. In 2010 he shared the prestigeous Wolf Prize in physics with John Clauser and Anton Zeilinger for their work on entangled quantum states. In 2013, he went to

---

[103] Richard Feynman to Alain Aspect, 28 Sep 1984. Richard P. Feynman Papers, Box 22, Folder 15, California Institute of Technology Archives. Aspect (pers. comm., 20 Apr 2005). Feynman's quotation, in Feynman (1982, p. 471), is: "It has not yet become obvious to me that there's no real problem. I cannot define the real problem, therefore I suspect there's no real problem, but I'm not sure there's no real problem. So that's why I like to investigate things." Before this fragment, Feynman had written, "Might I say immediately, so that you know where I really intend to go, that we always have had a great deal of difficulty in understanding the world view that quantum mechanics represents."

Copenhagen to receive the Bohr Medal, awarded by UNESCO, for his contribution to our understanding of entanglement.[104] It was a kind of closing of the circle. Reality of entanglement had been an insight of Einstein, which Einstein refused, in his dispute with Bohr. Bohm and Bell worked on Einstein's insight but were targets of criticism from Rosenfeld, the Bohr's longstanding assistant. Aspect went to Copenhagen as a protagonist in the setting of entanglement, a physical effect underpinning the burgeoning field of quantum information. The awarding of this Bohr Medal was thus telling of the changes in the status of foundations of quantum physics since the 1950s.

Conclusion

Hidden variables, considered a philosophical matter 30 years before, entered the optics laboratories, and occupied a place in mainstream physics. This is not to say that physicists reached a full consensus on the meaning of these experiments and that philosophical issues vanished.[105] Indeed, the compatibility between quantum physics and its non-locality, on one hand, and special relativity, on the other hand, remained a matter of dispute. There appeared demonstrations that quantum mechanics cannot be used to exchange superluminal messages. Bell, himself never felt comfortable with this kind of compatibility between these two major physical theories.[106] Other physicists did trust that detectors with higher efficiencies would lead to violations of quantum mechanics. Indeed, one year after the publication of Aspect's experiments, Marshall et al. (1983), published a paper entitled "Local realism has not been refuted by atomic cascade experiments," in which they built a hidden-variable model able to mimic quantum-mechanical predictions. Marshal, Santos, and Selleri had resumed a line of approach first suggested by Clauser and Horne (1974), by relaxing the fair sampling assumption adopted in the CHSH (1969) paper.

The most important historical lesson from the period analyzed, I think, is that the path from philosophy to physics required not only good theoretical ideas, experimental skills, and technological improvements, but also a change in the physics community's attitude about the status of the foundations of quantum mechanics as a subject for physics research. However, the path from

(continued)

[104] http://bohr2013.nbi.ku.dk/english/events_exhibitions/niels_bohr_medaljen/; http://www.unesco.org/new/en/media-services/single-view/news/unescos_niels_bohr_gold_medal_awarded_to_prominent_physicists_in_2013/#.UyN7jF4zjx4. Accessed on 14 March 2014.

[105] An example of a good physicist who did not grasp the full meaning of Bell's theorem even after Aspect's experiments is Abraham Pais, who later in his life became historian of physics. While writing Einstein's biography, Pais (1982, Chapter 25c) assessed the EPR paper had no bearing on physics and did not cite Bell's theorem as a development of this issue.

[106] Eberhard (1978), Ghirardi et al. (1980), Aspect (1981), Page (1982), Tausk (1967, 29-31), and Bell (2004a, b).

philosophy to physics was slow and sinuous, involved diverse factors, and not only the ones I discussed here. Even on the road I presented here, the perceptions of its protagonists about the recognition of these achievements evolved in different ways, according to their personal experiences and local contexts. The 1970s were a transitional decade for the research on Bell's theorem, in particular, and for the foundations of quantum mechanics, in general. The role of local contexts and personal stories can be measured if one considers that while Shimony met at the 1970 Varenna meeting a small but supportive environment for his research and Holt did not feel prejudices against the subject of his dissertation at Harvard in the early 1970s, Clauser had his academic career blocked and only found a similar environment to that found by Shimony at the 1976 Erice meeting, and Aspect only felt himself well accepted in the main milieu of the French Optics by 1978.

In parallel with the differences we have seen in this paper, some common features can be extracted from the biographical sketches of our characters. A rough collective biography of them can be drawn noting that many of them were after all, dissidents, or quantum dissidents, a theme I will elaborate further in the Chap. 9, while concluding this book. They fought against the dominant attitude among the physicists according to which foundational issues in quantum mechanics were already solved by the founding fathers of the discipline. Some of them, such as Bell, Clauser, and Shimony, were hard critics of the complementarity interpretation. The common ground of the quantum dissidents was minimal and focused just on the importance of the research into the foundations of quantum mechanics. They supported different interpretations of this physical theory and chose different approaches and issues in their research. While Clauser did not trust in the Copenhagen interpretation, Aspect had no strong philosophical qualms with this interpretation in itself; while in the 1950s Bohm tried to build models for mimicking quantum mechanics, Bell gave his attention to the critical analysis of its assumptions. The fact that their common platform was the critical analysis, both theoretical and experimental, of the foundations of quantum physics, instead of the development of just one alternative interpretation, or even the advocacy of their philosophical credo, was one of the sources of their strength. They had the benefit of a new professional environment, since the old generation of the founding fathers of quantum physics were no longer in the field of combat. They were also benefited by cultural changes in the late 1960s, which opened room for criticisms of science and criticisms within science, a topic we have discussed in the previous chapter. And yet, they did not just reflect their times but they also contributed to changing them. They were led to these issues not only by scientific motivations; philosophical, pedagogical, and even political factors were also influential, while varying from case to case. Their story is as a whole a story of success since

(continued)

foundations of quantum mechanics, or at least some research from this field, entered the mainstream of physics. Nevertheless, in each individual case, recognition did not always come in due time, and that may explain their different appreciation of the evolution of the field to which they dedicated their energy.

References

Aczel, A.D.: Entanglement: The Greatest Mystery in Physics. Four Walls Eight Windows, New York (2002)

Aspect, A.: Proposed experiment to test the nonseparability of quantum mechanics. Phys. Rev. D **14**(8), 1944–1951 (1976)

Aspect, A.: Experiences basées sur les inégalités de Bell. Journal de Physique – Colloque C **42**, 63–80 (1981)

Aspect, A.: Experimental tests of Bell's inequalities in atomic physics. In: Lindgren, I., Rosen, A., Svanbeg, S. (eds.) Atomic Physics, vol. 8, pp. 103–128. Plenum, New York (1983)

Aspect, A.: Bell's theorem: the naive view of an experimentalist. In: Bertlmann, R.A., Zeilinger, A. (eds.) Quantum (Un)Speakables from Bell to Quantum Information, pp. 119–153. Springer, Berlin (2002)

Aspect, A., Grangier, P., Roger, G.: Experimental tests of realistic local theories via Bell's theorem. Phys. Rev. Lett. **47**(7), 460–463 (1981)

Aspect, A., Dalibard, J., Roger, G.: Experimental test of Bell inequalities using time-varying analyzers. Phys. Rev. Lett. **49**(25), 1804–1807 (1982a)

Aspect, A., Grangier, P., Roger, G.: Experimental realization of Einstein-Podolsky-Rosen-Bohm Gedankenexperiment – a new violation of Bell inequalities. Phys. Rev. Lett. **49**(2), 91–94 (1982b)

Ballentine, L.E.: Resource letter IQM2: foundations of quantum mechanics since the Bell inequalities. Am. J. Phys. **55**, 785–792 (1987)

Bastin, T.: Quantum Theory and Beyond: Essays and Discussions Arising from a Colloquium. Cambridge University Press, Cambridge, UK (1971)

Belinfante, F.J.: A Survey of Hidden-Variables Theories. Pergamon, New York (1973)

Bell, J.S.: On the Einstein Podolsky Rosen paradox. Physics **1**, 195–200 (1964)

Bell, J.S.: On the problem of hidden variables in quantum mechanics. Rev. Mod. Phys. **38**(3), 447–452 (1966)

Bell, J.S.: Introduction to the hidden-variable question. In: d'Espagnat, B. (ed.) Foundations of Quantum Mechanics. Proceedings of the International School of Physics "Enrico Fermi", pp. 171–181. Academic, New York (1971)

Bell, J.S.: Quantum mechanical ideas. Science **177**, 880–881 (1972)

Bell, J.S.: Locality in quantum mechanics: reply to critics. Epistemol. Lett. **7**, 2–6 (1975)

Bell, J.S.: Testing quantum mechanics. Prog. Sci. Cult. Interdiscip. J. Ettore Majorana Centre **1**(4), 439–445 (1976)

Bell, J.S.: Atomic-cascade photons and quantum-mechanical nonlocality. Comments Atom. Mol. Phys. **9**, 121–126 (1980)

Bell, J.S.: On the impossible pilot wave. Found. Phys. **12**(10), 989–999 (1982)

Bell, J.S.: Beables for quantum field theory. In: Hiley, B.J., Peat, F.D. (eds.) Quantum Implications: Essays in Honour of David Bohm, pp. 227–234. Routledge & Kegan, London (1987)

Bell, J.S.: La nouvelle cuisine. In: Bell, J.S. (ed.) Speakable and Unspeakable in Quantum Mechanics—Collected Papers on Quantum Philosophy, pp. 232–248. Cambridge University Press, Cambridge (2004a)

Bell, J.S.: Speakable and Unspeakable in Quantum Mechanics: Collected Papers on Quantum Philosophy. With an Introduction by Alain Aspect. Cambridge University Press, Cambridge (2004b)

Bell, J.S., Nauenberg, M.: The moral aspect of quantum mechanics. In: De Shalit, A., Feshbach, H., Van Hove, L. (eds.) Preludes in Theoretical Physics, pp. 279–286. North Holland, Amsterdam (1966)

Bell, J.S., Shimony, A., Horne, M.A., Clauser, J.F.: An exchange on local beables. Dialectica **39** (2), 85–110 (1985)

Bell, J.S., Bell, M., Gottfried, K., Veltman, M.: Quantum Mechanics, High Energy Physics and Accelerators: Selected Papers of John S. Bell, with Commentary. World Scientific, River Edge, NJ (1995)

Bernstein, J.: Quantum Profiles. Princeton University Press, Princeton, NJ (1991)

Bertlmann, R.A., Zeilinger, A.: Quantum [Un]Speakables: From Bell to Quantum Information. Springer, Berlin (2002)

Bispo, W.F.d.O., David, D.F.G., Freire Jr., O.: As contribuições de John Clauser para o primeiro teste experimental do teorema de Bell: uma análise das técnicas e da cultura material. Revista Brasileira de Ensino de Física **35**, 3603 (2013)

Blumenthal, R.B., Ehn, D.C., Faissler, W.L., Joseph, P.M., Lanzerot, L.J., Pipkin, F.M., Stairs, D. G.: Deviation from simple quantum electrodynamics. Phys. Rev. Lett. **14**(16), 660–664 (1965)

Bohm, D.: Quantum Theory. Prentice-Hall, New York (1951)

Bohm, D.: A suggested interpretation of the quantum theory in terms of hidden variables – I & II. Phys. Rev. **85**(2), 166–179, 180–193 (1952)

Bohm, D.: Quantum theory as an indication of a new order in physics. Part A. The development of new order as shown through the history of physics. Found. Phys. **1**(4), 359–381 (1971)

Bohm, D., Aharonov, Y.: Discussion of experimental proof for the paradox of Einstein, Rosen, and Podolsky. Phys. Rev. **108**(4), 1070–1076 (1957)

Bohm, D., Bub, J.: A proposed solution of measurement problem in quantum mechanics by a hidden variable theory. Rev. Mod. Phys. **38**(3), 453–469 (1966)

Bohm, D.J., Hiley, B.J.: Intuitive understanding of nonlocality as implied by quantum-theory. Found. Phys. **5**(1), 93–109 (1975)

Broglie, L.d.: Sur la réfutation du thèoreme de Bell. Comptes Rendus de l'Académie des Sciences **278 Série B**, 721–722 (1974)

Bromberg, J.L.: The Laser in America, 1950-1970. MIT Press, Cambridge, MA (1991)

Bromberg, J.L.: Device physics vis-à-vis fundamental physics in Cold War America: the case of quantum optics. Isis **97**(2), 237–259 (2006)

Brown, H.R., Harré, R. (eds.): Philosophical Foundations of Quantum Field Theories. Oxford University Press, Oxford (1988)

Brush, S.G.: The chimerical cat: philosophy of quantum mechanics in historical perspective. Soc. Stud. Sci. **10**(4), 393–447 (1980)

Clark, P.M., Turner, J.E.: Experimental tests of quantum mechanics. Phys. Lett. A **26**, 447 (1968)

Clauser, J.F.: Proposed experiment to test local hidden-variable theories. Bull. Am. Phys. Soc. **14** (4), 578 (1969)

Clauser, J.F.: Von Neumann's informal hidden-variable argument. Am. J. Phys. **39**(9), 1095–1096 (1971a)

Clauser, J.F.: Reply to Dr. Wigner's objections. Am. J. Phys. **39**(9), 1098–1099 (1971b)

Clauser, J.F.: Experimental limitations to validity of semiclassical radiation theories. Phys. Rev. A **6**(1), 49–54 (1972)

Clauser, J.F.: Experimental distinction between quantum and classical field-theoretic predictions for photoelectric effect. Phys. Rev. D **9**(4), 853–860 (1974)

Clauser, J.F.: Experimental investigation of a polarization correlation anomaly. Phys. Rev. Lett. **36** (21), 1223–1226 (1976)

Clauser, J.F.: Early history of Bell's theory and experiment. In: Black, T.D. (ed.) Foundations of Quantum Mechanics, pp. 168–174. World Scientific, Singapore (1992)

Clauser, J.: Early history of Bell's theorem. In: Bertlmann, R.A., Zeilinger, A. (eds.) Quantum [Un]Speakables – From Bell to Quantum Information, pp. 61–98. Springer, Berlin (2002)
Clauser, J.F.: Early history of Bell's theorem. In: Bigelow, N.P. (ed.) Coherence and Quantum Optics VIII, pp. 19–43. Plenum & Kluwer, New York (2003)
Clauser, J.F., Horne, M.A.: Experimental consequences of objective local theories. Phys. Rev. D **10**(2), 526–535 (1974)
Clauser, J.F., Shimony, A.: Bell's theorem – experimental tests and implications. Rep. Prog. Phys. **41**(12), 1881–1927 (1978)
Clauser, J.F., Horne, M.A., Shimony, A., Holt, R.A.: Proposed experiment to test local hidden-variable theories. Phys. Rev. Lett. **23**(15), 880–884 (1969)
Cohen-Tannoudji, C., Diu, B., Laloë, F.: Mécanique quantique. Hermann, Paris (1973)
d'Espagnat, B.: Foundations of Quantum Mechanics – Proceedings of the International School of Physics "Enrico Fermi". Academic, New York (1971)
d'Espagnat, B.: The quantum theory and reality. Sci. Am. **241**(5), 158–181 (1979)
d'Espagnat, B.: Conceptual Foundations of Quantum Mechanics, 2nd edn. Addison-Wesley, Redwood City, CA (1989) [1st ed 1971]
Eberhard, P.H.: Bell's theorem and different concepts of locality. Nuovo Cimento B **46**(2), 392–419 (1978)
Einstein, A., Podolsky, B., Rosen, N.: Can quantum-mechanical description of physical reality be considered complete? Phys. Rev. **47**, 777–780 (1935)
Faraci, G., Gutkowski, D., Notarrigo, S., Pennisi, A.R.: Experimental test of EPR paradox. Lettere al Nuovo Cimento **9**(15), 607–611 (1974)
Feynman, R.P.: Simulating physics with computers. Int. J. Theor. Phys. **21**(6–7), 467–488 (1982)
Freedman, S.J., Clauser, J.F.: Experimental test of local hidden-variable theories. Phys. Rev. Lett. **28**(14), 938–941 (1972)
Freedman, S.J., Holt, R.A., Papaliolios, C.: Experimental status of hidden variable theories. In: Broglie, L.d., Flato, M. (eds.) Quantum Mechanics, Determinism, Causality and Particles, pp. 43–59. D. Reidel, Dordrecht (1976)
Freire Jr., O.: The historical roots of "foundations of quantum mechanics" as a field of research (1950–1970). Found. Phys. **34**(11), 1741–1760 (2004)
Freitas, F.H.A., Freire Jr., O.: Sobre o uso da Web of Science como fonte para a história da ciência. Revista da SBHC **1**(2), 129–147 (2003)
Fry, E.S.: Two-photon correlations in atomic physics. Phys. Rev. A **8**(3), 1219–1232 (1973)
Fry, E.S., Thompson, R.C.: Experimental test of local hidden-variable theories. Phys. Rev. Lett. **37** (8), 465–468 (1976)
Galison, P.: Image and Logic: A Material Culture of Microphysics. University of Chicago Press, Chicago, IL (1997)
Ghirardi, G.C., Rimini, A., Weber, T.: A general argument against superluminal transmission through quantum mechanical measurement process. Lettere al Nuovo Cimento **27**(10), 293–298 (1980)
Gilder, L.: The Age of Entanglement – When Quantum Physics Was Reborn. Knopf, New York (2008)
Gleason, A.M.: Measures on the closed subspaces of a Hilbert space. J. Math. Mech. **6**(6), 885–893 (1957)
Gordin, M.D.: The Pseudoscience Wars: Immanuel Velikovsky and the Birth of the Modern Fringe. The University of Chicago Press, Chicago, IL (2012)
Gottfried, K.: Quantum Mechanics. W.A. Benjamin, New York (1966)
Greca, I.M., Freire Jr., O.: Meeting the challenge: quantum physics in introductory physics courses. In: Matthews, M.R. (ed.) International Handbook of Research in History, Philosophy and Science Teaching, pp. 183–209. Springer, Dordrecht (2014)
Greenberger, D.: The history of the GHZ paper. In: Bertlmann, R.A., Zeilinger, A. (eds.) Quantum [Un]Speakable—From Bell to Quantum Information, pp. 281–286. Springer, Berlin (2002)

Harvey, B.: The effects of social context on the process of scientific investigation: experimental tests of quantum mechanics. In: Knorr, K.D., Krohn, R., Whitley, R. (eds.) The Social Process of Scientific Investigation, pp. 139–163. D. Reidel, Dordrecht (1980)
Harvey, B.: Plausibility and the evaluation of knowledge: a case-study of experimental quantum mechanics. Soc. Stud. Sci. **11**, 95–130 (1981)
Heisenberg, W.: Physics and Philosophy; the Revolution in Modern Science. Harper, New York (1958)
Holt, R.A.: Atomic cascade experiments. PhD dissertation, Harvard University (1973)
Holt, R.A., Pipkin, F.M.: Precision-measurement of lifetime of 7^3S_1 state of atomic mercury. Phys. Rev. A **9**(2), 581–584 (1974)
Holt, R.A., Pipkin, F.M.: Polarization correlation measurement on an atomic mercury cascade. Prog. Sci. Cult. Interdiscip. J. Ettore Majorana Centre **1**(4), 460 (1976)
Horne, M.A.: Experimental consequences of local hidden variable theories. PhD dissertation, Boston University (1970)
Horne, M.A.: On four decades of interaction with John Bell. In: Bertlmann, R.A., Zeilinger, A. (eds.) Quantum [Un]Speakable—From Bell to Quantum Information, pp. 99–102. Springer, Berlin (2002)
Horne, M.A., Shimony, A.: Local hidden-variable theories. Epistemol. Lett. **1**, 7–22 (1973)
Horne, M., Shimony, A., Zeilinger, A.: Down-conversion photon pairs: a new chapter in the history of quantum-mechanical entanglement. In: Anandan, J.S. (ed.) Quantum Coherence, pp. 356–372. World Scientific, Singapore (1990)
Jackiw, R., Shimony, A.: The depth and breadth of John Bell's physics. Phys. Perspect. **4**(1), 78–116 (2002)
Jackiw, R., Shimony, A.: Bell, John Stewart. In: Koertge, N. (ed.) New Dictionary of Scientific Biography, pp. 236–243. Thomson Gale, New York (2008)
Jammer, M.: The Philosophy of Quantum Mechanics – The Interpretations of Quantum Mechanics in Historical Perspective. Wiley, New York (1974)
Jauch, J.M., Piron, C.: Can hidden variables be excluded in quantum mechanics. Helvetica Physica Acta **36**(7), 827–837 (1963)
Kaiser, D. (ed.): Pedagogy and the Practice Of Science: Historical and Contemporary Perspectives. MIT Press, Cambridge, MA (2005)
Kaiser, D.: Turning physicists into quantum mechanics. Phys. World May:28–33 (2007)
Kasday, L.R.: Experimental test of quantum predictions for widely separated photons. In: d'Espagnat, B. (ed.) Foundations of Quantum Mechanics. Proceedings of the International School of Physics "Enrico Fermi", pp. 195–210. Academic, New York (1971)
Kasday, L.R., Ullman, J.D., Wu, C.S.: Angular-correlation of compton-scattered annihilation photons and hidden variables. Nuovo Cimento B **25**(2), 633–661 (1975)
Kevles, D.J.: The Physicists: The History of a Scientific Community in Modern America. Knopf, New York (1978)
Kocher, C.A., Commins, E.D.: Polarization correlation of photons emitted in an atomic cascade. Phys. Rev. Lett. **18**(15), 575–577 (1967)
Körner, S.: Observation and Interpretation – A Symposium of Philosophers and Physicists. Butterworths, London (1957)
Kragh, H.: An Introduction to the Historiography of Science. Cambridge University Press, Cambridge (1987)
Lamehi-Rachti, M., Mittig, W.: Quantum-mechanics and hidden variables – test of Bell's inequality by measurement of spin correlation in low-energy proton-proton scattering. Phys. Rev. D **14** (10), 2543–2555 (1976)
Lochak, G.: Parametres cachés et probabilités cachées. Epistemol. Lett. **6**, 41 (1975)
Lochak, G.: Thèses. Epistemological Letters **19**, 27–28 (1978)
Marshall, T.W., Santos, E., Selleri, F.: Local realism has not been refuted by atomic cascade experiments. Phys. Lett. A **98**(1–2), 5–9 (1983)

McGuire, J.H., Fry, E.S.: Restrictions on nonlocal hidden-variable theory. Phys. Rev. D **7**(2), 555–557 (1973)

Messiah, A.: Quantum Mechanics. North Holland, Amsterdam (1961)

Nussbaum, G.H., Pipkin, F.M.: Correlation of photons in cascade and coherence time of 6^3P_1 state of mercury. Phys. Rev. Lett. **19**(19), 1089–1092 (1967)

Page, D.N.: The Einstein-Podolsky-Rosen physical reality is completely described by quantum-mechanics. Phys. Lett. A **91**(2), 57–60 (1982)

Pais, A.: "Subtle Is the Lord": The Science and the Life of Albert Einstein. Oxford University Press, New York (1982)

Papaliolios, C.: Experimental test of a hidden-variable quantum theory. Phys. Rev. Lett. **18**(15), 622–625 (1967)

Paty, M.: The recent attempts to verify quantum mechanics. In: Lopes, J.L., Paty, M. (eds.) Quantum Mechanics, a Half Century Later, pp. 261–289. Reidel, Dordrecht (1977)

Pauli, W.: Remarques sur le problème des paramètres cachés dans la mécanique quantique et sur la théorie de l'onde pilote. In: George, A. (ed.) Louis de Broglie – physicien et penseur, pp. 33–42. Editions Albin Michel, Paris (1953)

Pauli, W., Meyenn, K. v.: Wissenschaftlicher Briefwechsel mit Bohr, Einstein, Heisenberg u. a. Band IV, Teil II 1953–1954. Springer, Berlin (1999)

Pinch, T.: What does a proof do if it does not prove? A study of the social conditions and metaphysical divisions leading to David Bohm and John von Neumann failing to communicate in quantum physics. In: Mendelsohn, E., Weingart, P., Whitley, R. (eds.) The Social Production of Scientific Knowledge, pp. 171–216. Reidel, Dordrecht (1977)

Pipkin, F.M.: Atomic physics tests of the basic concepts in quantum mechanics. Adv. Atom. Mol. Phys. **14**, 281–340 (1978)

Podlubny, I.: Comparison of scientific impact expressed by the number of citations in different fields of science. Scientometrics **64**(1), 95–99 (2005)

Rauch, H.: Quantum physics with neutrons: from Spinor symmetry to Kochen-Specker phenomena. Found. Phys. **42**(1), 153–172 (2012)

Rauch, H., Zeilinger, A., Badurek, G., Wilfing, A., Bauspiess, W., Bonse, U.: Verification of coherent Spinor rotation of fermions. Phys. Lett. A **54**(6), 425–427 (1975)

Redner, S.: Citation statistics from 110 years of physical review. Phys. Today **58**(6), 49–54 (2005)

Rosenfeld, L.: L'évidence de la complementarité. In: George, A. (ed.) Louis de Broglie – physicien et penseur, pp. 43–65. Editions Albin Michel, Paris (1953) [A slightly modified English version of this paper is Strife about complementarity, Science progress, 163 (1953), 1393–1410, reprinted in Robert Cohen and John Stachel (eds.). Selected papers of Léon Rosenfeld (Dordrecht, D. Reidel, 1979)]

Sachs, M.: Further comment on alternative to orthodox interpretation of quantum theory. Am. J. Phys. **37**(2), 228–229 (1969)

Schawlow, A.: Concluding remarks. In: Lindgren, I., Rosen, A., Svanbeg, S. (eds.) Atomic Physics, vol. 8, pp. 565–569. Plenum, New York (1983)

Schrödinger, E.: The present situation in quantum mechanics. In: Wheeler, J.A., Zurek, W.H. (eds.) Quantum Theory and Measurement, pp. 152–167. Princeton University Press, Princeton, NJ (1983) [Original publication in Naturwissenchaften 123, 807–812, 823–828, 844–849, 1935]

Shimony, A.: Role of observer in quantum theory. Am. J. Phys. **31**(10), 755–773 (1963)

Shimony, A.: Reply to Dr. Lochak. Epistemol. Lett. **8**, 1–6 (1976)

Shimony, A.: The point we have reached. Epistemol. Lett. **26**, 1–7 (1980)

Shimony, A.: Introduction. Dialectica **39**(2), 83–84 (1985)

Shimony, A.: Search for a Naturalistic World View (2 vols.). Cambridge University Press, Cambridge (1993)

Shimony, A.: John S. Bell: some reminiscences and reflections. In: Bertlmann, R.A., Zeilinger, A. (eds.) Quantum [Un]Speakable—From Bell to Quantum Information, pp. 51–60. Springer, Berlin (2002)

Siegel, A., Wiener, N.: Theory of measurement in differential-space quantum theory. Phys. Rev. **101**(1), 429–432 (1956)

Stapp, H.P.: S-Matrix interpretation of quantum theory. Phys. Rev. D **3**(6), 1303–1320 (1971)

Stöltzner, M.: Bell, Bohm, and von Neumann: some philosophical inequalities concerning no-go theorems and the axiomatic method. In: Placek, T., Butterfeld, J. (eds.) Non-locality and Modality, pp. 37–58. Kluwer, Dordrecht (2002)

Stone, L.: Prosopography. Daedalus **100**(1), 46–79 (1971)

Tausk, K.S.: A Medida na Mecânica Quântica. PhD dissertation, Universidade de São Paulo (1967)

Whitaker, A.: John Bell in Belfast: early years and education. In: Bertlmann, R.A., Zeilinger, A. (eds.) Quantum [Un]Speakable—From Bell to Quantum Information, pp. 7–20. Springer, Berlin (2002)

Wick, D.: The Infamous Boundary: Seven Decades of Controversy in Quantum Physics. Birkhäuser, Boston (1995)

Wiener, N., Siegel, A.: A new form for the statistical postulate of quantum mechanics. Phys. Rev. **91**(6), 1551–1560 (1953)

Wiener, N., Siegel, A.: The differential-space theory of quantum systems. Il Nuovo Cimento **2** (4 Suppl), 982–1003 (1955)

Wigner, E.P.: On hidden variables and quantum mechanical probabilities. Am. J. Phys. **38**(8), 1005–1009 (1970)

Wigner, E.P.: Rejoinder. Am. J. Phys. **39**(9), 1097–1098 (1971)

Wilson, A.R., Lowe, J., Butt, D.K.: Measurement of relative planes of polarization of annihilation quanta as a function of separation distance. J. Phys. G Nucl. Part. Phys. **2**(9), 613–624 (1976)

Wu, C.S., Shaknov, I.: The angular correlation of scattered annihilation radiation. Phys. Rev. **77** (1), 136 (1950)

Chapter 8
The 1980s and Early 1990s, Research on Foundations Takes Off

Abstract The 1980s were transitional times for foundations, moving from an early fringe position to mainstream physics. In the 1990s it would undergo more profound change, becoming part of the blossoming and promising field of quantum information. Different from previous times, the 1980s had a more diversified research agenda. New foundation experiments enabled by new techniques flourished and Bell's theorem experiments were resumed. New interpretations appeared and old ones were renewed. Theoretical breakthroughs were made in at least two different areas: the transition from the quantum description to the classical description through the idea of decoherence, and the early ideas related to the use of quantum physics to improve computers and computer science. Even without a thematic focus, however, the 1980s and the early 1990s were distinctive as being the years when technical advances made the implementation of different thought experiments possible. In the early 1990s, all the ingredients for the later emergence of quantum information were already present.

8.1 Introduction

In the late 1980s, the physicist Leslie E. Ballentine, a participant-observer in the quantum controversy and one of the proponents of the ensemble interpretation, which will be reported later, recorded his impressions on the professional status of the issues he was working on. "No longer can it be claimed that the interpretation of QM is a dull subject for endless debate by philosophers and armchair physicists. The foundations of QM has become what it always should have been, an exciting subject at the heart of mainstream physics" (Ballentine 1987, p. 787). This change in status had taken more than 30 years, from the early 1950s when Bohm's causal interpretation was considered only a matter for philosophical debates until the 1980s when the field gained its deserved recognition. The point of no return in this change may have been the announcement by Alain Aspect of his three experiments on Bell's theorem in the early 1980s, mentioned in Chap. 7. Indeed, Aspect's lecture tour reporting on his experiments and the coverage of this topic in philosophy and popular science books and magazines would hardly make his case

O. Freire Junior, *The Quantum Dissidents*, DOI 10.1007/978-3-662-44662-1_8

an illustration of the obstacles the quantum dissidents faced till the early 1980s (Deligeorges 1985; Cushing and McMullin 1989). Thus this decade was a transitional time for foundations moving from the early fringe position to mainstream physics. In the 1990s it would undergo further change becoming not only mainstream physics but also part of the flourishing field of quantum information.

In hindsight, the 1950s may be seen as the time when challenges to the dominant complementarity interpretation began to appear, through Bohm's causal interpretation and Everett's relative states presentation of quantum theory. The 1960s were dominated by the emergence of the quantum measurement issue as a problem, that is, the quantum measurement problem, with the dispute between Wigner and Rosenfeld marking the first clash inside the orthodoxy in quantum physics. The 1970s was the decade of the experiments on Bell's theorem and the early expectation of their results. In the 1980s the agenda diversified. New foundation experiments enabled by technical advances flourished even in the absence of Bell's theorem experiments, which were only resumed at the end of the 1980s. New interpretations appeared and the old ones were renewed. Theoretical breakthroughs were made in at least two different areas: the transition from the quantum description to the classical description through the idea of decoherence, and the early ideas related to the use of quantum physics to improve computers and computer science. However, periodization in decades is not useful here as the professional and intellectual climate of the 1980s lasted until it was superseded by and merged into the field of so-called quantum information in the mid-1990s.

Even without a thematic focus, however, the 1980s and the early 1990s had a common feature making it distinctive compared to previous times. They were the years when technical advances made the implementation of different thought experiments possible. The historian of physics Joan Bromberg, who works on the history of quantum optics in the US, aptly remarked that "one lead that historians have yet to pursue is constant reference that working physicists make to the role of new instrumentation" (Bromberg 2008, p. 327). In no other time in the history of research on the foundations of quantum mechanics had new technical possibilities played such a role. Bromberg cites a list of them: "interferometry with neutrons and atoms, masers that operated with a handful of atoms (micromasers), the production of pairs of entangled photons in nonlinear crystals (photon downconversion), the trapping of individual ions, ultra-fast optical and laser techniques, and new solid-state devices," and she correctly remarks that such apparatuses were not used in the early experiments on Bell's theorem (Bromberg 2008, p. 328). Studies published by Bromberg show a number of cases illustrating such a trend. For instance, she studied Marlan Scully's works and the quantum eraser (Bromberg 2006); Wheeler's proposition, and Carroll Alley's realization, of the delayed choice experiment; and Vigier & Helmut Rauch's involvement with neutron experiments with implications to the foundations of quantum theory (Bromberg 2008).

The plethora of new experiments testing quantum predictions in domains never touched before mobilized physicists at large, including many who were not interested in foundations of quantum mechanics in itself. Serge Haroche's case, which will be presented later, is emblematic of this new reality. Corporations also began to be interested in such subjects, even when applications were still far away from sight. In the early 1980s, Akira Tonomura (1942–2012), working for Hitachi—one of the leading Japanese electronics corporations—at its Central Research Laboratory, realized that there was a wide field for physics research which was being opened in the 1980s with experiments on foundations of quantum mechanics. He was one of the driving forces behind the series of conferences in Japan dedicated to foundational issues and experiments based on new technological advances. On Tonomura's initiative, Hitachi backed the conferences. A perusal of the proceedings of these conferences illustrates the mix of highly speculative theoretical proposals and empirically-controlled experiments, features of 1980s research in foundations (Nakajima et al. 1996). In the late 1980s, researchers at IBM used a scanning tunneling microscope to write the logo's company with 35 xenon atoms on a background of copper atoms, which called the attention of the media to a new scientific and technological domain: nanoscience and nanotechnology (Eigler and Schweizer 1990). Furthermore, at least in some of the cases studied by Bromberg, "the same scientists who worked on military devices simultaneously pursued fundamental and foundational topics," which led Bromberg to renew the historiographical debates on American physics during Cold War times (Bromberg 2006, p. 237). Thus the usual fabric of twentieth-century physics—an experimental science with relations to technological applications—was being woven around foundational issues. The reference to the usual fabric of physics requires, however, two clarifications. First, bringing corporations to the table does not mean a dramatic change in costs of foundational experiments, at least in the case of quantum optics. Indeed, while compared to other fields like high energy physics and astronomy, quantum optics still is a less expensive domain. Second, the threads of this fabric did not include the various interpretations of quantum physics. As we will see, in the 1980s, the proliferation of interpretations remained unrelated to experimental tests.

The main goal of this chapter is thus to illustrate how foundations had already become mainstream physics, undertaking experimental and theoretical activities as well as receiving due recognition. In addition, we will see that these were the times when all the ingredients necessary for the emergence of quantum information as a distinctive field were being put together. This chapter is organized exploiting the thematic diversity of the 1980s and early 1990s. We begin with Bell's theorem, which included theoretical and conceptual developments as well as technical advances. Then we discuss the theoretical and experimental achievements related to decoherence and the quantum-classical boundary. The next section is dedicated to the new techniques and new experiments in foundations of quantum physics. The following section gives a brief review of the proliferation of new interpretations and the renewal of the old ones. We conclude presenting the early achievements in the field of computer science, which would later merge with foundations to create the field of quantum information. This section ends spotting the emergence of quantum information, which is the end of the history covered by this book. Most of the biographical notices appear integrated in the narrative through the chapter;

however, I added an interlude to bring a biographical sketch of John Archibald Wheeler due to the singular role he played in this story.

8.2 The Fate of Bell's Theorem

Aspect's early 1980s experiments on Bell's theorem were so convincing that in subsequent years nobody bothered to replicate them. The reasons for this were related to the perceived unfeasibility of new experimental breakthroughs, as remarked by Aspect: "I do not see further meaningful progress can be made in the domain of Bell's inequalities, at least with our apparatus. We have exploited its maximal possibilities. Sure, an additional decimal could be obtained, but would it be worthy?" (*apud* Deligeorges 1985, p. 137). Aspect himself moved towards other rewarding topics of research. Invited by the French physicist and later Nobel Prize laureate Claude Cohen-Tannoudji, he began to work on the use of lasers to cool down atoms.[1]

Experiments were revived 5 years later due to the finding of a better source for pairs of photons with entangled polarizations. Instead of photons from atomic cascades, the new tests used parametric down-conversion (PDC), that is, pairs of photons created in the interaction between a laser beam and nonlinear optical crystals. In addition, as the use of this new source was improved, it began to attract the attention of other physicists for applications including the novelties of quantum cryptography and teleportation.

While the use of PDC in Bell's experiments was a new resource, the "creation of twin photon pairs"—"one of the most fascinating phenomena of quantum optics," according to Grynberg et al. (2010, p. 529), dates back to the early days of the laser. The prehistory of the twin photon pairs began as early as 1961, when Peter Franken and his colleagues at the University of Michigan, Ann Arbor, were able to focus a laser into a piece of quartz and generate the second harmonic with double frequency as output. The feat opened the field of nonlinear optics and illustrated the potential of the laser as a new technique (Brown and Pike 1995, p. 1427). However, while semi-classical approaches to light allowed physicists to deal with some of the non-linear effects, others required the full quantum treatment of light, that is, methods from quantum optics. This was the case with PDC, which is a case of parametric fluorescence. PDC is a physical effect which was first verified by David Burnham and Donald Weinberg in 1970. They observed "coincidences between photons emitted by an ammonium dihydrogen phosphate crystal pumped by a 325-nm He-Cd laser" (Burnham and Weinberg 1970). According to Brown and Pike (1995, p. 1435), Burnham and Weinberg "observed virtual simultaneity in the parametric production of optical-photon pairs, i.e. the splitting of a single photon

[1] A. Aspect, interviewed by the author and I. Silva, 16 Dec 2010 and 19 Jan 2011, American Institute of Physics, College Park, MD; AIP hereafter.

[after interaction with the non-linear crystal] ω_3 into two photons ω_1 and ω_2 of lower energy," in a process involving conservation laws, thus $\omega_3 = \omega_1 + \omega_2$, where ω is angular frequency. The use of PDC as a source for Bell's experiments in 1987 was not only the introduction of a different source. It also meant the introduction of new methods as the polarization correlation of a pair of photons could be treated in terms of two-photon interference, a phenomenon requiring methods from quantum optics, thus meshing experiments on entanglement with the wide field of quantum optics. Since the early 1960s, due to the works of Roy Glauber (Ou 2007; Horne et al. 1990; Silva 2013), optics had evolved to include the full quantum treatment of light as part of its theoretical toolkit.

In fact, quantum two-particle interference in physics arrived in physics through two different and independent paths. On the one hand, via people from quantum optics, and on the other, via people who were working on neutron interferometry, such as Anton Zeilinger and Michael Horne. An example from the first path, quantum optics physicists at Rochester, led by Leonard Mandel, were playing with pairs of photons produced from a PDC source and showed that two-photon interference is a strict quantum phenomenon, without a classical explanation, demonstrating what is now known as the Hong-Ou-Mandel effect (Ghosh et al. 1986; Hong et al. 1987). Unaware of these achievements, in 1985 Zeilinger and Horne tried to combine the interferometry experiments they were doing with Bell's theorem. Then they suggested a new experiment with Bell's theorem but instead of using correlation among polarizations (internal variables) they used linear momenta. They concluded that the quantum description of two-particle interferometry was completely analogous to the description of singlet spinstate used by Bell (Horne and Zeilinger 1985). However, they did not know how to produce such states in laboratories, because they "didn't know where to get a source that would emit pairs of particles in opposite directions." When Horne read the Ghosh-Mandel paper, they sent their paper to Mandel, who reacted saying: "This is so much simpler than the way we describe it, you should publish it." They called Horne's former supervisor, Abner Shimony, and wrote the "two-particle interferometry" paper explaining "the fundamental ideas of the recently opened field of two-particle interferometry, which employs spatially separated, quantum mechanically entangled two-particle states" (Horne et al. 1989).[2]

[2] The 1985 Horne and Zeilinger paper was prepared for a conference in Finland dedicated to the 50th anniversary of the EPR paper. This was the first Zeilinger paper to deal with Bell's theorem. Interview with Michael Horne, by Joan Bromberg, 12 Sep, 2002, AIP. Interview with Anton Zeilinger, by Olival Freire, 30 June 2014, AIP.

Picture 8.1 Arthur Zajonc, Anton Zeilinger, Mary Bell, and John Bell, at a conference in Amherst, MA, 1990. Courtesy of Anton Zeilinger

Interferometry experiments using PDC photon pairs were pioneered by two teams. According to Daniel Greenberger, Michael Horne, and Anton Zeilinger, who are key actors in this field, "real experiments commenced when Carroll Alley and Yanhua Shih at the University of Maryland first used down-conversion to produce an entangled state and when Ruba Ghosh and Leonard Mandel at the University of Rochester first produced two-particle fringes without using polarizers" (Greenberger et al. 1993, p. 22). PDC as a source for entangled photons entered Bell's experiments serendipitously, at least in the case of Yanhua Shih at the University of Maryland, College Park. The Soviet physicist Vladimir Braginsky was visiting the University of Maryland at College Park (UMCP) in the mid-1980s, as part of his interest in work on the detection of gravitational waves, an interest he shared with Carroll Alley.[3] Yanhua Shih, a Chinese physicist who was a teaching assistant at UMCP and a student of Alley's, was having difficulty calibrating the photodetectors as a certain experiment required measuring their absolute quantum efficiencies. They were thinking of sending them to the National Institute of Standards and Technology for calibration, which would have been expensive. This came up in a conversation among Shih, Alley, and Braginsky, and Braginsky suggested using PDC to calibrate the detectors, as David Klyshko was doing this in Moscow. This worked well. When it was time for Shih to begin his doctoral research, he and Alley came up with the idea of using PDC sources to do Bell's

[3] On Braginsky's biography and his collaboration with American physicists on the issue of detecting gravitational waves, see Braginsky, Vladimir B. Interview by Shirley K. Cohen. Pasadena, California, January 15, 1997. Oral History Project, California Institute of Technology Archives. Retrieved on April 12, 2014 from the World Wide Web: http://resolver.caltech.edu/CaltechOH:OH_Braginsky_V.

type experiments.[4] In all of this, chance did not play alone as the environment at UMCP was then very favorable to research on the quantum foundations. Alley was known for his distrust of the received view that foundational issues in quantum theory were already set and was involved in another experiment in foundations, the delayed choice experiment. Shih said he decided to work under the supervision of Alley because he was the only person at UMCP who really told him straightforwardly, "I don't know what is a photon."[5] Shih's choice of Alley reveals thus his interest in understanding, not only calculating, quantum physics.

In the first attempts to use PDC as source for experiments on Bell's theorem in the late 1980s, the results were meager compared to Aspect's previous results, the novelty just being the use of a new source. In 1988 Shih and Alley obtained violations of Bell's inequalities by three standard deviations and Zhe-Yu Ou and Leonard Mandel, at the University of Rochester, obtained about six standard deviations (Shih and Alley 1988; Ou and Mandel 1988). However, PDC sources also enabled tests with variables other than polarization and these wider possibilities would become instrumental for quantum information. This was the case with energy and time in experiments conducted at the Universität Frankfurt am Main by Brendel et al. (1992) and at Berkeley by Kwiat et al. (1993) following suggestions from Franson (1989).

Picture 8.2 Historian Joan Bromberg and physicist Yanhua Shih

[4] Interview of Yanhua Shih and Morton Rubin by Joan Bromberg on 14 May 2001, AIP. Interview with Carroll Alley by Joan Bromberg on 16 May 2006, AIP.

[5] Ibid. Yanhua Shih's professional career was twice affected by the turmoil in his home country, China. Due to the Cultural Revolution he went to university late and eventually graduated in physics from the Northwestern University, in Xi'an, and went to the University of Maryland to do his PhD. On completion, while preparing to return to China, the Tiananmen Square protests of 1989 took place. He decided not to return as he was informed he was on a blacklist due to his activities at the Chinese graduate student organization and looked for a job in the U.S. Shih was hired by the University of Maryland, Baltimore County, to reinforce a fresh doctoral program in physics, and he settled there for the rest of his scientific career. Interview of Yanhua Shih by Olival Freire and Joan Bromberg on 28 May 2014, AIP.

Picture 8.3 Drawing in the entrance of Shih's lab. The drawing was brought to Baltimore from Augusto Garuccio's lab in Bari

The use of PDC photons began to pay off. Shih and Morton Rubin, his colleague at University of Maryland, grasped the workings of the different types of conversion and got violations for the first time which were more meaningful than those obtained by Aspect in the early 1980s (Kiess et al. 1993). They obtained violations of 22 standard deviations, overtaking Aspect's 1981 experiment. Two years later another achievement came: a team comprising the converging efforts of Kwiat,

Zeilinger, K. Mattle and H. Weinfurter, from Innsbruck, and Sergienko and Shih, from Maryland, using a different pair of photons from PDC processes (Type-II noncollinear phase matching) obtained violations of Bell's inequality with over 100 standard deviations in less than 5 min (Kwiat et al. 1995). This result set PDC source as the standard resource for all the experiments with entangled photons from 1995 on.

Achievements related to the theoretical understanding of entanglement were also made. The most influential of these was the GHZ theorem.[6] In 1989, the Americans Daniel Greenberger and Michael Horne, who had been working on Bell's theorem since the late 1960s, jointly with Anton Zeilinger, introduced a novelty to the testing of entanglement, extending Bell's theorem in a different and interesting direction. According to Franck Laloë (2012, p. 100),

> For many years, everyone thought that Bell had basically exhausted the subject [... and early experiments ...] provided the most spectacular quantum violations of local realism. It therefore came as a surprise to many when in 1989 Greenberger, Horne, and Zeilinger (GHZ) showed that systems containing more than two correlated particles may actually exhibit even more dramatic violations of local realism.

The trio—Greenberger, Horne, and Zeilinger—analyzed Einstein's 1935 argument once again and were able to write what is now called the Greenberger-Horne-Zeilinger entangled states (GHZ) leading to conflicts between local realistic theories and quantum mechanics. However, unlike Bell's theorem, the conflict now is not of a statistical nature, as it was with Bell's theorem, insofar as each GHZ state leads to conflicting predictions with local realistic models. In principle, at least ideally, in this case, just one experimental run could pit quantum mechanics against local realism (Greenberger et al. 1989, 1990; Greenberger 2002).[7]

For Zeilinger, the GHZ theorem was a reward for a risky professional change. In the mid-1980s he had decided to leave neutron interferometry to build a research program in quantum and atomic optics from scratch. Zeilinger was supported by the Austrian Science Foundation and looked for the basics in the new field. Sometimes he did this through interaction with other teams, such as Leonard Mandel's in Rochester. Reasons for this choice were related to his understanding that these fields offered more opportunities than neutron interferometry. He was increasingly attracted by foundations of quantum physics and particularly entanglement. Curiously, neutron interferometry had been the common ground of collaboration among Zeilinger, Greenberger, and Horne but the GHZ theorem was a result that had implications far beyond their original interest.

---

[6] On the history of GHZ theorem, see Greenberger (2002); interview with Michael Horne, by Joan Bromberg, 12 September 12, 2002, AIP; and interview with Anton Zeilinger, by Olival Freire, 30 June 2014, AIP.

[7] The second paper contains a more detailed presentation of the GHZ theorem, its proof, and suggests possible experiments, including momentum and energy correlations among three and more photons produced through PDC.

Clifford Shull's laboratory at MIT had been the meeting point for the GHZ trio. Shull worked on neutron scattering, which gained him the 1994 Nobel Prize. Zeilinger was there as a postdoc student and later as a Visiting Professor. Horne's doctoral dissertation had been on Bell's theorem, under the supervision of Abner Shimony, as we have seen in Chap. 7. In the early 1970s Horne thought this subject was dead and turned to work with neutrons while teaching at the Stonehill College, MA, near Boston. Greenberger was at the City College of New York and had worked as a high-energy theorist and then moved to gravity. He became interested in neutron experiments looking for influences of gravity in quantum mechanics and met Zeilinger and Horne at a neutron conference in Grenoble (Greenberger 2002). Basic experiments with neutron interferometry were the bread and butter of the trio. The first intuition related to the GHZ theorem came from Greenberger, when he asked Zeilinger and Horne, "Do you think there would be something interesting with three particles that are entangled? Would there be any difference, something new to learn with a three-particle entanglement?" The work matured while he spent a sabbatical in Vienna working with Zeilinger. "I have a Bell's theorem without inequalities," was the manner in which he reported his results to Horne.[8]

The GHZ theorem was ready in around 1986 and was immediately well received when presented. N. David Mermin in the US (Mermin 1990), and Michael Redhead, in the UK, began to publicize it (Clifton et al. 1991), and Zeilinger recalls when he met Bell at a conference in Amherst, MA in 1990, how enthusiastic Bell was about the result. After this initial favorable reception they realized the subject deserved a better presentation than simply a conference paper.[9] They called Abner Shimony to join them in the writing of a more complete paper explaining the theorem and envisioning possible experiments. Bell's reaction, in particular, motivated Zeilinger to immediately think about taking this theorem to the laboratory benches. The road to experiments however, was dependent on both conceptual and experimental advances. Zeilinger considers it the most challenging experiment he had ever carried out. The quest for the required expertise, however, brought important preliminary results and spinoffs, such as the concept of "entanglement swapping" and the experiment on teleportation (Zukowski et al. 1993; Bouwmeester et al. 1997).[10] The twentieth century closed with Zeilinger successfully obtaining the experimental production of GHZ states which are in agreement with quantum theory and in disagreement with local realistic theories (Bouwmeester et al. 1999).

[8] Interview with Michael Horne, by Joan Bromberg, 12 Sep 2002, AIP. Interview with Anton Zeilinger, by Olival Freire, 30 June and 2 July 2014, AIP.

[9] In fact, there already was some competition around the most general proof of the theorem between on the one hand, Greenberger, Horne, and Zeilinger and on the other hand, Clifton, Redhead, and Butterfield. This competition is recorded in the paper by Clifton et al. (1991), on a "note added in prof", on p. 182.

[10] Anton Zeilinger, interviewed by Olival Freire, ibid.

8.2.1 The Ongoing Experiments with Entanglement

Let us continue a little beyond the time limits we set for our narrative as experiments with entanglement are a hallmark of what foundations of quantum physics has become as a field of research. In 1997 a team led by Zeilinger (Bouwmeester et al. 1997) reported the first quantum teleportation, that is, the use of entanglement for transmitting and rebuilding the state of a quantum system over arbitrary distances. Disputes ensued with Zeilinger's priority being challenged by a team including Francesco de Martini, Lucien Hardy, and Sandu Popescu, and bringing together physicists from Rome (La Sapienza), Oxford, Cambridge, and Bristol (Boschi et al. 1998).[11] In 1998 Zeilinger and his team improved on Aspect's 1982 experiment with time-varying analyzers. They reinforced the condition of locality by using a truly random setting of the analyzers, while in Aspect's case the use of a standing wave prevented a truly random change. The detectors were separated by 400 m. They got full agreement with quantum mechanics predictions and violations of Bell's inequality by over 30 standard deviations (Weihs et al. 1998). Let us remember that in 1982 Aspect had obtained violations of five standard deviations. In the same year Zeilinger managed to entangle photons without previous interactions, a procedure then christened "entanglement swapping" (Pan et al. 1998). In the same year in Geneva Nicolas Gisin obtained violations of Bell's inequalities, concerning energy-time, with pairs of photons separated by 10 km through optical fibers (Tittel et al. 1998).

Entering the twenty-first century, a team led by David J. Wineland from Boulder, Colorado, who went on to share the 2012 Nobel Prize in Physics, used massive entangled particles, beryllium ions, instead of photons to succeed in getting a violation of Bell's inequalities with higher efficiency detectors and closing the so-called detection loophole (Rowe et al. 2001). Before this experiment, in order to test Bell's theorem with optical photons and detectors with low detection efficiency, an additional assumption was required, the *fair sampling* assumption, admitting that results with a sample of pairs of photons would be independent of the results with all other pairs of photons. Plausible as this assumption was, and Bell was the first to consider that it was not the main loophole in the early experiments, it was indeed a lacuna reinforcing the point of view of physicists who were reluctant to accept the full implications of the experiments. Zeilinger, for his part, got more impressive results in Vienna. He obtained the entanglement of the orbital angular momentum states of photons (Mair et al. 2001); violations of Leggett inequalities, which were formulated in order to exhibit experimental contrast between quantum mechanics and even some classes of non-local realistic theories (Groblacher et al. 2007); and, more recently, he was able to make the breakthrough of getting quantum mechanics predictions and violations of Bell's inequalities for pairs of entangled photons separated by 144 km (Scheidl et al. 2010). This list of

[11] De Martini's challenge and Zeilinger's answer are in *Physics World*, "Teleportation: who was first?" 01 Mar 1998, pp. 23–24).

achievements is also an evidence of the leading role Zeilinger has been playing in this field, which is a return from the research conversion he had begun in the mid-1980s. Tantalizing experiments continue to be suggested, including the quest for quantum teleportation between ground and satellite systems. A more recent proposal involves the use of cosmic light from very distant quasars in such a way "that the quantum states of their light would be essentially unrelated," to close one of the last loopholes in the story of Bell's theorem. According to the proponents of this, Jason Gallicchio, Andrew Friedman, and David Kaiser, "in current experiments, with settings determined by quantum random number generators, only a small amount of correlation between detector settings and local hidden variables, established less than a millisecond before each experiment, would suffice to mimic the predictions of quantum mechanics" (Gallicchio et al. 2014).

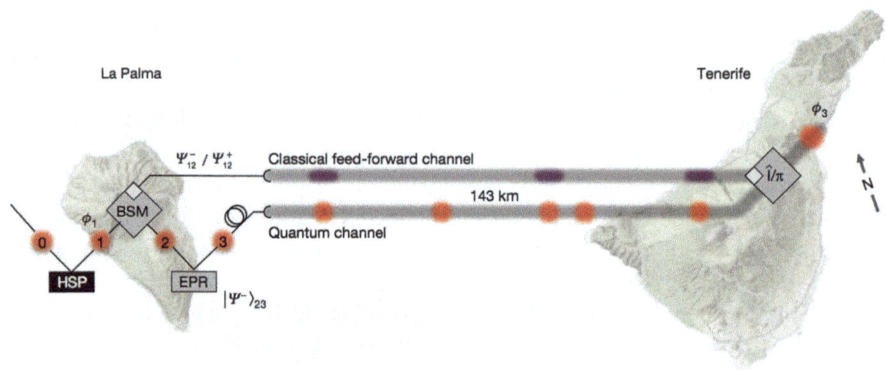

Picture 8.4 Quantum teleportation between the Canary Islands La Palma and Tenerife over both quantum and classical 143-km free-space channels. Ma, X.-S., et al.: Nature (2012), doi: 10.1038/nature11472. Courtesy Nature

Research on entanglement also led to its use in quantum cryptography. We postpone the introduction of its conceptual framework, initially suggested by Charles Bennett and Gilles Brassard with their BB84 protocol, to the section dedicated to works on computer science which led to the emergence of quantum information. Nicolas Gisin, a Swiss physicist, became the world leader in the domain of experiments with quantum cryptography. His review on this subject (Gisin et al. 2002) now has 2,659 citations. He also introduced the use of optical fibers in a series of experiments with entanglement, teleportation, and cryptography. Experiments with cryptography in optical fibers stoked the imagination of laymen, probably due to the legendary safety of Swiss banks, and brought wide press coverage. Gisin's case is an illustration of the changing profile of the research on foundations of quantum physics, becoming in the late 1980s and early 1990s a mix of basic science and potential applications. His thesis, under the supervision of Constantin Piron, was on quantum physics and statistical mechanics, more particularly on the derivation of non-linear dissipative equations analogous to

Schrödinger equation.[12] Later he showed that any "deterministic nonlinear Schrödinger equation would lead to signaling," that is, superluminal communication, thus leading to the conclusion that if these changes should be considered they "should be stochastic" (Gisin 1984). Despite this achievement he had difficulties in inserting himself into the Swiss academia, which expressed, however, conflicts inside researchers on foundations as Piron was also dedicated to this subject.

Piron had come from a tradition in theoretical physics in Geneva, which went back to Ernst Stueckelberg and Josef Jauch. As did Jauch, Piron worked on axiomatic quantum mechanics. Jauch and Piron's work became known as the "Geneva school" (Arthur 1981). As discussed in Chap. 7, Bell's works had challenged Jauch and Piron's attempts to reinforce von Neumann's proof against the possibility of hidden variables. As a matter of fact, Piron got upset with the success Bell obtained with these papers and reacted badly. Gisin, in contrast, had been fascinated by Bell's papers and attitudes since university. He recalls the first talk he attended given by Bell. Bell opened the talk saying "I am a quantum engineer but on Sundays I have principles" (Gisin 2002). Bell's theorem was a taboo subject in Piron's circle. Thus Gisin was not in line with the Geneva school and this schism had consequences. When he came back to Geneva from a postdoctoral stay at Rochester, he "was hoping Piron would take [him] in Geneva, helping to find connections in academy or industry [...] but Piron was not really helpful, not at all." Only in 1988, after 4 years working for a small company developing optical instrumentation for optical fibers and software, did he have an opportunity at the Group of Applied Physics at the University of Geneva.

Picture 8.5 Nicolas Gisin alongside the polarizer used in one of Aspect's early experiments

[12] Gisin's biographical information was drawn from Nicolas Gisin, interviewed by Olival Freire Jr, 2 Dec 2013, AIP. Bell's quantum engineer episode is also reported in Gisin (2002).

Picture 8.6 Thiago Barbosa, a student of Nicolas Gisin, at the Geneva lab

The list of places where the main experiments with Bell's theorem were carried out is also evidence of how widespread these kinds of physics experiments have become. The first generation was at Harvard, Berkeley, and Texas A&M University. The second was in Paris. The third generation emerged in such diverse places and institutions as Innsbruck, Vienna, Rome, Geneva, Caltech, Rochester, Maryland, Frankfurt, Oxford, Cambridge, Bristol, and Boulder. The work on Bell's theorem has received recognition through prizes and distinctions. More recently, the 2010 Wolf Prize, one of the most coveted physics prizes, was awarded to John Clauser, Alain Aspect, and Anton Zeilinger "for their fundamental conceptual and experimental contributions to the foundations of quantum physics, specifically an increasingly sophisticated series of tests of Bell's inequalities or extensions there of using entangled quantum states."[13] This prize acknowledged the role played by these physicists as leaders of the three successive generations of experiments with Bell's theorem.

We may conclude this section pondering the motivations behind this race for better entanglement experiments. What's the good for physics of this continued race? The popular dictum "what's good for the goose is good for the gander" may not be true for the foundations of physics and physics itself. Thus the ongoing series of Bell's experiments pose the following question: are more precise corroboration of quantum predictions and violations of Bell-type inequalities necessarily good for the health of physics or are they just more of the same? This question begs a two-level answer. First, physics has a tradition of improving measurement over

[13] http://www.wolffund.org.il/index.php?dir=site&page=winners&name=&prize=3016&year=2010&field=3008. Accessed on 18 Sep 2013.

time, the results of which are already known. This was the case with the speed of light, elementary electric charge, and other physical constants. However, the series of experiments with entanglement has other motivations. From a certain moment on, the motivation for these experiments also came from the perspective of harnessing entanglement for computing purposes. According to Yeang (2011, p. 331), the first to foresee this was David Deutsch, to whom we will come back later, in 1985 (Deutsch 1985a, b). The 1995 paper resulting from the collaboration between the teams led by Shih and by Zeilinger reported this string of promised applications in its first paragraph: "Recently, a whole wealth of curious and/or potentially useful applications of entangled states was proposed, from quantum communication, including cryptography and transfer of two bits of information in one photon, to quantum teleportation and 'entanglement swapping', to quantum computation" (Kwiat et al. 1995, p. 4337). Later, in the announcement of their 1999 experiment with GHZ states, Zeilinger's team (Bouwmeester et al. 1999) declared "such states of more than two entangled particles, known as Greenberger-Horne-Zeilinger (GHZ) states, play a crucial role in fundamental tests of quantum mechanics versus local realism and in many quantum information and quantum computation schemes." The possibility of these applications as reasons for these experiments was spotted by the historian Chen-Pang Yeang in his paper "Engineering Entanglement, Conceptualizing Quantum Information":

> ... the more recent expansion since the 1980s concerns not the foundations of quantum mechanics but the pragmatic and technological aspects of entanglement. Entanglement is not only strange but also useful, these researchers found. A lot of them even set aside the question of *why* quantum mechanics is so strange and rather focused on *how to utilize* the strange properties of quantum mechanics. Their answer gave rise to a new field known as quantum information. (Yeang 2011, p. 328)[14]

8.3 Theoretical and Experimental Breakthrough: Decoherence and the Quantum Classical Boundary

The 1980s saw the appearance of conceptual resources which together with entanglement would later become the theoretical foundations of quantum information: the decoherence effect and algorithms for the use of quantum theory in computer science. Decoherence is related to an old quantum problem, the transition from a quantum description to a classical description and the physical conditions which allow us to leave the former and adopt the latter. In other words, research on decoherence looked for the identification of the borders between the quantum and the classical world and the rules to successfully cross such a boundary. Nowadays the standard solution to this problem, in a nutshell, says that the coupling between the system under study and its environment makes the interference terms typical of

[14] Emphasis is in the original.

quantum descriptions disappear, in the limit of an infinite-sized environment. The assumption for this treatment is to consider quantum systems as open systems. The composite system, consisting of the quantum mechanical system plus the environment, is treated as a closed system, subject to Schrödinger's equation, and after its unitary evolution, the contribution of the environment is "traced out," resulting in the nonlinear evolution of the open quantum system. The same approach is able to give, in concrete cases, the time for the dilution of the interference terms. In mathematical terms, this coupling may be represented by the density matrix, where the terms concerning the quantum interferences appear in the off-diagonal terms. Tracing out this matrix leads to the master equations and then macroscopic variables may be obtained. This short description, however, veils the complex story behind it, as we will see.

Like most achievements in science, the lineage of the predecessors is always open to dispute. Some physicists and philosophers have referred to a "first period" of decoherence work, related to "closed systems" and developed by physicists such as the Dutch Nicolaas van Kampen, the Belgian Léon Van Hove, and the Italians Adriana Daneri, Angelo Loinger, and Giovanni Maria Prosperi. To suggest such a periodization, Castagnino et al. (2007) argued that these physicists "directed their attention to the emergence of classical macroscopic features from quantum microscopic descriptions." However, the "second period" was opened by a physicist—the Heidelberg physicist Heinz Dieter Zeh—who explicitly distanced himself from the work of the Italian physicists and criticized the assumption of quantum systems as closed systems. Indeed, one early insight into the solution of this problem was obtained by Zeh in a short paper written in 1967 and eventually published in 1970. Zeh criticized both "the arguments leading to inconsistencies in the description of quantum-mechanical measurement" and "those explaining the process of measurement by means of thermodynamical statistics." He was thus criticizing both Wigner's and Rosenfeld's stands as far as the measurement process was concerned. Zeh's main suggestion was that macroscopically different states "cannot be dynamically stable because of the significantly different interaction of their components with their environment," which was the seed of the current decoherence approach (Zeh 1970). The reception of this paper was a traumatic experience for him as it reflected the existing prejudices against the research in foundations at that time. We come back to this story later in this chapter. Zeh shelved the subject in the 1970s and only came back to it in the 1980s with the collaboration of Erich Joos, his German student. However, in the 1980s the race for understanding decoherence was being disputed with other teams, namely that of the Polish Wojciech H. Zurek, who went to the University of Texas at Austin to study and came under the influence of John Archibald Wheeler, and the Englishman Anthony Leggett, who counted on the collaboration of Amir Caldeira, his Brazilian student. The historical roads taken by Zeh and Zurek in their work on what we now call decoherence were analyzed by historian Kristian Camilleri (2009), while the path adopted by Leggett was studied by the historian Fábio Freitas.[15] In fact, historical studies show us that decoherence matured as a new physical effect along three different and independent routes. Let us examine their stories considering now the roles played by Zurek and Leggett.

From early on, Zurek gained visibility among the physicists interested in foundations of quantum mechanics. During his graduate studies, Wheeler invited him to work on a project to bring together and translate into English the corpus of the papers he thought researchers should be familiar with. This reflected the status foundations was obtaining. It was a procedure to manage a controversy already legitimated: putting at the disposal of the researchers the basic documents that constitute the corpus of the controversy. Since then, Wheeler and Zurek's *Quantum Theory and Measurements* (Wheeler and Zurek 1983) has maintained a mandatory presence in physics libraries throughout the world. On conceptual grounds, Zurek (1981) argued that "the form of the interaction Hamiltonian between the apparatus and its environment is sufficient to determine which observable of the measured quantum system can be considered 'recorded' by the apparatus." Moreover, he showed (Zurek 1982) that the interaction between the environment and the quantum apparatus entails a correlated system, which imposes "superselection rules" preventing the apparatus from "appearing in a superposition of states corresponding to different eigenvalues of the privileged *pointer observable*," that is, the macroscopic observables. Thus, he created the term einselection, short for environment-induced superselection. In addition to his own contributions to the building of decoherence as a physical effect (Zurek 1981, 1982; Unruh and Zurek 1989), Zurek contributed to publicizing research on decoherence with a paper—"Decoherence and the Transition from Quantum to Classical"—which would become a landmark in its popularization (Zurek 1991). The paper was one of the first to use the word decoherence and has so far been cited 1,350 times, which is unusual for a paper published in a non-technical journal such as *Physics Today*.[16] Indeed, it was with Zurek's paper that the term "decoherence" acquired wider currency among physicists. A survey on the Web of Science for papers with "decoherence" in their titles indicates the existence of about 10,000 papers. J. J. Halliwell (1989) was the first to use it in the context of quantum cosmology; D. Dieks (1989) was the third to use it and the first to use it in the context of the quantum measurement problem; while Zurek (1991) was the 24th, but this paper is by far the most cited paper among most of the first papers to use "decoherence" in their titles. Before 1989, the physics of decoherence appeared under other umbrellas. Zurek, for instance, used "pointer basis" and "superselection rules" in his first papers (Zurek 1981, 1982), only beginning to use "decoherence" from 1989 on (Zurek 1991).[17]

[15] Fábio Freitas, "Tony Leggett – Challenging Quantum Mechanics with Quantum Mechanics," unpublished paper, 2012.

[16] 1,353 citations on April 7, 2014. Source: *Web of Science.*

[17] All data from *Web of Science* were accessed on April 7, 2014.

Tony Leggett had been looking for a way to identify limits to the validity of quantum mechanics and his first venture into this domain involved dealing with a complex system in condensed matter physics, the superfluidity of Helium-3.[18] While he did not achieve this, he instead obtained a corroboration of quantum mechanics, this work gained him the Nobel Prize in 2003. Again tackling complex systems instead of simpler ones, Leggett went on to model macroscopic quantum tunneling in systems with Josephson junctions, which consist of two superconductors coupled by a barrier. Such a problem—quantum tunneling in macroscopic systems—was suggested by Leggett to Caldeira as a subject for his doctoral research in the late 1970s. More precisely, Caldeira and Leggett were interested in thermal fluctuations and how the coupling between a system and its environment would interfere with quantum tunneling in SQUIDS, which is an acronym for superconducting quantum interference devices. Caldeira was particularly qualified to deal with this problem as his Master dissertation in Brazil, under the supervision of Nicim Zagury, dealt with the dissipation of systems surrounded in a thermal bath. Caldeira's results showed that the coupling with the thermal bath would reduce the tunneling rate, and that the larger the system, the lesser the tunneling. Thus his results imply that the coupling between a system, initially described by a pure state, and its environment would destroy the quantum signature of the system leading the system to be described as a mixture. Furthermore, for concrete cases the time of such evolution of a system from quantum to classical descriptions, the time for damping, could be calculated. As these results had quantum theory as their assumption, they could appeal to experimental physics to check the validity of quantum theory for such systems. The work afforded Caldeira his PhD degree (Caldeira 1980) and the papers from the thesis (Caldeira and Leggett 1981, 1983a, b, 1985) skyrocketed in terms of citations.[19]

In the mid-1990s the experimenters took the decoherence challenge to the labs. Around the same time, two results were announced by teams led by Serge Haroche, in France, and David Wineland, in the US. They dealt with mesoscopic systems instead of macroscopic systems. Haroche and his colleagues, who came from an experimental tradition in quantum optics dealing with microwave cavities, sent a Rydberg atom through a superconducting microwave cavity in which there was a low-number photon field and were able to detect not only the quantum superposition but also its disappearance in the time predicted by the theoretical model (Brune et al. 1996). Haroche's experimental work was developed in collaboration with the Brazilian physics group led by Luiz Davidovich (Davidovich et al. 1996).[20] It could be compared to a Schrödinger's cat thought experiment fitted to test decoherence

[18] This paragraph is based on Freitas, 2012, op. cit. This paper draws Leggett's biographical information from his Nobel Prize biography and interviews with Leggett by Babak Ashraf on 25 March 2005 and Fábio Freitas, on 3 Aug 2011.

[19] 1,400; 2,466; 1,526; and 381 citations, respectively, till April 7, 2014, according to *Web of Science*.

[20] On the collaboration with Brazilian physicists, see Serge Haroche, interviewed by O. Freire, on 27 Feb 2004, AIP.

predictions. Wineland's team was able to make a similar comparison trapping ions with laser pulses and obtaining the quantum superposition of two spatially separated but localized positions (Monroe et al. 1996). There were conceptual differences between the two experiments, in addition to the difference in their materials. According to Zurek (2003, p. 766), while in the former the decoherence "is caused by entangling interactions with the quantum state of the environment," in the latter it "is simulated by classical noise in the observable complementary to the pointer." Haroche's and Wineland's feats were more than confirmation of the loss of coherence in mesoscopic systems, they were exemplars of the techniques developed in the 1980s and 1990s which enabled physicists to manipulate single photons and atoms. In 2012 the Nobel Prize in Physics was awarded in equal parts to them "for ground-breaking experimental methods that enable measuring and manipulation of individual quantum systems."[21] The magazine *Physics Today* (Smart 2012) reported the prize saying that "David Wineland used light to manipulate trapped atoms, and Serge Haroche used atoms to manipulate trapped light." Eventually, the macroscopic tunneling in SQUIDS, once envisaged by Leggett as a test for the validity of quantum mechanics, was in the labs corroborating quantum mechanics and decoherence further (Friedman et al. 2000; van der Wal et al. 2000; Gerry and Bruno 2013, pp. 157–166).

Let us turn to add the biographical background of some of the physicists who contributed, either theoretically or experimentally, to the understanding of decoherence.

8.3.1 Work on Decoherence: Zeh, Leggett, Zurek, and Haroche

"Dark ages" was the term H. D. Zeh used to describe the times between his first contact with the measurement problem in the late 1960s and the early 1980s when similar approaches were developed by W. H. Zurek, reaching a wider audience than before. Dark ages, however, may have been the consequence of the professional and cultural circumstances surrounding Zeh's shift of attention towards the foundations of quantum mechanics. By 1967 the theoretical nuclear physicist, then an assistant teacher (Privatdozent) at Heidelberg, turned his attention to the measurement problem, arriving at the conclusion that the interactions among macroscopic bodies and their environment prevent one from describing them as a closed system. Zeh's ideas were brought together in a paper sent to the Italian journal *Nuovo Cimento*. He did not work out the full implications of his claims, but afterwards noticed the connection between his ideas and the relative state interpretation advanced by Hugh

[21] http://www.nobelprize.org/nobel_prizes/physics/laureates/, accessed on April 7, 2014.

Everett in 1957, a connection which appeared in a second version of his paper eventually published in *Foundations of Physics* (Zeh 1970). The goings-on behind this paper highlight the obstacles faced by research on the foundations of quantum mechanics in the late 1960s and 1970s.[22]

According to Zeh the paper was turned down by several journals. *Nuovo Cimento*'s vice director, Franco Bassani, for instance, refused the paper based on a referee's report that stated: "the paper is completely senseless. It is clear that the author has not fully understood the problem and the previous contributions in this field," while *Die Naturwissenschaften*'s editor, F. Boschke, politely turned it down using the argument that some points were not clear for non-experts.[23] In addition to the refusal of a paper, Zeh began to face problems at his alma mater, Heidelberg University. As Zeh (2007) recalled, "it was absolutely impossible at that time to discuss these ideas with colleagues, or even to publish them. An influential Heidelberg Nobel Prize winner frankly informed me that any further activities on this subject would end my academic career!" The Nobel Prize winner was J. Hans D. Jensen, as Zeh later confirmed, adding that,

> When I wrote this paper which was published in 1970 [. . .] [Jensen] said he did not understand that, and he sent a copy unfortunately to Rosenfeld in Copenhagen. [. . .] He wrote a letter to Jensen which Jensen never showed me where he must have been very cynical about what I had said, and I remember that Jensen told that to some other colleagues, then when I noticed they were talking about them, they were chuckling. But he never told me precisely what was in this letter. [. . .] Then Jensen told me that I should not continue this work, and so then our relationship deteriorated.[24]

Eventually, this letter was unearthed and its content details the hardships faced by Zeh. It does not need further comments:

> I established a rule in my life never to step on anybody's toe, but a preprint written by a certain 'Toe' [Zeh, in German] from your institute that I have received makes me digress from that rule. I have all the reasons in the world to assume that such a concentrate of wildest nonsense is not being distributed around the world with your blessing, and I think to be of service to you by directing your attention to this misfortune.[25]

---

[22] The paragraphs on Zeh come from my paper "Quantum Dissidents . . .", *Studies in History and Philosophy of Modern Physics*, 40, 280–289, 2009.

[23] The original version, ["Probleme der Quantentheorie", in German], may be found at http://www.rzuser.uni-heidelberg.de/~as3/. Franco Bassani to H. D. Zeh, 3 Oct 1968, with an appended referee's report; Boschke to Zeh, 3 Oct 1968; I obtained copies of these letters due to the kindness of H. D. Zeh and F. Freitas.

[24] H. D. Zeh, interviewed by Fabio Freitas, 2008.

[25] "Ich mach es zu einer Lebensregel, so weit vermeidlich auf keinen Zeh zu treten, aber der Empfang eines von einem gewissen Dr. Zeh aus Ihrem Institut verfassten preprint veranlasst mich von dieser Regel abzuweichen. Ich habe allen Grund anzunehmen, dass ein solches Konzentrat wildesten Unsinnes nicht mit Ihrem Segen in die Welt verbreitet ist, und ich glaube Ihnen von Dienst zu sein, indem ich Ihre Aufmerksamkeit auf dieses Unglück richte." L. Rosenfeld to J. H. D. Jensen, 14 Feb 1968. Next Jensen tried to attenuate Rosenfeld's reaction, while fearing for its consequences: "I hope, that he does not quite have his reputation ruined," Jensen to Rosenfeld, 1 March 1968. Rosenfeld then considered Zeh's case in a "somewhat favorable light" though still

From the early 1970s on, Zeh practically abandoned nuclear physics, devoting himself full-time to the foundations of quantum mechanics. In hindsight, Zeh saw this move as a consequence of the professional obstacles he faced while approaching this subject:

> I concentrated on these issues because I had decided that my career was destroyed. Anyhow, I would never get a professorship because of these things already, and so I said, 'Now I can just do what I like and I don't have to try any more to get any position,' or something like that.[26]

Against such a background, it comes as no surprise that Zeh held what he considered to be the Copenhagen interpretation of quantum mechanics in no high esteem associated as it was to Rosenfeld, Bohr, Heisenberg, Pauli, and their connections to Jensen. Still in this context he expressed his resentment to John Archibald Wheeler, in 1986:

> I have always felt bitter about the way how Bohr's authority together with Pauli's sarcasm killed any discussion about the fundamental problems of the quantum. [...] I expect that the Copenhagen interpretation will some time be called the greatest sophism in the history of science, but I would consider it a terrible injustice if—when some day a solution should be found—some people claim that 'this is of course what Bohr always meant', only because he was sufficiently vague.[27]

A few years before, circa 1976, the historian of science David Edge had asked him, "Do you feel that physicists who hold unorthodox views about QM have any great difficulties in carrying out their work?" His answer, "exceptional difficulties." Asked "if a young physicist begins his professional life by working in this field, can his career be hampered?" he replied "definitely." Zeh also acknowledged that the number of physicists interested in the foundations of quantum mechanics was increasing and attributed this to the "decrease of the unjustified authority of N. Bohr, W. Pauli and other 'pragmatists'."[28]

In his early research into the foundations of quantum mechanics Zeh met not only adversaries (Zeh 2007). Indeed, the supreme irony is that the major support he found came from a physicist who shared the 1963 Nobel Prize in Physics with Jensen. We are speaking, of course, of Eugene Wigner and the manner in which he supported Zeh at the 1970 Varenna summer school and the publication of his paper in *Foundations of Physics*, episodes we analyzed in Chap. 4. A note about Zeh's relation to John Bell is also appropriate. In Varenna Zeh met Bell and found him

---

considering Zeh's paper "more like a possession claim of a monopoly of highest wisdom" than "as an invitation to a factual discussion," Rosenfeld to Jensen, 6 March 1968. The affair occupied three more letters between Rosenfeld and Jensen: Jensen to Rosenfeld, 10 Apr 1968; 9 May 1968; Rosenfeld to Jensen, 25 Apr 1968. Rosenfeld Papers, Niels Bohr Archive, Copenhagen. I am indebted to Anja Jacobsen and Felicity Pors for recovering these letters and Christian Joas for the German translation.

[26] H. D. Zeh, interviewed by Fabio Freitas, 2008, *op. cit.*

[27] H. D. Zeh to J. A. Wheeler, 30 Oct 1980, Wheeler Papers, Series II, Box Wo-Ze, folder Zeh, *WP*.

[28] Questionnaire sent to Zeh by David Edge, circa 1976. I am grateful to H. D. Zeh and F. Freitas for allowing me to consult this document.

supportive of his work despite not agreeing with it. In fact, at that time, Bell did not like Everett's interpretation, favoring Bohm's pilot wave, and he was totally excited with the theoretical and experimental debates on locality. Zeh, in turn, was not in tune with Bell as for him the superposition principle, which is the basis of quantum nonlocality, was always valid, while Bell expected to demonstrate through experiments the limits of quantum mechanics, as we have seen in Chap. 7. Disagreements apart, Zeh found Bell "very sympathetic, and he always asked the right questions. This was already a very great thing that there was somebody who was critical against the mainstream and put his finger on the right things. Only very few people did that."[29] Zeh also invited Bell twice to discuss foundational issues at Heidelberg, which diminished his own isolation at the university.

Anthony Leggett's education was very unusual for an outstanding physicist.[30] From secondary school, he was inclined and trained in the classics, except for a year of informal interaction with a Jesuit priest who introduced him to some advanced mathematics. He went to Balliol College, Oxford, and after graduating was about to do a doctorate in philosophy in order to become an academic. However, he doubted that philosophy, as it was being practiced, afforded objective criteria for deciding what is right or wrong, and he needed this. He began pondering the possibility of doing a second undergraduate course, this time in physics. The cultural changes in the West after the launching of Sputnik by the Soviets eased the way for him to get around the obstacles to do this. After getting his PhD in condensed matter under the supervision of Dirk ter Haar, he went to a postdoc at the University of Illinois, Urbana-Champaign, with David Pines, followed by a 1-year stay at Kyoto University in Japan to eventually land at the University of Sussex in 1967 where he would stay for 15 years. Only later did Leggett return to the University of Illinois.

This is not the place to describe Leggett's work on the superfluid liquid Helium-3, which led him, together with the Soviets Vitaly L. Ginzburg and Alexei A. Abrikosov, to Stockholm. Let me just say that Leggett's interests in foundations of quantum mechanics predates his seminal work on superfluidity. Indeed, around 1970, Leggett faced a second dilemma, considering the first when he left classics for a second degree in physics. According to his later recollections, "I found myself becoming increasingly bored with this area of research, and indeed with much of conventional physics; at the same time, thanks in part to a remarkable series of lectures delivered by my colleague Brian Easlea, I got more and more intrigued by the conceptual foundations of quantum mechanics."[31] Easlea was a nuclear

---

[29] H. D. Zeh, interviewed by Fabio Freitas, 2008, *op. cit.*

[30] The text on Leggett is entirely based on Fabio Freitas, "Tony Leggett - Challenging Quantum Mechanics with Quantum Mechanics," unpublished paper, 2012. This paper draws information on Leggett from his Nobel Prize biography (http://www.nobelprize.org/nobel_prizes/physics/laureates/2003/leggett-bio.html), his Nobel lecture (http://www.nobelprize.org/nobel_prizes/physics/laureates/2003/leggett-lecture.html), and interviews by Babak Ashraf on 25 March 2005 and Fábio Freitas, on 3 Aug 2011.

[31] See Leggett's Nobel lecture, on page 147, http://www.nobelprize.org/nobel_prizes/physics/laureates/2003/leggett-lecture.html, accessed on April 10, 2014.

physicist, a socialist, and a keen social critic of science, particularly of the nuclear race and male gender influence in the shaping of modern science (Easlea 1983). Leggett and Easlea became close friends. Back to the quantum, Leggett was particularly intrigued by the quantum measurement problem. Leggett obtained his first results in research on foundations in 1976, when he spent a time teaching in Ghana; dealing with the issue of hidden variables Leggett came upon Leggett's inequalities, one of the generalizations of Bell's theorem. This work was filed away for years but eventually found its way to be printed and tested in the lab (Leggett 2003; Groblacher et al. 2007). From 1980 on, Leggett became more and more involved in a program fully dedicated to the foundations of quantum mechanics. In Leggett's own words, this program may be defined as the quest for "theory of experiments to test whether the formation of quantum mechanics will continue to describe the physical world as we push it up from the atomic level towards that of everyday life (a program for which my shorthand is 'building Schrödinger's Cat in the laboratory')."[32] This definition was given in 2003 and Leggett recalled the evolution of this program, including the way it was received early on: "It is satisfying that this program, which when proposed twenty-five years ago met with considerable skepticism, seems in the last three or four years to have come to fruition, in the sense that several experiment[s] have [been] realized, which can be legitimately regarded as of the "Schrödinger's Cat" type."[33] Since the 1980s, through his papers, published in mainstream journals and proceedings as well as in journals such as *Foundations of Physics*, Leggett did not hide his expectations of exhibiting the limits of validity of quantum mechanics. In fact, even in the first days of his work on superfluid Helium-3, he initially envisioned the possibility of finding limits to the use of the quantum machinery, something he acknowledged in his Nobel lecture. So far this has not happened, but a lot has been learned on the way. The Brazilian historian of physics, Fábio Freitas, has asked why Leggett did not face professional obstacles while exposing such expectations since they conflict with the orthodoxy in quantum physics. And still, why has he not been widely portrayed among the researchers in foundations as one of these researchers? Freitas suggests a multifaceted answer. In a nutshell, Leggett's program and scientific style is to address foundational issues through complex cases, which demand a good deal of physics. Thus the physics community could accommodate itself with Leggett's heterodoxy insofar as he has always been pushing ahead the frontier of theoretical and experimental physics. Finally, his open distrust of the universal validity of quantum mechanics has left him apart from most researchers in foundations who have abandoned any doubts about universal validity after the blossoming of quantum information.[34]

---

[32] http://www.nobelprize.org/nobel_prizes/physics/laureates/2003/leggett-bio.html, accessed on April 10, 2014.

[33] Ibid.

[34] Freitas' point is far more substantiated than the summary I have done. See Freitas, ibid.

Wojciech H. Zurek is a Polish-born, naturalized American theoretical physicist. After undergraduate and master studies in physics in Poland he arrived at the University of Texas at Austin in 1975 to do a PhD degree. The following year Wheeler arrived at Austin and Zurek attended his courses while in parallel doing research in astrophysics and he became fascinated by the foundations of quantum mechanics. Over time he has exhibited a diversified interest in theoretical physics. With Tom Kibble he obtained a result in defect generation in non-equilibrium processes with wider applications, including astrophysics. He has worked mainly at Los Alamos National Laboratory where he became a Laboratory Fellow, which is a prestigious distinction in the frame of a U.S. National Laboratory researcher. We present later the non-cloning theorem, which was jointly created by Zurek and Wootters, and a result independently obtained by Dennis Dieks (Dieks 1982; Wootters and Zurek 1982). His works on decoherence in the 1980s (Zurek 1981, 1982, 1991, 1998; Unruh and Zurek 1989) contributed to set its research agenda. The 1991 *Physics Today* paper helped connect a wider audience in physics to the promises of these results. In addition, he edited with Wheeler the influential corpus book *Quantum Theory and Measurement* previously mentioned (Wheeler and Zurek 1983). More recently, his forays have been in the suggestion of an interpretation of quantum mechanics—the existential interpretation—centered on the concept of information and reminiscent of some of Bohr's thoughts. This suggested interpretation and results in decoherence merged into what Zurek christened Quantum Darwinism, which is a theoretical proposal for explaining the emergence of the classical world departing from the quantum world (Schlosshauer 2004, p. 1290, Zurek 2009).

Serge Haroche was born in Casablanca, Morocco, and after the end of the French protectorate he went to Paris. Haroche studied at the prestigious Ecole Normale Supérieure and knew he would work in science since he was a teenager. His training and research in physics was mostly done under the supervision of Claude Cohen-Tannoudji, who was awarded the Nobel Prize in Physics in 1997. After his PhD, in 1971, Haroche went to a postdoc at Stanford University working with Arthur Schawlow, who would also receive a Nobel Prize, in 1981. He has taught in a number of institutions, including Université de Paris VI, Ecole Polytéchnique, Harvard, Yale, and the Ecole Normale Supérieure, where he still has his lab. In 2001 he was selected to the chair of Quantum Physics at the traditional Collège de France. Coming from such a distinguished research milieu, Haroche had the burden of fulfilling the expectations around him. Indeed he early mastered a number of first-rate experimental techniques brought from atomic physics to quantum optics. He became a leader in fields such as laser spectroscopy, quantum beats, Rydberg atoms and QED cavities, and cavity-enhanced single-atom spontaneous emission. Through the novel use of some of the techniques deployed in such experiments came the groundbreaking 1996 experiment when a Schrödinger-cat type experiment spotted the loss of coherence of quantum states, namely decoherence.

Including Haroche in a collective biography of physicists dedicated to the research on the foundations of quantum physics would not do justice to his own perception of his scientific agenda. "I did not decide one day to look into the foundation of quantum mechanics. We just realized in our experiment that we were dealing with smaller and smaller systems. [...] then we were able to decrease the threshold until we had only one atom at a time in the cavity." This view is possibly related to his perceived connection between foundations and alternative interpretations of quantum mechanics as he explicitly tied one to the other, "You cannot find anything in my papers about foundations. I am always using standard quantum mechanics, and we are trying in our experiments to illustrate these standard quantum mechanics at work in specific situations."[35] For him standard quantum mechanics equals complementarity, thus Bohr's views. In a paper intended to convey to the wide physics audience the novelties of the Schrödinger-cat type experiments he and Wineland had carried out, he framed these results in the complementarity view. Indeed, he presented first the "orthodox Copenhagen interpretation" with instantaneous collapse, then the explanation of "modern decoherence theories" with a real physics process coupling the system and its environment, and concluded, "for all practical purposes, of course, the orthodox and decoherence points of view are equivalent" (Haroche 1998, p. 41). This does not mean, however, that Haroche was insensitive to competitors. In this same paper, interpreting his own experiments, he recorded, "attempts have been made to modify the quantum theory by adding subtle mechanisms that would 'explain' quantum choices in systems with macroscopic components," and cited Ghirardi's 1986 proposal of modifying the Schrödinger equation in order to get spontaneous collapses. This acknowledgment did not imply, however, any sympathies towards such proposals. According to him, "whether such theories will be successful and lead to testable experimental predictions remains dubious" (Haroche 1998, p. 41). Thus Serge Haroche is a committed Bohrian, who has given milestone contributions to the understanding of the foundations of quantum theory while not even considering himself a researcher in foundations.

8.4 New Techniques and New Experiments in Foundations of Quantum Physics

The 1980s and early 1990s were the times when dramatic technical advances were applied in experiments in foundations and a new array of experiments in this domain, most of them previously only thought experiments, were performed. Technical improvements—namely the ability to isolate, control, and observe single quantum systems such as electrons, photons, neutrons and atoms—were one of the factors the physicist Alain Aspect mobilized to coin the term "the second quantum revolution." Examples of novel real experiments we have already seen in this chapter are the improved Bell's type experiments, quantum cryptography and teleportation, and Schrödinger's cat type experiments. Let us now further illustrate the impact of the new techniques and present more examples of new thought-turned-real experiments.

[35] Serge Haroche, interviewed by Olival Freire, 27 Feb 2004, AIP.

8.4.1 Techniques

In 2008 the French physicist Serge Haroche wrote an essay entitled "Fifty Years of Atomic, Molecular and Optical Physics in Physical Review Letters". Haroche has been a leader in these fields and wrote this essay for the 50th anniversary of this prestigious journal. It provides an overview of the technical achievements we are looking for (Haroche 2008).[36] He recalls that in the inception of *PRL* "atomic physics was considered—at least by physicists working in other areas—as a mature field with a rather unpromising future," as the job was "to improve the resolution of spectroscopy" which was "testing elementary systems whose basic properties were fundamentally known." However, in the 1960s, optics was dramatically changing through the introduction of lasers and the theoretical developments of quantum optics. Still according to Haroche, "instead of being merely a probe of atomic spectra, light was becoming a tool to actively manipulate atoms, to force them to occupy some states out of the natural thermal equilibrium" and "atomic physicists were increasingly fascinated by this challenge and the perspectives it opened."

Haroche compares the early 1960s with the 2000s both in the scale of the laboratories as well as in the kind of system being studied, emphasizing the differences. He recalls, "even if AMO [atomic, molecular, and optics] physics has remained small scale when compared with other fields such as particle physics or experimental astrophysics, it has become much more complicated and sophisticated than the field was then." Talking of the present, he says,

> Hundreds of optical elements have to be aligned on meter-long tables, forming a maze of laser beams intersecting with precision on atoms or molecules localized in traps or propagating in well-controlled atomic beams. Sophisticated cameras observe the atomic evolution, and fast computers are required to control complex procedures and to measure correlation signals that would have been absolutely impossible to track in the pre-computer era. Two or three students must work together to manage the various aspects of the experiment.

As for the kind of systems under scrutiny, he remarks, "whereas physicists in the 1960s were studying collective atomic samples made of billions of particles, they can now juggle with single atoms or control small samples made of a few atoms." The same goes for light as "experimentalists now play with single photons or with fields made of a small controlled number of light quanta. They are able to build and reconstruct states of light with intrinsic quantum properties, unexplainable in classical terms." For the study of foundations of quantum mechanics, single quantum systems have been crucial, as observed by the physicist M. A. B. Whitaker, while reviewing experiments in the foundations of quantum theory, "only by such experiments studying single particles, and single correlated systems, may one move beyond the rather bland information obtainable from ensembles, and study the adequacy of standard approaches to quantum theory in these more theoretically challenging circumstances" (Whitaker 2000, p. 3).

---

[36] All Haroche's quotations in this section come from the same paper (Haroche 2008).

8.4.2 Experiments

From the long list of foundational experiments carried out in the 1980s let us dip in and select some which fire our imagination most: on one hand, the delayed-choice experiment, on the other hand, realizations with single systems, both for radiation and matter, of the old two-slit thought experiment, which had been envisioned to exemplify quantum wave-particle duality. As for the delayed-choice experiment, in 1978 Wheeler took an old idea from Bohr, fleshed it out, and christened it the delayed-choice experiment. The suggestion of this experiment and its realizations have been analyzed by the historian Joan Bromberg (2008), and I draw from her work here. Bromberg has also studied the cases of quantum eraser and which-path experiments (Bromberg 2006, 2008). The delayed-choice thought experiment concerns modifying the experimental setup after the experiment has begun—for instance, after the light emitted from a source has entered the interferometer—thus changing the whole phenomenon. Modifying the experimental setup may allow, for instance, an interference fringe—which is the wave signature of a phenomenon—to appear or disappear. I deliberately worded the thought experiment in Bohrian terms, that is, the phenomenon which is the object of quantum theory is the wholeness of systems under investigations plus devices required for such analysis. In these terms the experiment may seem straightforward. Now let us rephrase it in realistic terms including its time evolution along paths in the space-time. One can say that a later decision to change the device—for instance, blocking or unblocking the passage of radiation in a certain way—would change the early beginning of the system behavior causing it to take either one way or the other as particles, or alternatively to take both the ways as waves. More directly, "delayed choice suggests that an act of observation in the present can create the past. In particular, it suggests that by carrying out scientific observations, the human community can create its universe" (Bromberg 2008, p. 332). Wheeler deliberately played with these interpretation ambiguities in order to attract attention to the suggested experiment. In his very first presentation, he wrote "Can one choose whether the photon (or electron) *shall have* come through both of the slits, or only one of them, after it has already transversed this screen?" A few pages later, he attenuated the impact of his first words, using Bohr's terms: "Then let the general lesson of this apparent time inversion be drawn: 'no phenomenon is a phenomenon until it is an observed phenomenon,'" and then explained how a Bohrian could dissolve the paradox: "In other words, it is not a paradox that we choose what *shall* have happened after 'it has *already* happened.' It has not really happened, it is not a phenomenon, until it is an observed phenomenon" (Wheeler 1978, pp. 9–14).

Wheeler campaigned for this experiment to be carried out and Carroll Alley at Maryland University and Herbert Walther at the Max Planck Institute for Quantum Optics in Garching, West Germany, took up the challenge. Bromberg's analysis reveals, at least in the better-documented case of Alley's experiment, the dynamics driving this type of experiment in the 1980s. Alley had led two experiments with a diversity of applications. The first one was lunar ranging, using a laser for measurement of the distance between points on the earth and on the moon within centimeters, which required taking mirrors to the lunar surface. In addition to being a technical feat, such an experiment was useful to contrast different gravitation theories, for instance. The second was a time measurement with atomic clocks at different speeds and heights from the earth's surface. Such experiments, funded by the U.S. military, were instrumental in the building of the Department of Defense's Global Positioning System (GPS). As Bromberg says, "both of these projects demanded new technology." The invention of the laser had been crucial. Furthermore, the lunar ranging experiment required "detectors capable of registering single photons since the losses were so great that out of the 10^{19} photons sent out in each shot, 'one detects only one photoelectron per 10 to 20 shots!'" The experiment also "needed timers that could register the moment of detection with an accuracy of a few nanoseconds." The atomic clock measurement required improvements in the laser and the electronics, including improved Pockels cells. Again according to Bromberg, Alley and his colleagues "decided that a delayed-choice experiment was something they could do with the equipment they had from the lunar ranging and gravitational corrections experiments." Of no lesser importance, they also had the staff and students to do it. One of Bromberg's conclusions is that "the Maryland delayed-choice experiment is a classic example of research that piggybacked on major government financed projects." The results of the delayed choice experiments, both carried out by the American and the German teams, confirmed quantum predictions but their authors diverged as to how to interpret them. Since then, these experiments have become standard in popular science books as illustrations of quantum weirdness.

8.4.3 The Conspicuous Double Slit Experiment

The double slit experiment has been the most discussed thought experiment since the first days of quantum theory. Both the wave and particle behavior of the electron and photon can be exhibited depending on the designed setup. It was the pet model for Bohr in his debates with Einstein, later it became the standard ideal experiment for a conceptual introduction to quantum physics, and the bestselling *Feynman's Lectures on Physics* helped make it an icon in physics teaching. Feynman opened the volume dedicated to quantum mechanics with the frequently-quoted fragment saying he had chosen "to examine a phenomenon which is impossible, absolutely impossible, to explain in any classical way, and which has in it the heart of quantum mechanics. In reality, it contains the only mystery." And then he went on to discuss the double slit device with bullets, waves, and electrons (Feynman et al. 1963, vol III) . However, this had never been checked in the labs for single electrons or photons until the 1980s.

In 1979 the French physicist Claude Cohen-Tannoudji, who would share the 1997 Nobel Prize in Physics with Steven Chu and William Daniel Phillips for their research into methods of laser cooling and trapping of atoms, presented the syllabus for his course at the traditional Collège de France. "Taking into consideration recent experiments, let us try to answer the following question: could one get rid of the concept of photon, at least in the domain of optics?"[37] The question was central in the debates about the necessity of the full quantum treatment for light suggested by Roy Glauber, and the semi-classical approaches supported by Emil Wolf (Silva 2013; Silva Neto and Freire Jr. 2013). In the audience was Alain Aspect, who thought that the source of light he was using in Bell's type experiments was a kind of one-photon state Cohen-Tannoudji was explaining. The crucial point for him was that in all previous experiments with "single photons," which dated back to Taylor in 1909, the light impulsions could not be quantum mechanically described as single-photon states. The idea remained dormant as he was fully involved in the assembly of his experiments on Bell's theorem. Later on, he suggested using this source for a wave-particle experiment with single photons for Philippe Grangier's doctoral research. The source emits a pair of entangled photons from atomic decay and the idea was that by detecting one of the photons in one channel you could be sure you had a single photon in the other. Then the single photons may be used for an interference type experiment or a which-path type experiment.[38] Single photon states were also produced in 1986 in Rochester by Chung Ki Hong and Leonard Mandel (Hong and Mandel 1986; Grangier 2005).

Aspect and Grangier's 1986 experiment led to a milestone paper in the history of foundations of quantum physics not only because it reported the first experiment with wave particle duality for single photons but also due to the clear-cut conclusions drawn by its authors. After presenting his results, Aspect and colleagues (Grangier et al. 1986, p. 178) interpreted them in two different ways. Initially they used complementarity: "if we want to use classical concepts, or pictures, to interpret these experiments, we must use a particle picture for the first one, [...] on the contrary, we are compelled to use a wave picture, to interpret the second experiment. Of course, the two complementary descriptions correspond to mutually exclusive experimental set-ups." Aspect's inclination was towards another kind of explanation. It was an explanation based on a direct interpretation of the quantum mathematical formalism, without appealing to pictures, using concepts that had just emerged in quantum optics: "from the point of view of quantum optics, we will rather emphasize that we have demonstrated a situation with some properties of a 'single photon state'." Three years later, discussing the same results, the trio (Aspect et al. 1989, p. 128) went further in their epistemological choices. After presenting the explanation with complementary classical concepts, they added: "the logical conflict between these two pictures applied to the same light impulses constitutes one of most serious conceptual problems of quantum mechanics." Then they remarked that the experimental setups were incompatible and that this incompatibility was presented by Bohr as an element of coherence of quantum

[37] Cohen-Tannoudji's course is at http://www.phys.ens.fr/~Claude%20Cohen-Tannoudji/college-de-france/1979-80/cours1/cours1.pdf, accessed on 9 May 2014.

[38] Alain Aspect, interview with O. Freire & I. Silva, 16 Dec 2010 and 19 Jan 2011, AIP.

theory. And yet their choice was favorable to the kind of explanation which emphasizes the self-sufficiency of the quantum formalism. This formalism describes both experiments without appealing to pictures or classical concepts: "... if, on the contrary, one is restrained to the quantum mechanics formalism, the descriptions of the light impulses are the same. It is the same state vector (the same density matrix) that one must use for each experiment. The observable changes but not the description of light" (Aspect et al. 1989, p. 128). These choices are evidence of how physicists were increasingly depending on the very quantum mathematical formalism and dispensing with the use of pictures for their reasoning on quantum systems. Twenty years later, technical developments have been so dramatic that this experiment has become an educational tool for physics teaching using lasers, new materials and CCD cameras (Jacques et al. 2005).

Now, let's move from light to matter, or from photons to electrons. The wave behavior of electrons had already been tested and the workings of electronic microscopes, for instance, are based on this property. However, previous experiments had always dealt with a huge number of electrons. It was Akira Tonomura who was the first to test this property with single electrons. He worked first on electron holography (Tonomura 1987) and then moved to experiments to verify the existence of the Aharonov-Bohm effect, presented in Chap. 2 of this book. Until then the confirmation of this effect was unclear. After controversies, Tonomura settled the subject favorably to the existence of Aharonov-Bohm's effect (Tonomura et al. 1986). According to A. Howie (2012), Tonomura's "conclusive and elegant experiment of 1986 finally silenced critics, and was immediately recognized beyond the world of electron microscopy as a remarkable tour de force." In the late 1980s, Tonomura performed the double slit experiment with electrons sending them one by one towards the slits (Tonomura et al. 1989). The short movie they made has become part of the toolkit of physics teachers around the world, as has "an Internet video of his version of the classic 'double-slit experiment' which continues to demonstrate for many the central mystery of quantum mechanics. It shows how electrons travelling through a biprism arrive at a detector one by one, as particles, but over time build up a wave interference pattern" (Howie 2012).[39] In addition to illustrating this striking quantum feature, this experiment has contributed to the growth of a field dedicated to making optics with electrons, which involves the study of the wave features of electrons. However, it should be noted that in the same year that Tonomura passed away his precedence in this experiment was challenged by Rodolfo Rosa (2012), who argued that Italian physicists P. G. Merli, G. F. Missiroli, and G. Pozzi had carried out a similar experiment as early as 1972.

After presenting the delayed-choice experiment it is time for an interlude to make a brief biographical notice of Wheeler's involvement with foundations of quantum mechanics.

[39] The film is available at https://www.youtube.com/watch?v=oxknfn97vFE, accessed on 15 April 2014.

8.5 Interlude: Wheeler's Perennial Concern with the Quantum

John Archibald Wheeler was an American theoretical physicist with a PhD from Johns Hopkins University in the early 1930s and a postdoctoral stay in Europe, including Copenhagen. Back in the U.S., he collaborated with Bohr in the creation of the liquid drop model for nuclear fission in 1939. He actively participated in the Manhattan Project (Rhodes 1986). In the Cold War context, he was fully engaged in the US military endeavor, actively participating in the JASON Project (Aaserud 1995). For most of his academic career he was at Princeton University, with short exceptions in the 1930s, when he was at University of North Carolina at Chapel Hill and after his retirement when he went to the University of Texas, coming back to Princeton as *Emeritus Professor* in 1986. He was a protagonist in the revival of general relativity after World War II and a pioneer, jointly with DeWitt, in quantum gravity (Misner et al. 2009). He was a talented player in the physics word game, being the creator of the term "black hole." Wheeler garnered a string of awards and is considered an "Uncrowned Nobel Laureate," an accolade created by Robert Weber to describe those who, according to a poll among Nobel Laureates, "are the peers of prize winners in every respect save that of having the award" (Weber and Lenihan 1980, pp. 4–5).[40] He always considered himself a committed Bohrian. As we have shown in Chap. 3, motivated by Everett's dissertation work and by his own work in gravitation, he initially strongly supported Everett in the mid-1950s. Unfortunately, he decided to attempt to convince Niels Bohr of Everett's ideas before the approval of his thesis, and, indeed, went to Copenhagen to get this endorsement. As we extensively analyzed in Chap. 3, the only result from that enterprise was the Copenhagen physicist's dismissal of Everett's proposal. The event scarred Everett who left physics and academia for a successful career in mathematics related to U.S. defense, leaving Wheeler frustrated. Moreover, what we did not comment on in Chap. 3, was that Wheeler came out of it very wary of concerns with foundations of quantum theory. For more than a decade he kept his reflections on the quantum foundations, particularly on the relationship between the quantum and gravity, to himself, not coming back to the subject in the open.[41]

In the early 1970s the quantum foundations illness that had once inflicted Wheeler came back. The buzz around Bell's theorem experiments was brought to him by his colleague at Princeton, Wigner, and he got involved with Edward Fry's experiment (Misner et al. 2009, p. 45). Bohr, his former mentor on interpretation of

---

[40] I am thankful to Aurino Ribeiro Filho for this remark.

[41] In 1967, Wheeler wrote to Max Born: "You are one of the few persons who have contributed through your work and through your leadership to the elucidation of both general relativity and the quantum principles. [. . .]. Which of these two principles do you rank as the 'deepest'?" Born's answer was disappointing, ". . . I am afraid I am too old (85) to understand it." Wheeler to Born, 29 Sep 1967; Born to Wheeler, 17 Oct 1967. Wheeler Papers, Series I – Box Boh-Bu, Folder Born, M., American Philosophical Society, Philadelphia, PA.

quantum mechanics, had passed away more than a decade ago. Wheeler felt he needed to catch up on the developments in this subject. An invitation from Michel Paty and José Leite Lopes from Strasbourg to attend a conference dedicated to the 50th anniversary of de Broglie matter waves gave him the opportunity to do his homework. Thus, in the fall of 1973 he declined an invitation from A. O. Barut to give talks at the University of Colorado and the reason was "unhappily I can't accept [...] at this time because I expect to be going overseas for the spring semester to try to get up to speed in the realm of quantum mechanics."[42] In Bures-sur-Yvette, on the outskirts of Paris at the Orsay campus, he met Bernard d'Espagnat and John Bell. Five pages in his notebooks, on April 2, 1974, were dedicated to conversations with Bell on the interpretation of quantum physics.[43] In early May in Strasbourg he talked about Everett's proposal and its connection with cosmology, thus the relationship between quantum and gravity (Wheeler 1977; Lopes and Paty 1977). In Europe, political and ideological differences did not hamper fruitful exchanges with Lévy-Leblond, as we commented on in Chap. 6, both interested in Everett's dispensation with the role of observers in the usual quantum theory.

From the mid-1970s on, Wheeler became deeply engaged in the research on foundations of quantum physics and brought his prestige to this field, which badly needed professional recognition. He suggested the delayed-choice experiment (Wheeler 1978) and campaigned for its implementation; together with Zurek, he organized a volume bringing together the corpus of the quantum controversy (Wheeler and Zurek 1983); pushed for an interpretation of quantum mechanics centered on the notion of information (he coined the "it from bit"); and motivated students to work on foundational issues. Wheeler is also acknowledged for his abilities to identify and support good students, the most prominent among them being Richard Feynman. While at the University of Texas at Austin, which he transformed into a Mecca for quantum foundation devotees, he inspired at least three young physicists who would play key roles in the inception of quantum information: W. H. Zurek, David Deutsch, and William Wootters. Wheeler's enduring contributions to the research on foundations was recognized at the conference on "Fundamental Problems in Quantum Theory" held in 1995 under the auspices of the New York Academy of Sciences (Greenberger and Zeilinger 1995).

[42] Wheeler to A. O. Barut, 23 Oct 1973, Wheeler Papers, Series I – Box Ba-Bog, Folder Barut, A. Ibid.

[43] Wheeler Papers, Series V, Notebook June 1973–April 1974. Ibid.

8.6 The Proliferation of Interpretations

The most impressive feature regarding interpretations of quantum mechanics is the proliferation of interpretations of the same mathematical formalism. Numbers began in the 1950s, rose in the 1980s and mid-1990s, and continue to rise nowadays. The *Compendium of Quantum Physics* (Greenberger et al. 2009) provides short introductions to most of these interpretations, including the following: Bohm interpretation, Bohmian mechanics, complementarity principle, consistent histories, Copenhagen interpretation, GRW theory, hidden-variables models of quantum mechanics, Ithaca interpretation, many worlds interpretation, modal interpretations, orthodox interpretation, probabilistic interpretation, and transactional interpretation.[44] While there is some redundancy in this list, it is not comprehensive. It does not include, for instance, the stochastic interpretation and the ensemble interpretation.

This proliferation raises a question often asked by the non-experts in foundations of quantum physics. Why is there this ongoing proliferation if predictions coming from standard quantum theory have been so widely confirmed in recent decades? We can obtain a philosophical answer following Max Jammer (1974, pp. 1–20) in his suggestion of a distinction between formalism and interpretation. While this distinction is far from being unproblematic even among philosophers of science, it was important as Jammer's book was widely read and played a role into this controversy itself. Indeed, Jammer was saying that different interpretations could be accommodated in the same formalism and that corroboration of formalism does not equal corroboration of one specific interpretation. In comparison, let us recall that in the 1930s Wolfgang Pauli had tried unsuccessfully to christen quantum theory as "complementarity theory" by analogy with relativity theory (Enz 2002, p. 249) and in the 1950s Léon Rosenfeld strongly reacted to Heisenberg's use of "Copenhagen interpretation" as this term could induce the existence of other interpretations, as we commented in Chap. 3.

However, the increasing diversity of interpretations did not only result from the logical possibility of this diversity, it came mainly from acute conceptual problems in the foundations of quantum theory as the narrative constructed throughout this book has attempted to show. The most influential conceptual problems have been the quantum measurement problem, and the related problem of the transition from quantum to classical descriptions, and the compatibility between quantum theory and general relativity, which concerns the quest for a theory of quantum gravity. In addition, traditional interpretations, both the complementarity view and von Neumann's approach, have an unequivocal instrumentalist flavor. In the case of complementarity, this was presented as a huge epistemological lesson. On the other hand, realistic views demand alternative interpretations of quantum mechanics and the increase of these views in the second half of the twentieth century has

[44] For some of these interpretations, larger introductions can be found at the Stanford Encyclopedia of Philosophy, at http://plato.stanford.edu/.

been noted by different scholars (Brush 1980; Elkana 1984).[45] Thus the increase of realism among physicists and philosophers involved with foundations favored the appearance of realistic alternative interpretations. Add to these conceptual and philosophical aspects the legitimacy obtained by the quantum controversy as a scientific controversy with philosophical implications (Freire Jr. 2003) and you have a clue to explaining the current proliferation. This was the fuel both for the birth of new interpretations (consistent histories, spontaneous collapse) and the revival of old ones (Bohm-de Broglie's and Everett's approaches).

The frame of possible interpretations for quantum physics in the 1980s thus reflected the rising concern with the quantum measurement problem. This issue was present in the inception of this theory but it became fully explicit only in the 1960s, as discussed in Chap. 4. The holy grail of physicists involved in research on foundations has become to describe physically and mathematically when and how a quantum measurement occurs, or its related issue, when the quantum description should be replaced by the classical one. As we have seen, it motivated research programs which ultimately led to the understanding of decoherence. As for new interpretations, the two newest and most influential interpretations came out of directly dealing with this issue: the consistent history and the collapse theories.

The consistent history interpretation was born between 1984 and 1990, and its founding fathers were Robert Griffiths, Roland Omnès, and Murray Gell-Mann and James Hartle (Freire Jr. 2013).[46] Robert Griffiths is a prominent statistical physicist working at Carnegie Mellon University in Pittsburgh, who turned his attention to research on the interpretation of quantum mechanics in the early 1980s. In his seminal paper "Consistent histories and the interpretation of quantum mechanics," published in 1984, he suggested mathematical criteria using classical rules of probability to produce conditional probabilities for sequences of events at different times and showed that such criteria could be applied to systems described by the usual quantum mechanical formalism. He called these criteria a "consistent history approach" because they were able to identify sequences of events, now called consistent histories, which were meaningful in a quantum mechanical treatment. These criteria constitute, for him, a regulatory principle to adopt in quantum theory. For Griffiths (1984, p. 219), the main advantage of his approach was that it could be applied to closed (isolated) quantum systems between successive measurements thus without taking measurement as a central process for quantum theory. Therefore, one can speak about the physical meaning of a quantum state even in the

[45] Historian and philosopher of science Yehuda Elkana (1984, p. 503) remarked, with irony: "An open-minded, fair-thinking, egalitarian, liberal philosopher will generally tend to designate himself a realist or a scientific realist. This is 'a good thing' to be. Idealist attitudes like positivism, operationalism, behaviorism are nowadays mostly rejected by philosophers of science and are contraposited to realism. Relativism, though not necessarily an idealist position, is also considered to be the opposite of realism, and is generally talked of as 'the threat.'"

[46] I draw from this paper to present the consistent history approach. It analyzes the extent to which this approach can be considered a new orthodoxy.

absence of measurement processes, which is an advantage for a philosophical approach to quantum physics in terms of realism.

While his approach differed from the traditional interpretations, Griffiths did not see it as an alternative interpretation. He saw it "as an extension and (we hope) clarification of what is, by now, a 'standard' approach to quantum probabilities." Griffiths saw it as a part of "an extended controversy which is far from being resolved" about the "physical interpretation to the solutions (including boundary and initial conditions)" of the Schrödinger equation (Griffiths 1984, p. 221).

Roland Omnès is a theoretical physicist from the Université de Paris XI in Orsay who worked on particle and field physics before moving to foundations of quantum mechanics. In an answer to a reviewer of the first major publication of his proposal, he highlighted his own contribution to the consistent history approach. Asked about "what is common and what is different in [his] approach with Griffith's [sic] history description," he replied that "as far as mathematical techniques are concerned, Griffith's [sic] construction is used," and added that "the conceptual foundations are different because what is proposed here is a revision of the logical foundation of quantum mechanics" (Omnès 1987, p. 172). Omnès revealed in this answer his intellectual heritage, that of modern axiomatization which comes from the mathematicians David Hilbert and Henri Cartan, to whom Omnès acknowledges influence through Cartan's teachings (Omnès 1988a, p. 931). In fact, in a three-paper follow up (Omnès 1988a–c), he developed the logical and theoretical machinery that allowed him "to construct consistent Boolean logics describing the history of a system, following essentially Griffiths' proposal" (Omnès 1988a, p. 893).

Murray Gell-Mann and James Hartle came from very different backgrounds. It was the quantization of gravitation which led them to foundations of quantum physics, as they acknowledged in their first joint paper: "we will discuss the implications of quantum cosmology for one of the subjects of this conference—the interpretation of quantum mechanics" (Gell-Mann and Hartle 1989, p. 322). Previously Hartle had worked out what is now known as the Hartle-Hawking wave function of the universe in collaboration with Stephen Hawking, a solution of the Wheeler-DeWitt equation for quantizing gravitation. From the University of California, Santa Barbara, Hartle and his former PhD supervisor, the particle physics 1969 Nobel Prize winner Gell-Mann of Caltech, approached the issue of interpreting quantum mechanics in the late 1980s. The main merit of their contribution was to associate the attribution of classical probabilities in quantum systems as preached by Griffiths and Omnès with decoherence, an understanding of which was just emerging. The connection was that "decoherence requires a sufficiently coarse-grained description of alternative histories of the universe" (Gell-Mann and Hartle 1989, 1990). According to Gell-Mann (1994, p. 144), "coarse graining typically means following only certain things at certain times and only to a certain level of detail." While the first papers they jointly published were more programmatic, they eventually published a more technical piece in which "the connections among decoherence, noise, dissipation, and the amount of coarse graining necessary to achieve classical predictability are investigated quantitatively" (Gell-Mann and Hartle 1993).

As for affiliations, Gell-Mann and Hartle departed from the point of view that all standard interpretations, Copenhagen included, which presuppose a classical domain or an external observer, are inadequate for cosmology because "measurements and observers cannot be fundamental notions in a theory that seeks to discuss the early universe when neither existed." They acknowledged Everett as the first to suggest "how to generalize the Copenhagen framework so as to apply quantum mechanics to cosmology." However, they considered Everett's work incomplete as Everett was not able to "adequately explain the origin of the classical domain or the meaning of the 'branching' that replaced the notion of measurement." Thus, Gell-Mann and Hartle considered the works of Zeh, Zurek, and Joos and Zeh with regard to decoherence as a "post-Everett" stage, and identified their own proposal as part of this trend, along with Griffiths' and Omnès' (Gell-Mann and Hartle 1990).

We begin our comment on collapse theory through the introduction of its main creator, the Italian physicist GianCarlo Ghirardi. He did his PhD in theoretical physics at University of Milan in 1959. Later, in 1963, he moved to the University of Trieste to take up a full-time teaching position. In Trieste, Ghirardi was a key figure animating the close relations between the university and the International Centre for Theoretical Physics (ICTP) created by Abdus Salam in the early 1960s. On the occasion of his 70th birthday, he was honored with a conference at ICTP entitled "Are There Quantum Jumps? On the Present Status of Quantum Mechanics," a gathering which illustrated his prestige among Italian physicists and abroad. Ghirardi began his career working in high energy physics, with some works on phenomenology in this subject, dealing with scattering theory and symmetries. In the late 1970s, he shifted his scientific interests towards foundations of quantum physics; indeed he became more and more involved with foundations, reacting to Bell's theorem and its implications. However, his interest in foundations was present from his early career, as he has vivid reminiscences of the impact on him of a talk given by Prosperi in the 1960s on his work with Daneri and Loinger (Ghirardi 2007). In 1980, motivated by the experimental results confirming quantum mechanical predictions and violating Bell's inequalities, he and his colleagues Rimini and Weber were among those who argued for the compatibility between quantum mechanics and special relativity, showing that quantum nonlocality does not allow the superluminal sending of messages (Ghirardi et al. 1980).[47] In his capacity as a journal referee, he analyzed the paper by Nick Herbert (1982) suggesting superluminal signaling, and elaborated a no-cloning theorem analogous to the one by Wootters and Zurek (1982). The proof slept as a referee report and was only published later (Ghirardi and Weber 1983; Ghirardi 2007, p. 2893; Kaiser 2012, pp. 195–235).

In the mid-1960s Ghirardi's work on foundations paid off with the proposal, co-authored with Rimini and Weber, of a changed Schrödinger equation, later referred to as the GRW from the initials of its authors, but called by him the

[47] This issue is presented in Chap. 7.

"Dynamical Reduction Program." The change meant the addition of non-linear and stochastic terms to that equation (Ghirardi et al. 1986). Due to this, the GRW proposal should not be considered in strict terms an interpretation of quantum mechanics. Indeed it is a modified theory. This stochastic term should not lead to different predictions from standard quantum theory for microsystems with few degrees of freedom. By the same token it should explain the absence of superposition of eigenstates in the description of macroscopic systems. The contrived mathematical apparatus thus constructed implies that "predictions of GRW Theory coincide almost always with those of standard QM," but "there are domains in which the two theories do not yield the same predictions, but these are (so far) beyond the reach of experimental test" (Frigg 2009).[48]

According to a later synthesis, they were looking for "new dynamics [...] characterized by the feature of not contradicting any known fact about microsystems and of accounting, on the basis of a unique, universal dynamical principle, for wavepacket reduction and for the classical behavior of macroscopic systems" (Bassi and Ghirardi 2003, p. 257). It is thus a proposal for fixing quantum mechanics by restricting its weird features to microsystems. The proposal was well-accepted by John Bell, who strongly supported it, and brought Philip Pearle, who was working along parallel lines, in contact with Ghirardi (Ghirardi 2002, 2007, p. 2905; Ghirardi et al. 1990). In fact, Pearle, from Hamilton College in the U.S., had suggested a first proposal in this direction as early as 1976 (Pearle 1976) and had been interested in interpretations of this theory since at least 1970 (Bloch et al. 1968; Ballentine et al. 1971). As in Ghirardi's model, the disappearance of quantum interferences depends on the size of the systems, or the number of degrees of freedom, Bell summarized Ghirardi's proposal saying that in the GRW theory "the [Schrödinger's] cat is not both dead and alive for more than a split second" (Bell 2004, p. 204). Since then this proposal has been framed as a true research program with attempts to solve its problems, such as the need of relativistic generalization and the promise of phenomena where this theory and the usual quantum mechanics depart from each other. Indeed, this is crucial for the future of GRW, as it is not an interpretation of quantum mechanics but rather a different theory.

In terms of the number of citations, works on foundations of quantum mechanics brought more prestige to Ghirardi than his earlier work on theoretical physics, which was already well-regarded. However, as somebody who experienced the changes in the professional status of foundations, he holds the following memories from the earlier times (Ghirardi 2007, p. 2895):

> Actually, I remember very well that the shared attitude was more or less the following: these problems are nonscientific problems and might interest exclusively philosophically minded people, but are of no relevance for the scientific enterprise. The practical

[48] For a detailed presentation of the GRW theory, see Ghirardi, GianCarlo, "Collapse Theories", The Stanford Encyclopedia of Philosophy (Winter 2011 Edition), Edward N. Zalta (ed.), URL: http://plato.stanford.edu/archives/win2011/entries/qm-collapse/, accessed on 16 April 2014.

> counterpart of this position was that it made extremely difficult, even for people deeply involved in the subject, to be taken seriously and to get the due academic recognitions. In the subsequent years (1965–1990) the situation changed, slowly but continuously ...

The GRW theory and its modified versions are seen with affection by many of the researchers in the field of foundations both for its daring assumptions and for the possibility, even in a distant future, of an experimental contrast with standard quantum mechanics. One should take into account that changing the Schrödinger equation into a nonlinear equation had been envisioned by many—including Louis de Broglie and Eugene Wigner—unsuccessfully. The research program on collapse theories thus fills a lacuna in the broad spectrum of possible solutions for the concerns many physicists have with the standard quantum theory. This is one of the strengths of this proposal. As the GRW theories may lead to different predictions from quantum theory, this is also considered by many one of their strengths. Adrian Kent, for instance, supports the collapse theories but he acknowledges their shortcomings, considering that "the mathematics of collapse seem a little ad hoc and utilitarian," and sees them "at best only a step in roughly the right direction." His interest in these theories derives from the possibility of revealing limits of the validity of quantum theory, which he, maybe optimistically, expects to happen in the next two decades (Kent 2014). A detailed analysis of the proposals aiming to change Schrödinger's equation to dispense with collapse was done by Franck Laloë (2001).

In addition to these new interpretations, old ones also underwent change. Bohm's original model led to variety of changes, as analyzed in Chap. 2. Summarizing what we have discussed there, Bohm's ideas were revived in different ways. Chris Philippidis and Chris Dewdney, together with Basil Hiley used computers to obtain graphs of trajectories from Bohm's model (Philippidis et al. 1979) and this drew the attention of many researchers. Then Detlef Dürr, Sheldon Goldstein, and Nino Zanghi coined the term Bohmian mechanics to describe their work on Bohm's original model. They constructed their approach adopting just two premises: the state which describes quantum systems evolves according to Schrödinger's equation and particles move, that is, they have a speed in the configuration space. With this approach, without referring to the quantum potential and the difficult problem of its physical interpretation, they derived the same results one gets both with standard quantum mechanics and with Bohm's original approach for nonrelativistic phenomena (Dürr et al. 1992, 2009). This approach has been useful for discussing quantum chaos, and for this reason it has received acceptance well beyond physicists just interested in the foundations of quantum mechanics. In another direction, and more recently, one of the supporters of Bohm-de Broglie's causal interpretation, Antony Valentini, has extended it in order to lead to empirically distinct predictions, at least in the cosmological domain (Valentini 2007, 2010). Valentini achieved this by making a different derivation for the quantum equilibrium hypothesis, which Bohm had assumed in the 1950s. This derivation has been challenged by the supporters of Bohmian mechanics, thus bringing controversy among Bohm's intellectual heirs.

Everett's many-worlds also split into many variations. It was influential in the early work of Zurek, on Gell-Mann and Hartle's work on consistent histories, and on Deustch's work on quantum computation, as we will see. In the twenty-first century it has gained a stronghold among physicists and philosophers, which was reflected in the commemoration of the 50th birthday of Everett's thesis through colloquia and a cover page of the prestigious journal *Nature* on 5 July 2007. Everett's supporters however, continue to deal with an intractable problem, how to obtain statistical laws from the Everettian framework where there is no ingredient of randomness. According to critics, such as Kent (2014), "the key scientific question is why the experimental evidence for quantum theory justifies a belief in many worlds in the first place." Kent acknowledges the work Everettians have done, "but think they have all failed."

Other interpretations have also survived and are alive and well. The old stochastic interpretation was boosted with the development of full stochastic electrodynamics in the hands of Emilio Santos, Trevor Marshall, Luis de la Peña, Ana Cetto, and Miguel Ferrero, among others (Peña and Cetto 1996). An updated presentation of this approach is (Peña, Cetto, and Valdes-Hernandez 2015). Ferrero got his PhD degree under the supervision of Santos and went on to organize a successful series of conferences in Oviedo, Spain, dedicated to foundational issues. When quantum information emerged, however, Ferrero moved to a philosophical position which tries to obtain epistemological lessons from the standard formalism of quantum theory (Ferrero et al. 2013; Ferrero and Sánchez-Gómez 2014). The ensemble interpretation, systematized by Leslie Ballentine in the early 1970s, has survived without major changes (Home and Whitaker 1992; Ballentine 1998), while more recently it has been challenged due to potential conflicts with quantum mechanics' predictions (Pusey et al. 2012), and its acceptance among scholars working on foundations has declined as a consequence of the current practice of experiments with quantum single systems (Schlosshauer et al. 2013, p. A52). An updated view of some of the distinct interpretations for quantum mechanics, including current supporters of Bohr's, may be obtained in the interviews-turned-book organized by the philosopher Maximilian Schlosshauer (2011). A fine analysis of the various interpretations is presented in Franck Laloë's recent book (Laloë 2012).

Let us conclude this section with a few lessons from the proliferation of alternative interpretations of quantum mechanics. All of them present varying unsatisfactory features, which explain why none of them have obtained unqualified support among physicists and philosophers. Possibilities of different predictions continue to keep expectations high, as testified by Valentini's and Ghirardi's proposals and by Kent's comments. Meanwhile, with currently available evidence, they are empirically equivalent. This illustrates the philosophical thesis of the underdetermination of scientific theories by empirical data, at least in some of its versions, a thesis of Pierre Duhem and Willard Van Orman Quine (Harding 1976; Cushing 1994; Schlosshauer 2011).[49] Furthermore, Jeffrey Bub (2005) has played

[49] Maximilian Schlosshauer (2011, pp. 63–64) does not cite the Duhem-Quine thesis, but his analysis suggests its very content: "Such an irreducible plurality of interpretations would tell us

with an analogy between the history of Einstein's principle of relativity and the quantum mechanics case to argue that Bohmian and no-collapse theories are doomed to have no excess empirical content over quantum mechanics, if the information-theoretic constraints are considered.[50] This case highlights the possibility that some interpretations of quantum theory may be intrinsically underdetermined by empirical data. Finally, the diversity of interpretations has been useful for the development of physics. Bell's theorem was inspired by Bohm's interpretation, and Everett's interpretation has been influential in works in the birth of decoherence and quantum computation. We may extend further this final conclusion by saying that tolerance towards diversity may be more helpful to science than strict adhesion to the dominant views. Thus, at least in science, a Hundred Flowers policy may fare better than a Nonproliferation treaty.

I am not alone in the defense of diversity of interpretations as a fecund resource in quantum physics. More recently, Schlosshauer et al. (2013, p. A53) argued that "many ideas in physics didn't come out of thin air, but came about when someone started looking at a problem with a certain philosophical disposition. In this sense, we may think of the plurality of views in quantum mechanics as productive." A similar case has been made in other contexts. I want to exemplify with one statement by physicist Victor Weisskopf where you may fill the following blanks with any of the alternative interpretations of quantum mechanics and Weisskopf's statement would still be meaningful.[51]

> [...]'s talk has special significance. His approach differs from the one that is used by most theorists. I believe that the content and the results are the same, but he uses a very different terminology and a different way of reasoning. In some instances it brings out certain physical features of the theory that are hidden in the customary approach. But [...]'s formulations are of great value just because they are so different. In poetry, art and music we value highly new ways of expressing the same contents. In theoretical physics there is not enough variety of presentation. Most of the theorists stick to the generally employed ways of arguing and of calculating. This brings about too much uniformity although it helps to understand the papers of those authors. We must be grateful to [...] for showing us another way and we should devote more efforts to understand it. Perhaps the physical content is not so different but some of the problems of the orthodox approach appear in a new light.

In the factual quotation, Weisskopf was reacting to Julian Schwinger's presentation of his "source theory," which was formulated as an alternative to quantum

---

that we're free to embellish—some may say encumber—the formalism with entities of our choice, if such a maneuver helps us visualize what's going on, but that in doing so we'll be crossing into strictly metaphysical terrain. And if we follow such a reading to its logical (if radical) conclusion, then quantum theory might even contain a lesson about the task of physics: that the search for 'what the world is made of,' for a unique, definitive, fundamental ontology at the heart of everything, may be ultimately misguided."

[50] These constraints are derived from the Clifton–Bub–Halvorson theorem, see Bub (2005).

[51] Weisskopf's statement is published in the proceedings of the International Colloquium on the History of Particle Physics, 21–23 July 1982, Paris, *Journal de Physique*, Colloque 8, Suppl. 12, 1982, on p. 422. I am thankful to Thiago Hartz for bringing this quotation to my attention.

field theory. In fact, eventually, this proposal did not attract many adherents and faded due to a lack of supporters. The same may happen in the future with some of the alternative interpretations of quantum mechanics, but meanwhile physics gains from the existence of such a diversity of interpretations.

8.7 Early Quantum Information Achievements[52]

By the mid-1990s all the ingredients which would merge into quantum information were already present. Among these ingredients, we may list trust in entanglement as a physical fact, theory and experiments about decoherence, technical advances allowing the manipulation of single quantum systems, results from computer science, interest in the subject from the military and major corporations, and a cohort of researchers, many of them from the ranks of research in the foundations of quantum theory. We have already seen some of these factors, but we need to see closer the results from computer science and the evidences of the blossoming of quantum information as a new field of research in physics.

The experimental corroboration of entanglement in EPR experiments in the early 1980s sparked the imagination of some for its use in superluminal transmission of signals. Among them was the physicist Nick Herbert, who was part of an informal ring of people, mostly scientists, based at Berkeley and interested in foundations of physics as well as their implications in the new age world view. Herbert wrote a paper meaningfully entitled "Flash—a Superluminal Communicator Based Upon a New Kind of Quantum Measurement" (Herbert 1982), and even before its publication, as a result of the refereeing process and its circulation as a preprint, some physicists were inspired to refute it. From this refutation emerged the no-cloning theorem, a type of no-go theorem, which says you cannot replicate a quantum state from an arbitrary unknown quantum state. Such a result poses a limitation on the use of entanglement to convey information. In fact, the standard procedure for preventing errors through the sending of a message is its multiplication, that is, redundancy. The refutation was published by Dennis Dieks (1982) from Utrecht, William K. Wootters and Zurek (1982), both closely connected to Wheeler at the University of Texas at Austin, and Ghirardi and Weber (1983). Wooters and Zurek initially considered it an easily derivable result and intended to publish it in a low impact journal. Wheeler disagreed, suggesting the title and its submission to *Nature*. Wheeler was right about the value of the paper; it now has more than 1,600 citations and is considered a seminal paper on quantum information.[53] However, only in hindsight can we consider these papers forerunners of the quantum information boom as at the time of their publication they were considered

[52] In this section we draw from Yeang (2011) and from our paper: Freire and Greca, Informação e teoria quântica. *Scientiae Studia*, 1(1), pp. 11–33, 2013.

[53] Web of Science, consulted on 17 April 2014.

only in relation to debates on foundational issues. This story is well described by David Kaiser (2012, pp. 195–235) in the book *How the hippies saved physics*. This case is the best illustration of how even Californian New Agers contributed to the development of investigations on the foundations of quantum theory.

Quantum information researchers, however, would prefer to date the inception of quantum information to a different event at about the same time (Nielsen and Chuang 2010, p. 7). This was a conference which brought to the forefront a physicist with a greater recognition than Herbert; none less than Richard Feynman, an icon in American theoretical physics in the twentieth century. This 1981 conference at MIT was dedicated to "Physics and Computation." Among the issues dealt with was the reduction in the size of microprocessors and Moore's Law (Nielsen and Chuang 2010, pp. 7, 39–40, passim), which would lead to a scale at which quantum effects should be taken into account. Paul Benioff (1982) suggested a computer model based on the Turing machine but using quantum dynamics. The Turing machine was a mathematical model which forms the basis of the current programmable computer; thus Benioff was cautious suggesting a combination of usual computer science with quantum physics. Feynman (1982) went further suggesting that only a computer entirely based on quantum mechanics could cope with calculation of quantum systems as they require higher computational capabilities than any foreseeable classical computer. Computational strength is related to the concept of computational complexity and the quest for increasing strength had led computer scientists to generalize the original Turing machine. The stronger version of this thesis is the called the Church-Turing thesis and it says "Any algorithmic process can be simulated efficiently using a probabilistic Turing machine." Notwithstanding, one could ask, "might it not turn out at some later date that yet another model of computation allows one to efficiently solve problems that are not efficiently soluble within Turing's model of computation?" (Nielsen and Chuang 2010, pp. 5–7).

David Deutsch faced this problem and asked "whether it is possible for a quantum computer to efficiently solve computational problems which have no efficient solution on a classical computer, even a probabilistic Turing machine" (Nielsen and Chuang 2010). Deutsch had spent a few years at the University of Texas at Austin, where he was introduced to Everett's interpretation through Wheeler and DeWitt and claims it was instrumental in the algorithm he developed. According to the historian Chen-Pang Yeang, DeWitt played an especially important role. "He was the one who introduced me to Everett's many-worlds interpretation of quantum mechanics, and to the wider implications of quantum field theory, and it was because of his take on both the formalism and interpretation of quantum mechanics that I got interested in quantum computers" (Deutsch 2000, quoted in Yeang 2011, p. 332). Deutsch's first result was to formulate a type of quantum Turing machine and an algorithm to run on a quantum computer (Deutsch 1985a). By doing this he was the first to explicitly suggest exploiting quantum superposition, and entanglement implied in Bell's theorem, to make a computing machine (Yeang 2011, p. 331). However, Deutsch's proposal had its limitations. First, quantum superposition is not easy to manipulate as it quickly disappears through decoherence. Second, in this proposal Deutsch could not recover all the output values at the same time. It was therefore just a first step. He kept working on the subject publishing a number of results including a joint paper with Richard Jozsa which showed the superiority of a quantum computer for the calculation of some problems (Deutsch and Jozsa 1992). According to Yeang (2011, p. 335), "Deutsch started his inquiries into quantum computing with broad philosophical questions concerning the physicality of universal computing, [...] and the interpretation of quantum mechanics. Yet, he ended up discovering a potential application of such philosophical exercises."

However, practical applications were not yet foreseeable. Despite having become an "advocate and devotee of quantum computation," Deutsch was aware that the true potentialities of quantum computers were not yet set. At the end of the day, one should ask, "what was the quantum computer useful for?" Deutsch "had observed [that] a quantum computer must have impressive algorithms with practical potential" (Yeang 2011, p. 334). In the early 1990s a major breakthrough in the building of quantum algorithms superior to the classical ones was made. It came from a person totally unrelated to physics, not to say to the foundations of physics, namely Peter Shor.

Peter Shor was a "mathematical prodigy" who "received his BS from Caltech in mathematics and PhD from MIT in applied mathematics" (Yeang 2011, p. 336). He showed that two problems, intractable till then, could be efficiently solved on a quantum computer. Shor (1994) argued "that he could use a quantum algorithm to solve the so-called 'order-finding' problem with a significantly lower time complexity than the traditional approaches" (Yeang 2011, p. 336). While this may seem an abstract problem without mundane applications, according to Yeang, Shor showed that such a strategy could be used to deal with problems with huge practical implications, among them factorizing a large integer number. We should recall that all the cryptography used for financial transactions on the Internet in the mid-1990s was based, and it still is, on the fact that computers are slow to factorize huge integer numbers. Technically the safety of such transactions is assured by the RSA cryptosystem, which had been described by Ron Rivest, Adi Shamir, and Leonard Adleman in 1977. Shor was telling the world that a quantum computer could break all the current cryptography schemes in a shorter time. His seminal work had immediate follow-ups, already in the frame of the booming field of quantum information (Shor 1995, 1996).

From cryptography came the other early achievement in quantum information. Charles Bennett was trained in chemistry at Brandeis University, where he got his BS in 1964, and Harvard University, where he got his PhD in 1970. He then spent a time at Argonne Laboratory and got a position at the IBM Research Center in Yorktown Heights, New York. Here, under the influence of Rolf Landauer, he moved to the physics of computation and directed his attention to the relationship between physics and information. In the early 1980s, under the influence of the experiments confirming quantum mechanics' predictions and violating Bell's inequalities, particularly the experiments conducted by Alain Aspect, he came to reflect on how to use entanglement for cryptography. Jointly with Gilles Brassard from Université de Montréal, they developed the BB84, the first protocol for quantum cryptography (Bennett and Brassard 1984). Historian Chen-Pang Yeang explains BB84's appeal as a safe cryptography procedure: "the key principle of this scheme is that the quantum state of a particle is changed permanently after a measurement;" thus if "a person sends a message coded into, say, the polarized state of a photon, to another person," and "an eavesdropper is trying to tap this message, then he has to make a measurement of the photon's state, which changes it permanently" (Yeang 2011, p. 341). Later, Artur Ekert (1991) developed a system of quantum cryptography explicitly suggesting a key distribution based on quantum entanglement. Bennett extended the possibilities of using entanglement's weirdness for information purposes further. In collaboration with Brassard and C. Crepeau, R. Jozsa, A. Peres, and W. K. Wootters, he suggested the possibility of quantum teleportation (Bennett et al. 1993), which was soon achieved in the labs, as we have seen at the beginning of this chapter.

While part of the emerging discipline of quantum information, quantum cryptography became a field in itself, probably due to its first practical implementations. Some scientometric data may illuminate the strength of this field. A survey on the Web of Science database with the topic "quantum cryptography" finds 1,736 papers, the first one in 1988. At least two of these papers—(Ekert 1991; Gisin et al. 2002)—have more citations than the most cited paper under the topic "quantum information," which confirms the intrinsic and autonomous interest the subject has awoken.[54]

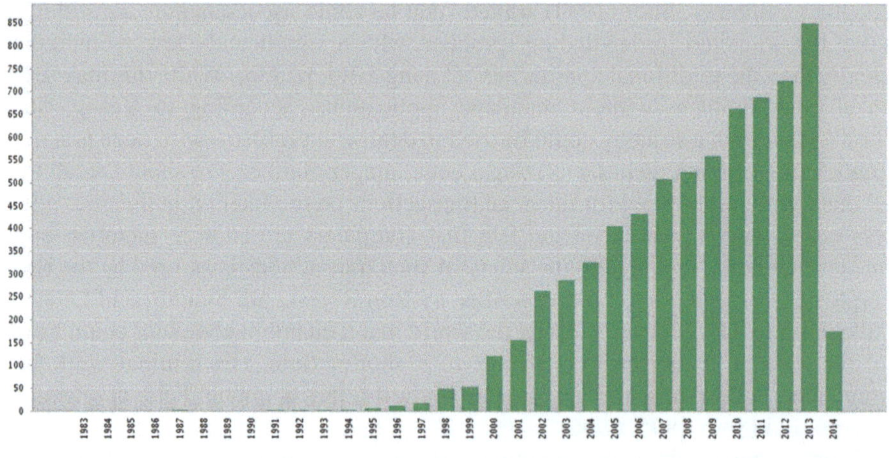

Picture 8.7 Number of papers using "quantum information" as topic. Source: Web of Science, May 2014

[54] Source: Web of Science, accessed on 21 April 2014. This survey shows one of the caveats to take into account while taking data from this source. The 1984 paper by Bennett and Brassard, which opened the subject for research, is not listed in this survey probably because it was presented at a conference. Publishing original papers at conferences is usual in some fields, including computer science. The Web of Science is trying to fix the issue but its coverage still has shortcomings like this.

We may then date the beginning of the blossoming research on quantum information to the mid-1990s.[55] A survey on the Web of Science database with the topic "quantum information" brings up 6,850 papers; the first published in 1983, 4 papers till 1991 and 15 till 1994, then 7 only in 1995, 12 in 1996, 18 in 1997, and 49 in 1998. The full number of papers is portrayed in the graph in Picture 8.7, showing the number of papers year by year.[56] Not by chance, conferences had their titles changed to include quantum information and many researchers in quantum optics began to add quantum information to their lab names. The Oviedo conferences, held in Spain in 1993, 1996, and 2003, were titled, respectively, "Fundamental problems in quantum physics," "New developments on fundamental problems in quantum physics," and "International conference on quantum information. Conceptual foundations, developments and perspectives" (Ferrero and Van der Merwe 1995, 1997; Ferrero 2003). In 2003, the Austrian Academy of Science created the "Institute for Quantum Optics and Quantum Information," based in Vienna and Innsbruck, reflecting the leadership Zeilinger had acquired in these fields. Thus the story told in this book comes to its end. Other stories, however, including the relationship between quantum information and research in the foundations of quantum physics, begin.

Historians dealing with the history of quantum information will need to take into consideration Sam Schweber's statement, which was formulated dialoguing with Paul Forman's criticism to postmodernity in science: "People working in [...] quantum computers, ... are principally concerned with the creation of novelty—of entities or effects that did not previously exist in the world—[...] and are no longer concerned with establishing the foundational theory that governs the interactions and determines the evolution of the structures that populate that domain" (Forman 2012; Schweber 2014).

References

Aaserud, F.: Sputnik and the Princeton 3 – the National-Security Laboratory that was not to be. Hist. Stud. Phys. Biol. Sci. **25**, 185–239 (1995)

Arthur, R.T.W.: Quantum-mechanics, a half century later – Lopes, JL, Paty, M. Philos. Sci. **48**(1), 156–161 (1981)

[55] This presentation of the early works in quantum information is far from comprehensive. We have not presented, for instance, the contributions by Lov Grover, Benjamin Schumacher and Stephen Wiesner, all from the mid-1990s, the end of the period we are analyzing. We suggest the paper written by Chen-Pang Yeang (2011), which is the first historical account of the emergence of this field, meaningfully titled "Engineering Entanglement, Conceptualizing Quantum Information," and the book by Nielsen and Chuang (2010), which is a conceptual introduction to the subject.

[56] Ibid. Still, another caveat about the accuracy of scientometric data. The 1993 quantum teleportation paper, written by Bennett and colleagues, has collected till now more than 5,646 citations, but it did not appear in this survey, probably because it did not explicitly use the term "quantum information" in its abstract.

Aspect, A., et al.: Wave particle duality for a single photon. J Optic Nouvelle Revue D Optique **20** (3), 119–129 (1989)
Ballentine, L., et al.: Quantum-mechanics debate. Phys. Today **24**(4), 36–44 (1971)
Ballentine, L.E.: Resource letter IQM2: foundations of quantum mechanics since the Bell inequalities. Am. J. Phys. **55**, 785–792 (1987)
Ballentine, L.E.: Quantum Mechanics: A Modern Development. World Scientific, River Edge, NJ (1998)
Bassi, A., Ghirardi, G.C.: Dynamical reduction models. Phys. Rep. **379**(5–6), 257–426 (2003)
Bell, J.S.: Speakable and Unspeakable in Quantum Mechanics: Collected Papers on Quantum Philosophy. With an Introduction by Alain Aspect. Cambridge University Press, Cambridge (2004)
Benioff, P.A.: Quantum-mechanical Hamiltonian models of discrete processes that erase their own histories – application to turing-machines. Int. J. Theor. Phys. **21**(3–4), 177–201 (1982)
Bennett, C.H., Brassard, G.: Quantum cryptography: public key distribution and coin tossing. In: International Conference on Computers, Systems & Signal Processing, pp. 175–179 (1984)
Bennett, C.H., et al.: Teleporting an unknown quantum state via dual classical and Einstein-Podolsky-Rosen channels. Phys. Rev. Lett. **70**(13), 1895–1899 (1993)
Bloch, I., et al.: Comment on – Alternative to orthodox interpretation of quantum theory. Am. J. Phys. **36**(5), 462–463 (1968)
Boschi, D., et al.: Experimental realization of teleporting an unknown pure quantum state via dual classical and Einstein-Podolsky-Rosen channels. Phys. Rev. Lett. **80**(6), 1121–1125 (1998)
Bouwmeester, D., et al.: Experimental quantum teleportation. Nature **390**(6660), 575–579 (1997)
Bouwmeester, D., et al.: Observation of three-photon Greenberger-Horne-Zeilinger entanglement. Phys. Rev. Lett. **82**(7), 1345–1349 (1999)
Brendel, J., et al.: Experimental test of Bell inequality for energy and time. Europhys. Lett. **20**(7), 575–580 (1992)
Bromberg, J.L.: Device physics vis-à-vis fundamental physics in Cold War America: the case of quantum optics. Isis **97**(2), 237–259 (2006)
Bromberg, J.L.: New instruments and the meaning of quantum mechanics. Hist. Stud. Nat. Sci. **38** (3), 325–352 (2008)
Brown, R.G.W., Pike, E.R.: A history of optical and optoelectronic physics in the twentieth century. In: Brown, L.M., Pais, A., Pippard, B. (eds.) Twentieth Century Physics, vol. III, pp. 1385–1504. IOP and AIP Press, New York (1995)
Brune, M., et al.: Observing the progressive decoherence of the "meter" in a quantum measurement. Phys. Rev. Lett. **77**(24), 4887–4890 (1996)
Brush, S.G.: The chimerical cat: philosophy of quantum mechanics in historical perspective. Soc. Stud. Sci. **10**(4), 393–447 (1980)
Bub, J.: Quantum mechanics is about quantum information. Found. Phys. **35**(4), 541–560 (2005)
Burnham, D.C., Weinberg, D.L.: Observation of simultaneity in parametric production of optical photon pairs. Phys. Rev. Lett. **25**, 84–87 (1970)
Caldeira, A.O.: Macroscopic Quantum Tunnelling and Related Topics, University of Sussex (1980)
Caldeira, A.O., Leggett, A.J.: Influence of dissipation on quantum tunneling in macroscopic systems. Phys. Rev. Lett. **46**, 211–214 (1981)
Caldeira, A.O., Leggett, A.J.: Quantum tunnelling in a dissipative system. Ann. Phys. **149**(2), 374–456 (1983a)
Caldeira, A.O., Leggett, A.J.: Path integral approach to quantum Brownian-motion. Phys. A **121** (3), 587–616 (1983b)
Caldeira, A.O., Leggett, A.J.: Influence of damping on quantum interference – an exactly soluble model. Phys. Rev. A **31**(2), 1059–1066 (1985)
Camilleri, K.: A history of entanglement: decoherence and the interpretation problem. Stud. Hist. Philos. Mod. Phys. **40**, 290–302 (2009)

Castagnino, M., et al.: A general conceptual framework for decoherence in closed and open systems. Philos. Sci. **74**, 968–980 (2007)
Clifton, R.K., et al.: Generalization of the Greenberger Horne Zeilinger algebraic proof of nonlocality. Found. Phys. **21**(2), 149–184 (1991)
Cushing, J.: Quantum Mechanics – Historical Contingency and the Copenhagen Hegemony. The University of Chicago Press, Chicago, IL (1994)
Cushing, J.T., McMullin, E.: Philosophical Consequences of Quantum Theory: Reflections on Bell's Theorem. University of Notre Dame Press, Notre Dame, IN (1989)
Davidovich, L., et al.: Mesoscopic quantum coherences in cavity QED: preparation and decoherence monitoring schemes. Phys. Rev. A **53**(3), 1295–1309 (1996)
Deligeorges, S.: Le Monde quantique. Éditions du Seuil, Paris (1985)
Deutsch, D.: Quantum-theory, the Church-Turing principle and the universal quantum computer. Proc. R. Soc. Lond. Ser. A Math. Phys. Eng. Sci. **400**(1818), 97–117 (1985a)
Deutsch, D.: Quantum-theory as a universal physical theory. Int. J. Theor. Phys. **24**(1), 1–41 (1985b)
Deutsch, D., Jozsa, R.: Rapid solution of problems by quantum computation. Proc. R. Soc. Lond. Ser. A Math. Phys. Eng. Sci. **439**(1907), 553–558 (1992)
Dieks, D.: Communication by EPR devices. Phys. Lett. A **92**(6), 271–272 (1982)
Dieks, D.: Resolution of the measurement problem through decoherence of the quantum state. Phys. Lett. A **142**(8–9), 439–446 (1989)
Dürr, D., et al.: Bohmian mechanics. In: Greenberger, D., Hentschel, K., Weinert, F. (eds.) Compendium of Quantum Physics – Concepts, Experiments, History and Philosophy, pp. 47–55. Springer, Berlin (2009)
Dürr, D., et al.: Quantum chaos, classical randomness, and Bohmian mechanics. J. Stat. Phys. **68** (1–2), 259–270 (1992)
Easlea, B.: Fathering the Unthinkable: Masculinity, Scientists and the Nuclear Arms Race. Pluto, London (1983)
Eigler, D.M., Schweizer, E.K.: Positioning single atoms with a scanning tunnelling microscope. Nature **344**, 524–526 (1990)
Ekert, A.K.: Quantum cryptography based on Bell's theorem. Phys. Rev. Lett. **67**, 661–663 (1991)
Elkana, Y.: Transformations in realist philosophy of science from Victorian Baconianism to the present day. In: Mendelsohn, E. (ed.) Transformation and Tradition in the Sciences: Essays in Honor of I. Bernard Cohen, pp. 487–511. Cambridge University Press, Cambridge (1984)
Enz, C.P.: No Time to Be Brief: A Scientific Biography of Wolfgang Pauli. Oxford University Press, Oxford (2002)
Ferrero, M.: Special issue: International conference on quantum information. Conceptual foundations, developments and perspectives, 13–18 July 2002. J. Mod. Phys. **50**, 6–7 (2003)
Ferrero, M., Sánchez-Gómez, J.L.: Coming from material reality. Found. Sci. (2014) Online First, DOI 10.1007/s10699-014-9362-2.
Ferrero, M., Van der Merwe, A.: Fundamental Problems in Quantum Physics. Kluwer, Dordrecht (1995)
Ferrero, M., Van der Merwe, A.: New Developments on Fundamental Problems Quantum Physics. Kluwer, Dordrecht (1997)
Ferrero, M., et al.: A further review of the incompatibility between classical principles and quantum postulates. Found. Sci. **18**(1), 125–138 (2013)
Feynman, R.P.: Simulating physics with computers. Int. J. Theor. Phys. **21**(6–7), 467–488 (1982)
Feynman, R.P., et al.: The Feynman Lectures on Physics. Addison-Wesley, Reading, MA (1963)
Forman, P.: Production and curation: modernity entailed disciplinarity, postmodernity entails antidisciplinarity. Osiris **27**(1), 56–97 (2012)
Franson, J.D.: Bell inequality for position and time. Phys. Rev. Lett. **62**(19), 2205–2208 (1989)
Freire Jr., O.: A story without an ending: the quantum physics controversy 1950-1970. Sci. Educ. **12**(5–6), 573–586 (2003)

Freire Jr., O.: Orthodoxies on the interpretation of quantum theory: the case of the consistent history approach. In: Katzir, S., Lehner, C., Renn, J. (eds.) Traditions and Transformations in the History of Quantum Physics, pp. 293–307. Edition Open Access, Berlin (2013)
Friedman, J.R., et al.: Quantum superposition of distinct macroscopic states. Nature **406**(6791), 43–46 (2000)
Frigg, R.: GRW theory (Ghirardi, Rimini, Weber model of quantum mechanics). In: Greenberger, D., Hentschel, K., Weinert, F. (eds.) Compendium of Quantum Physics Concepts, Experiments, History and Philosophy, pp. 266–270. Springer, Berlin (2009)
Gallicchio, J., et al.: Testing Bell's inequality with cosmic photons: closing the setting-independence loophole. Phys. Rev. Lett. **112**, 110405 (2014)
Gell-Mann, M.: The Quark and the Jaguar: Adventures in the Simple and the Complex. W.H. Freeman, New York (1994)
Gell-Mann, M., Hartle, J.B.: Quantum mechanics in the light of quantum cosmology. In: Proceedings of the 3rd International Symposium on the Foundations of Quantum Mechanics, Tokyo, pp. 321–343 (1989)
Gell-Mann, M., Hartle, J.B.: Quantum mechanics in the light of quantum cosmology. In: Zurek, W.H. (ed.) Complexity, Entropy, and the Physics of Information, pp. 425–458. Addison-Wesley, Redwood City, CA (1990)
Gell-Mann, M., Hartle, J.B.: Classical equations for quantum-systems. Phys. Rev. D **47**(8), 3345–3382 (1993)
Gerry, C.C., Bruno, K.M.: The Quantum Divide: Why Schrödinger's Cat Is Either Dead or Alive. Oxford University Press, Oxford (2013)
Ghirardi, G.C.: John Stewart Bell and the dynamical reduction program. In: Bertlmann, R.A., Zeilinger, A. (eds.) Quantum [Un]speakables – From Bell to Quantum Information, pp. 287–305. Springer, Berlin (2002)
Ghirardi, G.C.: Some reflections inspired by my research activity in quantum mechanics. J. Phys. A Math. Theor. **40**, 2891–2917 (2007)
Ghirardi, G.C., Weber, T.: Quantum-mechanics and faster-than-light communication – methodological considerations. Nuovo Cimento Della Societa Italiana Di Fisica B Gen Phys Relat. Astron. Math. Phys. Methods **78**(1), 9–20 (1983)
Ghirardi, G.C., et al.: A general argument against superluminal transmission through quantum mechanical measurement process. Lettere al Nuovo Cimento **27**(10), 293–298 (1980)
Ghirardi, G.C., et al.: Unified dynamics for microscopic and macroscopic systems. Phys. Rev. D **34**(2), 470–491 (1986)
Ghirardi, G.C., et al.: Markov-processes in Hilbert-space and continuous spontaneous localization of systems of identical particles. Phys. Rev. A **42**(1), 78–89 (1990)
Ghosh, R., et al.: Interference of 2 photons in parametric down conversion. Phys. Rev. A **34**(5), 3962–3968 (1986)
Gisin, N.: Quantum measurements and stochastic-processes. Phys. Rev. Lett. **52**(19), 1657–1660 (1984)
Gisin, N.: Sundays in a quantum engineer's life. In: Bertlmann, R.A., Zeilinger, A. (eds.) Quantum [Un]speakable – From Bell to Quantum Information, pp. 199–207. Springer, Berlin (2002)
Gisin, N., et al.: Quantum cryptography. Rev. Mod. Phys. **74**(1), 145–195 (2002)
Grangier, P.: Experiments with single photons. Séminaire Poincaré http://www.bourbaphy.fr/grangier.pdf%5D (2005)
Grangier, P., et al.: Experimental-evidence for a photon anticorrelation effect on a beam splitter – a new light on single-photon interferences. Europhys. Lett. **1**(4), 173–179 (1986)
Greenberger, D.: The history of the GHZ paper. In: Bertlmann, R.A., Zeilinger, A. (eds.) Quantum [Un]speakable – From Bell to Quantum Information, pp. 281–286. Springer, Berlin (2002)
Greenberger, D.M., Zeilinger, A.: Fundamental Problems in Quantum Theory: A Conference Held in Honor of Professor John A. Wheeler. New York Academy of Sciences, New York (1995)
Greenberger, D., et al.: Going beyond Bell's theorem. In: Kafatos, M.C. (ed.) Bell's Theorem, Quantum Theory and Conceptions of the Universe, pp. 69–72. Kluwer, Dordrecht (1989)

Greenberger, D.M., et al.: Bell theorem without inequalities. Am. J. Phys. **58**(12), 1131–1143 (1990)

Greenberger, D.M., et al.: Multiparticle interferometry and the superposition principle. Phys. Today **46**(8), 22–29 (1993)

Greenberger, D.M., et al.: Compendium of Quantum Physics Concepts, Experiments, History and Philosophy. Springer, Berlin (2009)

Griffiths, R.B.: Consistent histories and the interpretation of quantum-mechanics. J. Stat. Phys. **36** (1–2), 219–272 (1984)

Groblacher, S., et al.: An experimental test of non-local realism. Nature **446**(7138), 871–875 (2007)

Grynberg, G., et al.: Introduction to Quantum Optics – From the Semi-classical Approach to Quantized Light. Cambridge University Press, New York (2010)

Halliwell, J.J.: Decoherence in quantum cosmology. Phys. Rev. D **39**(10), 2912–2923 (1989)

Harding, S.G. (ed.): Can Theories Be Refuted? Essays on the Duhem-Quine Thesis. Synthese Library, vol. 81. D. Reidel, Dordrecht (1976)

Haroche, S.: Entanglement, decoherence and the quantum/classical boundary. Phys. Today **51**(7), 36–42 (1998)

Haroche, S.: Essay: fifty years of atomic, molecular and optical physics in physical review letters. Phys. Rev. Lett. **101**, 160001 (2008)

Herbert, N.: Flash – a superluminal communicator based upon a new kind of quantum measurement. Found. Phys. **12**(12), 1171–1179 (1982)

Home, D., Whitaker, M.A.B.: Ensemble interpretations of quantum-mechanics – a modern perspective. Phys. Rep. **210**(4), 223–317 (1992)

Hong, C.K., Mandel, L.: Experimental realization of a localized one-photon state. Phys. Rev. Lett. **56**(1), 58–60 (1986)

Hong, C.K., et al.: Measurement of subpicosecond time intervals between 2 photons by interference. Phys. Rev. Lett. **59**(18), 2044–2046 (1987)

Horne, M.A., Zeilinger, A.: A Bell-type EPR experiment using linear momenta. In: Lahti, P., Mittelstaedt, P. (eds.) Symposium on the Foundations of Modern Physics – 50 Years of the Einstein-Podolsky-Rosen Gedanken Experiment, pp. 435–439. World Scientific, Singapore (1985)

Horne, M.A., et al.: Two-particle interferometry. Phys. Rev. Lett. **62**(19), 2209–2212 (1989)

Horne, M., et al.: Down-conversion photon pairs: a new chapter in the history of quantum-mechanical entanglement. In: Anandan, J.S. (ed.) Quantum Coherence, pp. 356–372. World Scientific, Singapore (1990)

Howie, A.: Akira Tonomura (1942–2012). Nature **486**, 324 (2012)

Jacques, V., et al.: Single-photon wavefront-splitting interference – an illustration of the light quantum in action. Eur. Phys. J. D **35**(3), 561–565 (2005)

Jammer, M.: The Philosophy of Quantum Mechanics – The Interpretations of Quantum Mechanics in Historical Perspective. Wiley, New York (1974)

Kaiser, D.: How the Hippies Saved Physics: Science, Counterculture, and the Quantum Revival. W. W. Norton, New York (2012)

Kent, A.: Our quantum problem – When the deepest theory we have seems to undermine science itself, some kind of collapse looks inevitable. Aeon. London, Aeon Magazine. http://aeon.co/magazine/nature-and-cosmos/our-quantum-reality-problem/ (2014). Accessed 7 May 2014

Kiess, T.E., et al.: Einstein-Podolsky-Rosen-Bohm experiment using pairs of light quanta produced by type-II parametric down-conversion. Phys. Rev. Lett. **71**(24), 3893–3897 (1993)

Kwiat, P.G., et al.: High-visibility interference in a Bell-inequality experiment for energy and time. Phys. Rev. A **47**(4), R2472–R2475 (1993)

Kwiat, P.G., et al.: New high-intensity source of polarization-entangled photon pairs. Phys. Rev. Lett. **75**(24), 4337–4341 (1995)

Laloë, F.: Do we really understand quantum mechanics? Strange correlations, paradoxes, and theorems. Am. J. Phys. **69**(6), 655–701 (2001)

Laloë, F.: Do We Really Understand Quantum Mechanics? Cambridge University Press, New York (2012)
Leggett, A.J.: Nonlocal hidden-variable theories and quantum mechanics: an incompatibility theorem. Found. Phys. **33**(10), 1469–1493 (2003)
Lopes, J.L., Paty, M.: Quantum Mechanics: A Half Century Later. Holland/D. Reidel, Dordrecht/ Boston, MA (1977)
Mair, A., et al.: Entanglement of the orbital angular momentum states of photons. Nature **412** (6844), 313–316 (2001)
Mermin, N.D.: What's wrong with these elements of reality? Phys. Today **43**(6), 9–11 (1990)
Misner, C.W., et al.: John Wheeler, relativity, and quantum information. Phys. Today **62**(4), 40–46 (2009)
Monroe, C., et al.: A "Schrodinger cat" superposition state of an atom. Science **272**(5265), 1131–1136 (1996)
Nakajima, S., et al. (eds.): Foundations of Quantum Mechanics in the Light of New Technologies [Selected Papers from the Proceedings of the First Through Fourth International Symposia]. World Scientific, Singapore (1996)
Nielsen, M.A., Chuang, I.L.: Quantum Computation and Quantum Information. Cambridge University Press, Cambridge (2010)
Omnès, R.: Interpretation of quantum-mechanics. Phys. Lett. A **125**(4), 169–172 (1987)
Omnès, R.: Logical reformulation of quantum-mechanics. 1. Foundations. J. Stat. Phys. **53**(3–4), 893–932 (1988a)
Omnès, R.: Logical reformulation of quantum-mechanics. 2. Interferences and the Einstein-Podolsky-Rosen experiment. J. Stat. Phys. **53**(3–4), 933–955 (1988b)
Omnès, R.: Logical reformulation of quantum-mechanics. 3. Classical limit and irreversibility. J. Stat. Phys. **53**(3–4), 957–975 (1988c)
Ou, Z.-Y.J.: Multi-Photon Quantum Interference. Springer, New York (2007)
Ou, Z.Y., Mandel, L.: Violation of Bells-inequality and classical probability in a 2-photon correlation experiment. Phys. Rev. Lett. **61**(1), 50–53 (1988)
Pan, J.W., et al.: Experimental entanglement swapping: entangling photons that never interacted. Phys. Rev. Lett. **80**(18), 3891–3894 (1998)
Pearle, P.: Reduction of state vector by a nonlinear Schrödinger equation. Phys. Rev. D **13**(4), 857–868 (1976)
Peña, L., Cetto, A.M.: The Quantum Dice: An Introduction to Stochastic Electrodynamics. Kluwer, Dordrecht (1996)
Peña, L., Cetto, A.M., Valdes-Hernandez, A.: The Emerging Quantum—The Physics Behind Quantum Mechanics. Springer, Heidelberg (2015)
Philippidis, C., et al.: Quantum interference and the quantum potential. Nuovo Cimento Della Societa Italiana Di Fisica B Gen. Phys. Relat. Astron. Math. Phys. Methods **52**(1), 15–28 (1979)
Pusey, M.F., et al.: On the reality of the quantum state. Nat. Phys. **8**(6), 475–478 (2012)
Rhodes, R.: The Making of the Atomic Bomb. Simon & Schuster, New York (1986)
Rosa, R.: The Merli–Missiroli–Pozzi two-slit electron-interference experiment. Phys. Perspect. **14**, 178–195 (2012)
Rowe, M.A., et al.: Experimental violation of a Bell's inequality with efficient detection. Nature **409**(6822), 791–794 (2001)
Scheidl, T., et al.: Violation of local realism with freedom of choice. Proc. Natl. Acad. Sci. U. S. A. **107**(46), 19708–19713 (2010)
Schlosshauer, M.: Decoherence, the measurement problem, and interpretations of quantum mechanics. Rev. Mod. Phys. **76**, 1267–1305 (2004)
Schlosshauer, M. (ed.): Elegance and Enigma – The Quantum Interviews. Springer, Heidelberg (2011)
Schlosshauer, M., et al.: The interpretation of quantum mechanics: from disagreement to consensus? Annalen Der Physik **525**(4), A51–A54 (2013)

Schweber, S.S.: Writing the biography of Hans Bethe: contextual history and Paul Forman. Phys. Perspect. **6**, 179–217 (2014)

Shih, Y.H., Alley, C.O.: New type of Einstein-Podolsky-Rosen-Bohm experiment using pairs of light quanta produced by optical parametric down conversion. Phys. Rev. Lett. **61**(26), 2921–2924 (1988)

Shor, P.W.: Algorithms for quantum computation – discrete logarithms and factoring. In: Proceedings of the 35th Annual Symposium on Foundations of Computer Science, pp. 124–134 (1994)

Shor, P.W.: Scheme for reducing decoherence in quantum computer memory. Phys. Rev. A **52**(4), R2493–R2496 (1995)

Shor, P.W.: Fault-tolerant quantum computation. In: Proceedings of the 37th Annual Symposium on Foundations of Computer Science, pp 56–65 (1996)

Silva, I.: Uma história do conceito de fóton na segunda metade do século XX: para além de histórias do modelo bola de bilhar. Salvador, Universidade Federal da Bahia and Universidade Estadual de Feira de Santana. PhD dissertation (2013)

Silva Neto, C.P., Freire Jr., O.: Herch Moysés Nussenzveig e a Ótica Quântica: consolidando disciplinas através de escolas de verão e livros-texto. Revista Brasileira De Ensino De Fisica **35** (2), 2601–2611 (2013)

Smart, A.G.: Physics Nobel honors pioneers in quantum optics. Phys. Today **65**(12), 16–18 (2012)

Tittel, W., et al.: Violation of Bell inequalities by photons more than 10 km apart. Phys. Rev. Lett. **81**(17), 3563–3566 (1998)

Tonomura, A.: Applications of electron holography. Rev. Mod. Phys. **59**(3), 639–669 (1987)

Tonomura, A., et al.: Evidence for Aharonov-Bohm effect with magnetic-field completely shielded from electron wave. Phys. Rev. Lett. **56**(8), 792–795 (1986)

Tonomura, A., et al.: Demonstration of single-electron buildup of an interference pattern. Am. J. Phys. **57**(2), 117–120 (1989)

Unruh, W.G., Zurek, W.H.: Reduction of a wave packet in quantum Brownian-motion. Phys. Rev. D **40**(4), 1071–1094 (1989)

Valentini, A.: Astrophysical and cosmological tests of quantum theory. J. Phys. A Math. Theor. **40** (12), 3285–3303 (2007)

Valentini, A.: Inflationary cosmology as a probe of primordial quantum mechanics. Phys. Rev. D **82**, 063513 (2010)

van der Wal, C.H., et al.: Quantum superposition of macroscopic persistent-current states. Science **290**(5492), 773–777 (2000)

Weber, R.L., Lenihan, J.M.A.: Pioneers of Science: Nobel Prize Winners in Physics. Institute of Physics, Bristol (1980)

Weihs, G., et al.: Violation of Bell's inequality under strict Einstein locality conditions. Phys. Rev. Lett. **81**(23), 5039–5043 (1998)

Wheeler, J.A.: Include the observer in the wave function? In: Lopes, J.L., Paty, M. (eds.) Quantum Mechanics: A Half Century Later, pp. 1–18. D. Reidel, Dordrecht (1977)

Wheeler, J.A.: The "past" and the "delayed-choice" double-slit experiment. In: Marlow, A.R. (ed.) Mathematical Foundations of Quantum Theory, pp. 9–48. Academic, New York (1978)

Wheeler, J.A., Zurek, W.H.: Quantum Theory and Measurement. Princeton University Press, Princeton, NJ (1983)

Whitaker, M.A.B.: Theory and experiment in the foundations of quantum theory. Prog. Quant. Electron. **24**, 1–106 (2000)

Wootters, W.K., Zurek, W.H.: A single quantum cannot be cloned. Nature **299**(5886), 802–803 (1982)

Yeang, C.-P.: Engineering entanglement, conceptualizing quantum information. Ann. Sci. **68**(3), 325–350 (2011)

Zeh, H.D.: On the interpretations of measurement in quantum theory. Found. Phys. **1**, 69–76 (1970) [Reprinted in Wheeler and Zurek, Quantum Theory and Measurement, pp. 342–349]

Zeh, H.D.: Roots and fruits of decoherence. Quant. Decoherence Poincare Semin. **2005**(48), 151–175 (2007)

Zukowski, M., et al.: "Event-ready-detectors" Bell experiment via entanglement swapping. Phys. Rev. Lett. **71**, 4287–4290 (1993)

Zurek, W.H.: Pointer basis of quantum apparatus – into what mixture does the wave packet collapse. Phys. Rev. D **24**(6), 1516–1525 (1981)

Zurek, W.H.: Environment-induced super-selection rules. Phys. Rev. D **26**(8), 1862–1880 (1982)

Zurek, W.H.: Decoherence and the transition from quantum to classical. Phys. Today **44**(10), 36–44 (1991)

Zurek, W.H.: Decoherence, einselection and the existential interpretation (the rough guide). Philos. Trans. Roy. Soc. a Math. Phys. Eng. Sci. **356**(1743), 1793–1821 (1998)

Zurek, W.H.: Decoherence, einselection, and the quantum origins of the classical. Rev. Mod. Phys. **75**, 715–775 (2003)

Zurek, W.H.: Quantum Darwinism. Nat. Phys. **5**, 181–188 (2009)

Chapter 9
Coda: Quantum Dissidents - A Collective Biographical Profile

Abstract This chapter draws a collective biographical profile of a sample of physicists who were protagonists in the research on the foundations of quantum physics between the 1950s and the early 1990s. We have studied the cases of Bohm, Vigier, Everett, Zeh, Bell, Clauser, Shimony, Horne, Wigner, Rosenfeld, d'Espagnat, Selleri, DeWitt, Aspect, Bub, Tausk, Leggett, Wheeler, Zurek, Ghirardi, Haroche, Greenberger, Zeilinger, Gisin, and Shih. We analyze their training and early career, their achievements, their qualms with quantum mechanics, their motivations for such research, professional obstacles they faced, their attitude towards the Copenhagen interpretation, and their success and failures. Most of them were dissidents, fighting against the dominant attitude among physicists at the time when foundational issues were considered to be already solved by the founding fathers of the discipline. Theirs is a story of success as the foundations of quantum mechanics finally entered the physics mainstream, despite the fact that their hope to set limits of validity for quantum mechanics was not fulfilled.

9.1 Introduction

Through a diverse number of case studies involving different actors, places, and issues, we have seen that the flourishing field of research on the foundations of quantum mechanics at the end of the twentieth century was neither a consequence of a gradual linear evolution from the inception of quantum mechanics in the mid-1920s nor a direct result of new techniques which enabled the execution of many *Gedankenexperiments*. Indeed, diverse factors have played their roles in the evolving controversy over the foundations of this physical theory. These factors were diverse enough to include conceptual, philosophical and ideological issues, professional biases, generational, political and cultural changes, and the diversity of the social and professional climates in which physics was practiced throughout the century. In addition to this, there were conceptual and theoretical breakthroughs, technical innovations, *Gedankenexperiments* and factual experimental feats as well as technological expectations. However, not all of them prevailed at the same time; in fact in each diachronic slice of this history the workings of only a few could be

O. Freire Junior, *The Quantum Dissidents*, DOI 10.1007/978-3-662-44662-1_9

found. Our job as historian was thus to disentangle the roles played by each factor in each local and temporal context. I hope this has been achieved. Now it is time to conclude and I would like to use a different approach. Instead of examining each case, with its own chronology, place, issues, and contexts, I want to consider the ensemble of physicists who dealt with foundations of quantum mechanics between the early 1950s and the early 1990s. In order to gain an insight into this, my strategy is to make a collective biographical profile of the physicists who were the protagonists in this area. This strategy was inspired by what professional historians call prosopography (Stone 1971; Kragh 1987); however, I use qualitative instead of quantitative methods to do it.[1]

In order to build this collective biography, I raised the following questions about these people: What was their training and early career before approaching the foundations of quantum mechanics? What were their achievements in the foundations of quantum mechanics and in other fields of physics? When and where did their qualms with quantum mechanics arise? Why were they attracted to the foundations of quantum mechanics? Did they face professional obstacles while working on the foundations of quantum mechanics? Were they critical of what they perceived as the Copenhagen interpretation? How did they appreciate Bohr's thoughts on the epistemological questions of quantum mechanics? Did they succeed in their careers while working on the foundations of quantum mechanics? What kind of network did they develop especially while working on the foundations of quantum mechanics? To answer such questions I turned to oral histories, biographies, and archival materials.

As the sample of physicists analyzed is not comprehensive, there remains an element of arbitrariness in the choices I have made. I hope the remaining arbitrariness will be less important than the insights we can obtain from such a sample. I began with the young generation who criticized the standard view of quantum physics in the early 1950s, considering people such as David Bohm (1917–1992), Jean-Pierre Vigier (1920–2004), and Hugh Everett (1930–1982). They challenged the received wisdom that the foundational issues were already solved by the founding fathers of quantum mechanics. Thus, for a question of balance, I also considered the case of the founding fathers who, in some way or other, got involved in the debate in the 1950s. I considered both supporters of the standard view, namely Niels Bohr (1885–1962), Wolfgang Pauli (1900–1958), Werner Heisenberg (1901–1976), John von Neumann (1903–1957), and Max Born (1882–1970), as well as its critics, such as Albert Einstein (1879–1955), Erwin Schrödinger (1887–1961), and Louis de Broglie (1892–1987). Then I turned my attention to physicists who moved the subject of foundations on the physics agenda from a fringe position in the early 1960s to a blossoming field in physics in the early 1980s. They were Eugene Wigner (1902–1995), Léon Rosenfeld (1904–1974), John Bell (1928–1990), John Clauser (1942–), Abner Shimony (1928–), Michael Horne (1943–),

[1] Some parts of this chapter were drawn from my paper "Quantum dissidents: Research on the foundations of quantum theory circa 1970" (Freire Jr. 2009).

Heinz Dieter Zeh (1932–), Bernard d'Espagnat (1921–), Franco Selleri (1936–2013), Bryce DeWitt (1923–2004), Jeffrey Bub (1942–), Klaus Tausk (1927–2012), Basil Hiley (1935–), Alain Aspect (1947–), and Anton Zeilinger (1945–). Finally, I examined a few cases of physicists who began to work on these issues in the 1980s, when the mix of groundbreaking experiments and theoretical developments consolidated research in foundations as mainstream in physics. The physicists considered were Anthony Leggett (1938–), Wojciech H. Zurek (1951–), John A. Wheeler (1911–2008), GianCarlo Ghirardi (1935–), Daniel Greenberger (1933–), Nicolas Gisin (1952–), Yanhua Shih (1949–), and Serge Haroche (1944–). Thus roughly 30 individuals were considered for building this biographical profile. To prevent repetition, I do not come back to the biographical information which was presented through the book, focusing instead on the analysis of this information.

My main argument is that a rough collective biography of these figures can be drawn noting that most of them were dissidents, or rather quantum dissidents. Pushing the foundations of quantum physics into common physics required not only good theoretical ideas, experimental skills, and technological improvements, but also a change in the physics community's attitude to the status of the foundations of quantum mechanics as a subject for physics research. These physicists fought against the dominant attitude among physicists at the time according to which foundational issues in quantum mechanics had already been solved by the founding fathers of the discipline. Thus, they challenged the bias against the research on the foundations, and many of them were hard critics of what they recognized as the complementarity interpretation. Their common ground, however, was minimal and focused solely on the importance of the research into the foundations of quantum mechanics. Critical of each other's work, they supported different interpretations of this physical theory and chose different approaches and issues in their research. The fact that their common platform was the critical analysis, both theoretical and experimental, of the foundations of quantum physics, rather than the development of just one alternative interpretation or even the advocacy of their philosophical credo, was one of the sources of their strength. Their story is on the whole a story of success as the foundations of quantum mechanics, or at least some research from this field, entered the physics mainstream. However, many of these quantum dissidents aimed to break standard quantum mechanics, that is, to reveal its limits. Such expectations were confounded; quantum mechanics predictions have been further corroborated in all tests in recent decades. Indeed, physical effects that form the basis of the current quantum information boom, entanglement and decoherence, are implications of standard quantum mechanics. Thus, not all of the dreams of the quantum dissidents came true.

9.2 Achievements

The scientific achievements of our protagonists in the period under study are truly impressive. This can be witnessed through scientometric data, as some of their papers are highly cited. More convincing, however, is the list of their achievements. Major breakthroughs were obtained by Bell, Clauser, Shimony, Aspect, Horne, Zeilinger, Greenberger, Shih, and Gisin, who contrasted local realism with quantum mechanics and contributed to set entanglement as a new quantum physical effect. Research on decoherence coalesced later. In our sample, Zeh, Leggett, Zurek, and Haroche were the people who brought the subject from theoretical insights to experimental corroboration. Everett, Wheeler, Wigner, d'Espagnat, Zeh, Shimony, Rosenfeld, Bub, Tausk, and Ghirardi broadened the understanding of what is at stake when quantum measurement is concerned. They did not agree on the details. Some of them, such as Bohr and Rosenfeld, barely acknowledged that there was a quantum measurement problem, and others, Ghirardi, for instance, suggested changes in the very quantum theory to cope with the problem. DeWitt and Wheeler helped to bring the quantization of gravitation to the forefront of physics, while this challenge remains unsolved. Selleri fed the rush for better experiments and brought the quantum controversy to a wider audience. Diversity of interpretations for quantum theory has been a driving force fomenting the research on foundations, and Bohm and Everett, as well as Bell and Ghirardi, are responsible for the main share in this sense. From the old guard, with the advantage of hindsight, we can better evaluate Einstein's contribution. He saw quantum theory implied entanglement or nonseparability, but he considered it a sign of the incompleteness of quantum theory. Schrödinger saw it as the quantum fingerprint and christened it entanglement. Bohr's interpretation, while philosophically controversial, stands the proof of time. However, as quantum physics was, and continues to be, a field for daring theoretical speculations, at many times very divergent, the litmus test for some of them is not yet set. Let us consider two examples. Wheeler took quantum physics for granted, suggested impressive new experiments sure that they would corroborate quantum physics, and tried to see what quantum physics was telling us about the deepest truths. His "it from bit" encapsulates his boldest conjecture. Ghirardi's approach was the other way around. He wanted to fix quantum theory in order to reconcile it with the intuition derived from our everyday experience. However, he did not do it only through mere hopes. He built what a few others only thought, a non-linear Schrödinger equation, in order to physically explain measurements and collapses. In fact, our understanding of the workings of quantum mechanics has widened through the work of these protagonists.

9.3 Synopsis of the Quantum Controversy Dynamics

This biographical sample, when considered jointly with the cases analyzed throughout this book, provides us with a synopsis of the factors driving the dynamics that research on foundations underwent between the early 1950s and 1990s, from a marginal place to the mainstream of physics.

First of all, there were unsolved and pressing theoretical issues, such as the compatibility between quantum mechanics and local hidden variables, the challenge of meshing gravity and quantum theories, the quantum measurement problem, and the boundary between classical and quantum descriptions. However, these problems were not posed as such in advance. They were being formulated as such, and developed from the early 1950s on through the minds and hands of the characters we have analyzed in this book. Thus the acknowledgment of their relevance was fruit of the controversy over the quanta. Second, there are individuals who stood up and used their intellectual skills and professional reputation to push these theoretical issues forward, as well as their experimental implications, which began to appear in the early 1970s, on the mainstream physics agenda. Thus they made a high professional bid or, in Bourdieusian terms, they adopted a strategy of subversion, they risked their reputation to reconfigure the discipline, to change what was considered to be good physics (Bourdieu 1975). Many of the physicists portrayed in this collective biography held this role, but I would single out some of them: David Bohm, Hugh Everett, Jeffrey Bub, Klaus Tausk, John Clauser, Abner Shimony, Michael Horne, Alain Aspect, Anton Zeilinger, Heinz Dieter Zeh, Nicolas Gisin, and Wojciech H. Zurek. Framing otherwise, they had the moral courage to stand up and tell their fellow physicists that foundations were not solved problems, instead these were problems worthy of investigation in spite of the dominant view that such problems had already been solved. The strategy of subversion was also adopted by some physicists who were not new entrants in the discipline. Still keeping Bourdieu's terms, they had enough symbolic capital to endeavor a risky path. John Bell, Eugene Wigner, Bernard d'Espagnat, Bryce DeWitt, Franco Selleri, Anthony Leggett, and GianCarlo Ghirardi fit this portrait. Together, they were the moral force behind the growing network of physicists working on the foundations of quantum physics.

Third, in the late 1960s and early 1970s the cultural and political unrest in American and European universities had an unforeseen payoff. The unrest helped open the invisible but strong walls of the physics discipline—I am here using Lenoir's metaphor for the development of scientific disciplines (Lenoir 1997)—to the entrance of themes not highly valued till then, namely foundations of quantum mechanics. Fourth, there were the experimental physicists, who did their jobs well, particularly those who carried out very sensitive experiments even before the general improvement of techniques in atomic and quantum optics. I am thinking here, for instance, of experiments such as those carried out by Clauser, Fry, and Aspect. Fifth, from the mid-1980s on, there appeared an impressive number of cutting-edge experiments such as those conducted by Haroche, Alley, Gisin,

Zeilinger, and Shih. Finally, there were contributions linking quantum physics to computer science, as the works by Bennett, Brassard, Deutsch, and Shor presented in the previous chapter. Some of these contributions came from scientists who were not involved in the research on the foundations of quantum physics. Blended, they fuelled the appearance of quantum information, in the mid-1990s. This synopsis is probably oversimplified, and yet it gives us a glimpse of the complex plethora of factors driving foundations of quantum mechanics from the margins of the discipline to its mainstream.

9.4 Training, Professional Losses, Philosophical Trends, and Interpretations

Most of our protagonists were not trained in research on the foundations of quantum mechanics. However, exceptions grew through the time. Everett, Bub, Tausk, Horne, Aspect, Gisin, and Shih worked on foundations during their doctoral training. Clauser, Shimony, and Zurek went to foundations soon after their PhDs. The others switched to foundations at a later stage in their careers. Thus they managed and minimized possible damages to their careers. They were attracted to foundations from very different domains, including particle physics, nuclear physics, quantum field theory, and relativity. Some of them, such as Zeh, Leggett, and Selleri, acknowledged the existence of unsolved problems in foundations while switching to this field; others, such as Bell and Clauser, had been concerned with such questions since their early training. For a few of them, such as Clauser, Selleri, and Leggett, to a certain extent, the political and cultural climate of the late 1960s and early 1970s influenced their decision to undertake research into foundations.

Some of these physicists suffered professional setbacks while dedicating themselves to foundations of quantum theory. The most prominent cases are Everett's frustration after the discussion of his thesis on relative states in Copenhagen, Tausk's career truncated after his preprint on the measurement problem at Trieste and his PhD examiner's board in São Paulo, Zeh's career stymied at Heidelberg as a consequence of a paper which is now considered a forerunner of decoherence, and Clauser's difficulty getting a job at an American university due to his experiments on hidden variables. Bohm's case is a more complex one as there was an overlapping between the poor reception of his causal interpretation among his fellow physicists and the non-renewal of his contract at Princeton during the McCarthy times. However, even in this case, one can wonder if, in the late 1950s, when Bohm was looking for a job in Europe, whether he would have obtained a better position if he had continued to work on plasma and collective variables instead of the causal interpretation. Prejudices were attenuated but were not fully eliminated in the 1980s. Zeilinger, for instance, recalls that working on foundations did not help him when looking for a position in European universities at the time. This list is not exhaustive and testimonies from Bell, Shimony, d'Espagnat, Ghirardi, and Aspect, among others, reveal how they perceived contemporary prejudices against the research on foundations. Only a fine-grained analysis, in each case, may determine how effective this bias was.

As for the philosophical background of our characters and their attitudes towards the complementarity interpretation, we can say that most of them were committed to certain brands of scientific realism and were critics of what they realized as being Bohr's complementarity. This seems to have been a major change compared to the philosophical trends prevalent at the inception of quantum mechanics. In an earlier paper Brush (1980) tried to explain this change claiming the existence of cyclic oscillations between "Romantic" and "Realist" periods in Western culture and science. We leave this renaissance of realism in current physics as an open question to further investigation. As for the attitudes towards the complementarity interpretation, it seems necessary to note how they evolved over time. In the early 1950s, most of our characters, particularly among the founding fathers of this theory, supported it. This scene dramatically changed later as almost all of the physicists portrayed here who entered this field in the late 1960s and the 1970s were critics of Bohr's view. And yet from the 1980s a different scene may be drawn. Some of our characters openly supported complementarity, but now it was an acknowledged interpretation among others, neither superior nor inferior. Recently, a poll conducted among attendees of a conference dedicated to the foundations of quantum physics included the question "What is your favorite interpretation of quantum mechanics?"[2] Each person could give more than one answer. The results revealed 42 % support "Copenhagen," 24 % favor "Information-based/information-theoretical," 18 % agree with "Everett" in its diverse versions, 9 % support "Objective collapse" like GRW, 6 % favor "Quantum Bayesianism," 6 % favor "Relational quantum mechanics," and null support to "De Broglie-Bohm," "Modal interpretation," "Statistical (ensemble) interpretation," and "Transactional interpretation." 12 % declare "Other," the same tell "I have no preferred interpretation." While the authors of the poll were cautious to present their result as a "snapshot" instead of a representative sample, it is indicative of the current diversity of views. Furthermore, we may conclude that such a diversity of interpretations ultimately played a positive role in our understanding of what quantum mechanics is. Suffice to recall that Bell's work came from his engagement with alternative interpretations of quantum mechanics. This could be taken as a lesson for scientific controversies; tolerance with heterodoxies may be more fruitful than uncritical adhesion to received views. As I have suggested in the previous chapter, at least in this case a Hundred Flowers policy fares better than a Non-Proliferation treaty.

---

[2] Schlosshauer et al. (2013, p. 225).

9.5 The Quantum Dissidents

Considering such a diversity of personal and professional profiles, one may ask what these characters had in common, in addition to the theme of research they chose. My conclusion, which was formulated earlier, is that a family portrait of this full group would portray them as dissidents, quantum dissidents. Most of our protagonists were critical of what they perceived as the complementarity interpretation, the Copenhagen interpretation, or the usual interpretation of quantum mechanics, which they saw as the widespread interpretation of this theory. This stands despite the current difficulty in identifying exactly what the Copenhagen interpretation was (Howard 2004; Camilleri 2009). However, they did not share a unique alternative interpretation of quantum mechanics. They shared the professional and intellectual attitude that issues in foundations of quantum mechanics were worthy enough to be pursued as part of a professional career in physics, and that denying this was a dogmatic attitude. This was the main feature of their dissidence, as most physicists at the time disagreed with this. Indeed contemporary wisdom in physics until the 1980s was that problems in foundations had already been solved by the founding fathers of quantum physics.

Picture 9.1 Conference at Amherst College, MA, August 1990. Some of the quantum dissidents are portrayed. L-R are: first row: Leggett, [Zajonc], [?], Bell; second row: Pearle, [?], Jarrett, Mermin, Weisskopf; third row: Horne, [?], [?], Gottfried; fourth row: Bernstein, Claudia Tesch[e], [?], Greenberger; fifth row: [?], Gould, Shimony, Greenstein, and Zeilinger. AIP Emilio Segre Visual Archives, Gift of Abner Shimony

The literature on quantum mechanics invites the metaphorical use of the term dissident, a common term in politics and religion. Here I am using it in the context of a scientific controversy to portray most of the characters in the sample of physicists we have just analyzed. Analogous terms have already been widely used. Indeed, Heilbron (2001) wrote on the "missionaries of the Copenhagen spirit," Wheeler used "heresy" when speaking about Everett's views (Osnaghi et al. 2009, footnote 240), Wigner (1963) presented his own views as the orthodoxy in quantum mechanics, Popper portrayed Einstein, Schrödinger, Bohm and de Broglie as "dissenters" (Popper and Bartley 1982, p. 100), DeWitt referred to the "rigid Copenhagen doctrine" (Freire Jr. 2009, p. 287), and Jammer (1974, p. 250) wrote about the "unchallenged monocracy of the Copenhagen school in the philosophy of quantum mechanics" while analyzing debates on interpretation in the early 1950s, just to cite a few. The metaphor is appealing and may be extended. Like many dissidents in the second half of the twentieth century, such as Nelson Mandela, Luís Inácio Lula da Silva, or Martin Luther King, they or the cause they embraced won, at least in the medium term. The foundations of quantum mechanics eventually became a respectable field of research. However, the metaphor does not always work. The expectations of breaking down quantum mechanics or revealing its limits have not been fulfilled. Bell's Hamletian dictum—"quantum mechanics is rotten" (Gottfried 1991)—has yet to be confirmed. Standard quantum mechanics has entered the twenty-first century more experimentally corroborated than ever.

A last word on the usefulness of this metaphor. Is it a useful concept in the history of quantum mechanics for describing all physicists who have worked on these topics, except of course for those who supported the standard interpretation, that is the complementarity view? While it may be used to describe protagonists from the early 1950s, such as Bohm and those who entered this field in the late 1960s, such as Clauser and Selleri, it is clear that the metaphor runs dry when applied to most physicists who entered the field in the late 1980s. An illuminating case is Alain Aspect's. While he had entered the field in the mid-1970s, suffering from contemporary prejudices, he ended as the iconic figure of what foundations ultimately became, namely mainstream physics. Indeed his is the most illustrative case to trace the transition in the value of the research in foundations, from the fringe to the mainstream. Another interesting case is that of Zeilinger. He has been a supporter of Bohr's views as well an active participant in current research on the foundations of quantum physics.

Such a biographical profile begs as many questions as it answers. I will just address a couple of them. The history of science has been enriched by taking into account the role of local contexts in the production of science. Our cases are evidence that institutions in Paris, Berkeley, Heidelberg, Vienna, Italy or the US have played different roles in this story. The very existence of a friendly environment towards foundations in some of the US physics departments begs an explanation, as it challenges available historical literature which emphasizes pragmatism and the Cold War as factors pitting American physics against philosophically-loaded physical research and as favorable to applied research. As Bromberg (2006) has shown "device" physics is not contradictory to "fundamental" physics in the case of quantum optics in America.[3] Kaiser (2007) has framed such a context, and its change, in term of the pedagogical constraints on American physics throughout its development in the twentieth century and also relating it to cultural changes of the times (Kaiser 2012). Finally, the coverage of our survey left lacunas that may only be filled with further historical research. Two of them—foundations of quantum physics in Soviet and Japanese physics—are conspicuous. Throughout this book, and also from other sources, we have seen evidence of the interest in this issue among physicists in these countries. Their interest may not have been influential enough to shape the history we describe in this book differently, that is, the move of foundations from the fringe to the mainstream physics. Notwithstanding, they are of intrinsic interest to the history of physics in the twentieth century.

References

Bourdieu, P.: Specificity of scientific field and social conditions of progress of reason. Soc. Sci. Inf. (Information Sur Les Sciences Sociales) **14**, 19–47 (1975)

Bromberg, J.L.: Device physics vis-à-vis fundamental physics in Cold War America: the case of quantum optics. Isis **97**, 237–259 (2006)

Brush, S.G.: The chimerical cat: philosophy of quantum mechanics in historical perspective. Soc. Stud. Sci. **10**, 393–447 (1980)

Camilleri, K.: Constructing the myth of the Copenhagen interpretation. Perspect. Sci. **17**, 26–57 (2009)

Freire Jr., O.: Quantum dissidents: research on the foundations of quantum mechanics circa 1970. Stud. Hist. Philos. Mod. Phys. **40**, 280–289 (2009)

Gottfried, K.: Does quantum mechanics carry the seeds of its own destruction? Phys. World 34–40 (1991)

Heilbron, J.: The earliest missionaries of the Copenhagen spirit. In: Galison, P., Gordin, M., Kaiser, D. (eds.) Science and Society – The History of Modern Physical Science in the Twentieth Century. Routledge, New York (2001)

Howard, D.: Who invented the "Copenhagen interpretation"? A study in mythology. Philos. Sci. **71**, 669–682 (2004)

Jammer, M.: The Philosophy of Quantum Mechanics – The Interpretations of Quantum Mechanics in Historical Perspective. Wiley, New York (1974)

Kaiser, D.: Turning physicists into quantum mechanics. Phys. World May, 28–33 (2007)

Kaiser, D.: How the Hippies Saved Physics: Science, Counterculture, and the Quantum Revival. W. W. Norton, New York (2012)

Kragh, H.: An Introduction to the Historiography of Science. Cambridge University Press, Cambridge (1987)

Lenoir, T.: Instituting Science: The Cultural Production of Scientific Disciplines. Stanford University Press, Stanford, CA (1997)

Osnaghi, S., Freitas, F., Freire Jr., O.: The origin of the Everettian heresy. Stud. Hist. Philos. Mod. Phys. **40**, 97–123 (2009)

[3] See Bromberg (2006) for a discussion of this issue.

Popper, K.R., Bartley, W.W.: Quantum Theory and the Schism in Physics. Rowan and Littlefield, Totowa, NJ (1982)
Schlosshauer, M., Kofler, J., Zeilinger, A.: A snapshot of foundational attitudes toward quantum mechanics. Stud. Hist. Philos. Mod. Phys. **44**, 222–230 (2013)
Stone, L.: Prosopography. Daedalus 46–79 (1971)
Wigner, E.P.: Problem of measurement. Am. J. Phys. **31**, 6–15 (1963)

Index

O. Freire Junior, *The Quantum Dissidents*, DOI 10.1007/978-3-662-44662-1

MIX
Papier aus verantwortungsvollen Quellen
Paper from responsible sources
FSC® C105338

If you have any concerns about our products,
you can contact us on
ProductSafety@springernature.com

In case Publisher is established outside the EU,
the EU authorized representative is:
Springer Nature Customer Service Center GmbH
Europaplatz 3, 69115 Heidelberg, Germany

Printed by Libri Plureos GmbH
in Hamburg, Germany